Structural Steelwork Calculations and Detailing

Structural Steelwork Calculations and Detailing

T. J. MacGINLEY,
BE, ME, C Eng, MI Struct E, MIE Aust, M Weld I
Senior Lecturer, Department of Civil Engineering, Sunderland Polytechnic

LONDON BUTTERWORTHS

THE BUTTERWORTH GROUP

ENGLAND
Butterworth & Co (Publishers) Ltd
London: 88 Kingsway, WC2B 6AB

AUSTRALIA
Butterworths Pty Ltd
Sydney: 586 Pacific Highway, NSW 2067
Melbourne: 343 Little Collins Street, 3000
Brisbane: 240 Queen Street, 4000

CANADA
Butterworth & Co (Canada) Ltd
Toronto: 14 Curity Avenue, 374

NEW ZEALAND
Butterworths of New Zealand Ltd
Wellington: 26–28 Waring Taylor Street, 1

SOUTH AFRICA
Butterworth & Co (South Africa) (Pty) Ltd
Durban: 152–154 Gale Street

First published in 1973

ISBN 0 408 70469 1

Printed in Hungary

Preface

The purpose of the book is to present the structural steelwork calculations involved in the design of elements in simple steel-frame buildings.

The book is written for students beginning courses in design for the Higher National Certificate, Higher National Diploma and for the first two years of degree courses in Civil and Structural Engineering. It may also be of assistance to students taking professional examinations and it could find a place in structural steel design offices as a reference book for junior staff members.

It is assumed that the student will be doing an elementary course in theory of structures at the same time or will have completed such a course. Where necessary, reference should be made to suitable text books on theory of structures. A knowledge of the analysis of indeterminate structures is not required.

The book is designed to introduce the student to the British Standard BS449: *The Use of Structural Steel in Building*. The book should be read in conjunction with this code. Emphasis is placed on modern methods of construction using shop welding and bolted site joints.

Sketches are an important part of the design process and most worked examples conclude with sketches giving the information required for detailing. This leads to the final chapter of the book which sets out recommendations for detailing and which can be read independently.

Problems are given at the end of each chapter and if the student works through some of these he should then be able to apply in practice the principles set out in the chapter. All problems are worked in SI units.

Acknowledgements

Extracts from

BS 449, Part 2, 1969, *The use of structural steel in building*
CP3, Part 1, 1967, Chapter V, *Loading: Dead and imposed loads*
CP3, Part 2, 1970, Chapter V, *Loading: Wind loads*
BS 499, Part 2, 1965, *Welding terms and symbols*
BS 4190, 1967, *ISO Metric black hexagon bolts, screws and nuts*
BS 4395, Part 1, 1969, *High-strength friction-grip bolts and associated nuts and washers for structural engineering. Metric series.* Part 1. *General grade*
BS 1865, 1964, *General requirements for the metal-arc welding of mild steel*
CP114, Part 2, 1969, *The structural use of reinforced concrete in buildings*

are reproduced by permission of the British Standards Institution 2 Park Street, London WIA 2BS, from whom copies of the complete standards may be obtained.

The author gratefully acknowledges permission to use material for this book from the literature on structural hollow sections published by The British Steel Corporation, Tubes Division.

A general acknowledgement is due to the British Constructional Steelwork Association, as the author has been greatly influenced by their publications and has used their methods to suit the requirements of this book.

Contents

1

Introduction

1.1 STEEL-FRAME BUILDINGS AND STRUCTURAL ELEMENTS

Steel-frame buildings consist of a skeletal framework which carries all the loads to which the building is subjected. The steel members are used to carry lateral loads when acting as beams and girders, and axial loads when acting as stanchions, struts and ties. Internally the structure is covered with floor plates and slabs to carry floor loads. Externally, sheeting and glazing form the cladding for the framework and carry wind and snow loads on the roof and walls. Internal partitions in this type of building are generally non-load-bearing. The plate elements collect the imposed loads from the floors, and the wind and snow loads from the roof and walls, and transmit them to the steel frame. The steel frame is made up of the following load-bearing structural elements:

beams, girders, trusses, stanchions and bracing.

Connections join individual members to form some of the elements as well as joining the elements together to form the frame. Stanchion bases transmit the building loads to the foundations.

The structural elements are fabricated in shops where working conditions can be carefully controlled. The frame is broken down into parts of suitable size for the method of transport and erection to be used. The members stack together occupying little volume for transport to site. The designer must have an appreciation of the methods of fabrication and erection, as well as the problems of transport, in order to set the size of the separate members and design and detail suitable connections.

The elements are made up of rolled and finished sections and compound and built-up sections as shown in Figure 1.1. To provide exa-

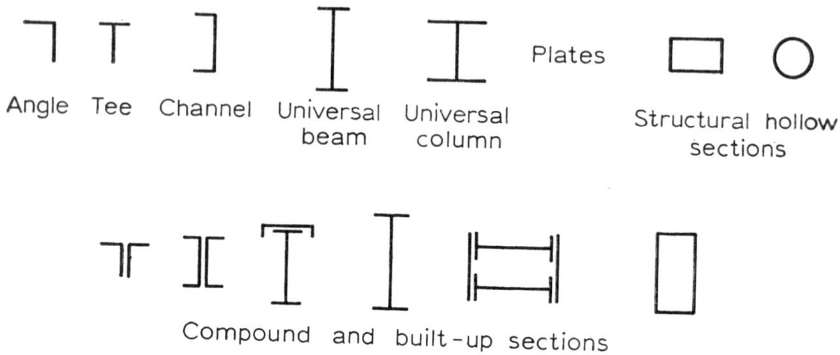

Figure 1.1 Structural elements

mples for design calculation and detailing, structural elements, connections and bases are chosen from the steel frame buildings shown in the following figures. Figure 1.2: Industrial building—trusses, stanchions, bracing, stanchion bases; Figure 1.3: Industrial building—trusses, crane girders, stanchions, bracing, stanchion bases; Figure 1.4: Multi-storey office building—beams, girders, stanchions, stanchion bases. An index to the calculation problems is given in the figures.

1.2 SCOPE

An architect draws up plans for a building to meet the client's requirements. The structural engineer may propose various framing arrangements. Several schemes may require investigation to determine which is the most economical. For a given arrangement of the frame, the problem in structural design consists of:

(a) Estimation of loading
(b) Analyses of main frames, trusses, floor systems, bracing, etc. Analyses refer to finding the loads, shears and bending moments at all points or at recognisable critical points on every member. These values are used in the next step in design
(c) Design of the elements using the data from (b)
(d) Production of the steel arrangement and detail drawings from the designer's calculations and sketches.

This book is limited to design and detailing of structural elements. A complete building is not treated as one design problem. However, all the elements are designed in the appropriate chapters.

The structural frames chosen are primarily of simple pin-jointed design, and elements, joints, etc., are chosen from these for the various

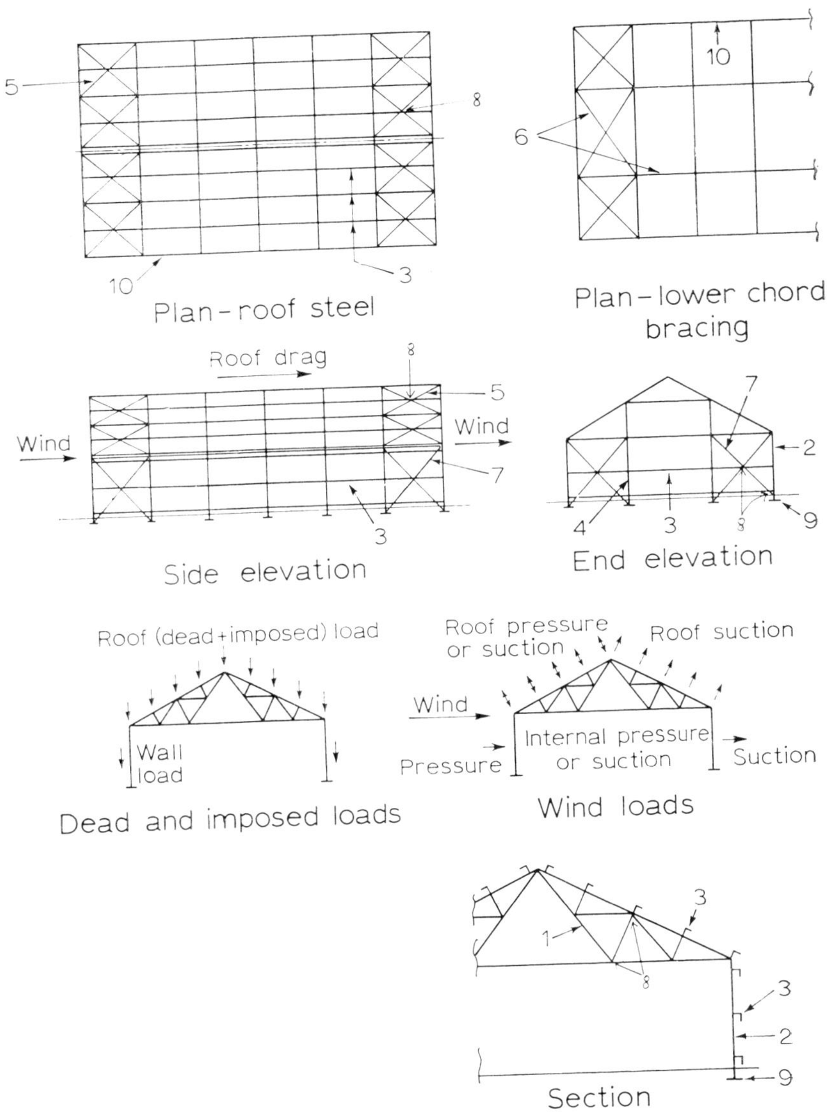

Element	Chapter	Element	Chapter
1 Roof truss	11	6 Lower chord bracing, struts	11
2 Stanchion	7	7 Wall bracing	12
3 Purlins and sheeting rails	10	8 Truss and bracing connections	2
4 Gable stanchions	7	9 Stanchion base	9
5 Roof bracing	12	10 Eaves strut	7, 11

Figure 1.2 Industrial building

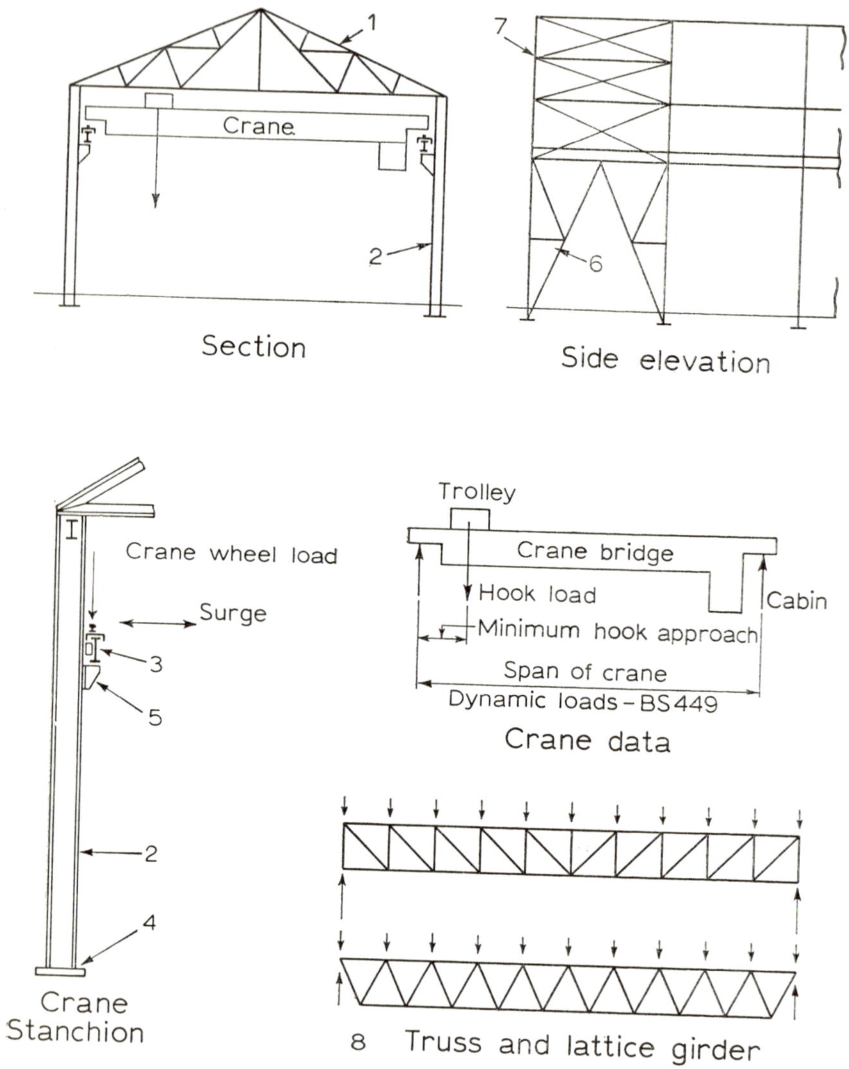

Element	*Chapter*	*Element*	*Chapter*
1 Roof truss	11	5 Bracket	2
2 Crane stanchion	8	6 Wall bracing	12
3 Crane girder	5	7 Roof bracing	12
4 Stanchion base	9	8 Truss and lattice girder	11

Figure 1.3 Building with cranes. Truss and lattice girder

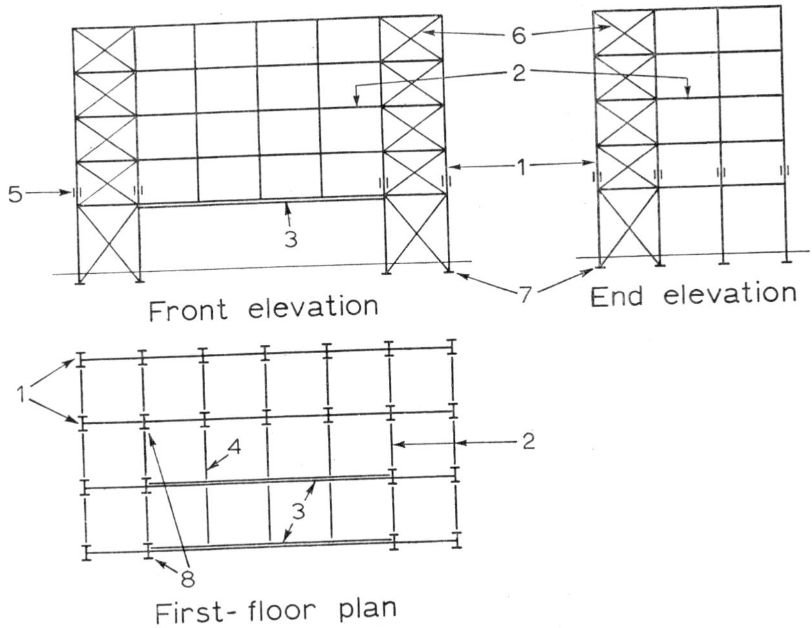

Loading
Dead loads — steel, floor slab, screed and finish, ceiling, fire protection partitions, services machinery
Imposed load — depends on use—Refer to CP3: Chapter V: Part 1
Wind load — external pressure and suction, roof drag and internal pressure and suction. Refer to CP3: Chapter V: Part 2

Element	*Chapter*
1 Stanchions	6, 7
2 Floor beams	3
3 Plate girder	4
4 Connections—beams to girders	2
5 Column splice	6
6 Bracing	12
7 Stanchion base	9
8 Connections—beam to stanchion	2

Figure 1.4 Multi-storey office building

problems discussed in the book. In problems, the method of obtaining the load, the design principles, detail and relevant code requirements are brought out. The calculations are longer than would be required in an actual design office. Some of the aspects of rigid-frame design and the moment connections needed for this type of construction are discussed. Welding and high-strength friction-grip bolts find important applications here.

All design is in accordance with BS 449, 1969, Part 2: *The Use of Structural Steel in Building*.

Grade 43 mild steel is used in all design problems. The allowable stresses for the higher-strength steels, grades 50 and 55, are given in the BS 449 code. All calculations are in SI units. Code requirements relevant to the worked problems are noted and discussed. Reference should be made to the *Handbook on Structural Steelwork*, published jointly by The British Constructional Steelwork Association Ltd and The Constructional Steel Research and Development Organisation, for metric properties of sections, safe-load tables, etc. Safe-load tables for bolts and welds are given in the Appendix.

Design and detail using structural hollow sections has been introduced in parallel with that for rolled steel sections. The use of these sections is very important in modern construction. Special methods for their use have been developed by the British Steel Corporation, Tubes Division, and these methods are followed here.

1.3 PRINCIPLES OF DESIGN

The aim of structural design is to produce a safe and economical structure that fulfils its required purpose. Theoretical knowledge alone is not sufficient since a theoretical analysis is based on an idealised structure, for example, representing members by their centrelines and assuming perfect pin and fixed joints. Thus calculated forces and deflections are only approximations to the actual ones. The actual structure is subjected to forces that cannot always be accurately estimated, for example, imposed loads are set down in CP3, Chapter V, Part 1, to represent the probable loading on a floor. Unsuitable details can lead to high stress concentrations and early failure.

The designer, uses his knowledge of structural mechanics, the codes of practice and practical experience to produce a safe solution to a given problem. There are, of course, various solutions to any given problem and alternatives must be investigated to determine the most economical answer.

Codes of practice (CPs) are drawn up by the British Standards Institution with the guidance of panels of experts from the professional institutions, including engineers from the Universitites, colleges, professions, government authorities and industry. These codes give allowable stresses and safe practices for design and construction. From an analysis of the structural frame, the shears, axial loads and moments on the structural elements are determined. The size of element is chosen so that the allowable stresses, deflections, etc., given in the codes are not exceeded and so that any other relevant requirements such as minimum thickness of material or maximum width/thickness ratios are satisfied.

1.4 LOADING

In general, representative loading has been chosen for the member under consideration. For example, for a plate girder the design can be carried out for a given set of loads. The problem of obtaining the loading could be treated separately from the design of the girder.

In certain cases it has been considered desirable to derive the actual design loads on the member. Here separate load systems may require investigation, and the allowable stresses may change for different load combinations. Examples are the dead, imposed and wind loading on the roof trusses and side stanchions of a building and the dynamic loading on crane girders.

Typical dead-load values for slabs, floor steel, sheeting, partitions, etc., are given in the various problems. For dead, imposed and wind loads, reference should be made to

CP3, Chapter V, Part I, 1967: *Dead and imposed loads*
CP3, Chapter V, Part 2, 1970: *Wind loads*

published by the British Standards Institution.

Dynamic loading for crane girders is given in Part 1 of the code, as well as imposed loads for all types of floors and roofs. Dynamic loading is also given in Chapter 3 of BS 449 1969, Part 2. The method of deriving the wind loading is given in Part 2 and this has been used in some of the worked problems.

1.5 DETAILING

The final chapter of this book deals with the detailing of structural steelwork.

In the earlier chapters, sketches in design problems are made to show loading and features pertinent to the design. At the end of most worked problems a sketch is given showing the basic design information such as the overall size, the span of the member, the sizes of sections or plates, the stiffener spacing, drilling and welding data, etc. These sketches are used by the draughtsmen to produce the working drawings. The detail drawings show all the information for fabrication and, with the general arrangement and marking plans, give the information for erection.

The draughtsman must know the acceptable methods of showing steelwork, the scales to be used and the methods of specifying members, plates, bolts, welding, etc. He must also be able to draw standard details and he must have a knowledge of the methods of fabrication and erection.

2

Connections

Connections join individual elements together to form the structural frame, for example, floor beams to stanchions in a building. Bolts and welding are used to make connections. Riveted connections are virtually obsolete.

2.1 BLACK BOLTS AND CLOSE-TOLERANCE BOLTS

2.1.1 Direct-Shear and Tension Joints

Bolts are arranged in single-shear or double-shear joints as shown in Figure 2.1(a), (b) and (c). Figure 2.1(d) shows a splice joint connecting two double angles where one group of bolts is in single shear and the other is in double shear. A joint with the bolts in tension is shown in Figure 2.2.

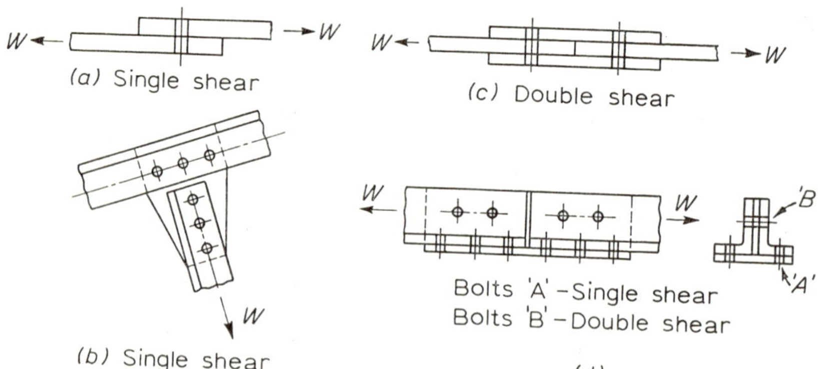

Figure 2.1 Bolts in single- and double-shear joints

A shear joint can fail in the following four ways:

1 Shearing of the bolt shank
2 Bearing failure of plate or member connected
3 Tension failure of plate or member connected
4 Shearing of plate ends due to insufficient edge distance

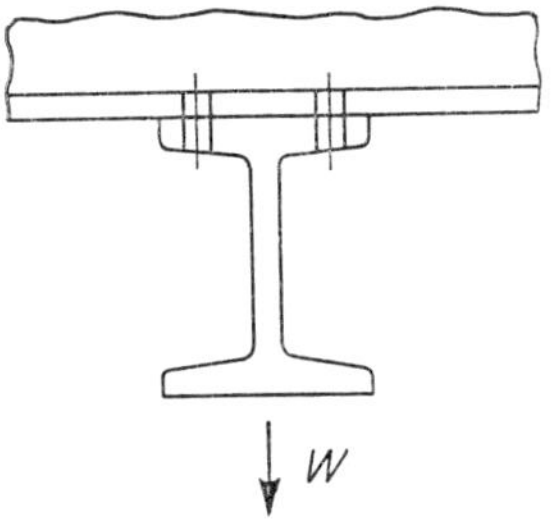

Figure 2.2 Bolts in a tension joint

The provision of sufficient bolts of a suitable diameter prevents failure of the joint by modes 1 and 2. To prevent failure by mode 3, tension members are designed for the net section after deducting holes. A table of minimum edge distances is given in BS 449 and mode-4 failure will not occur if these distances are used.

The shear stress in the bolts is given by

$$\text{Single shear:} \qquad \text{Shear stress} = \frac{\text{Load } W}{\text{No. of bolts} \times \text{Area of shank}}$$

$$\text{Double shear:} \qquad \text{Shear stress} = \frac{\text{Load } W}{2 \times \text{No. of bolts} \times \text{Area of shank}}$$

The shank area for bolts is determined from the following considerations.

Black bolts. These are manufactured from rolled bars and are used in holes of 2 mm clearance. The nominal diameter of the bolt is used in design calculations.

Close-tolerance bolts. These are accurately manufactured and used in holes drilled to a tolerance of +0.15 mm and −0 mm. The bolt diameter is used in design. Higher stresses are permitted than with black bolts.

Tolerances for holes are given in clause 59 of BS 449.

The bearing stress is given by

$$\text{Bearing stress} = \frac{\text{Load } W}{\text{Area of contact of the plate} \times \text{No. of bolts}}$$

where

Area of contact = Diameter of bolt × Plate thickness

When the bolts are in double shear, the bearing is termed enclosed bearing. The allowable stress in single bearing is 20% lower than the allowable enclosed bearing stress.

For bolts in tension, the stress is calculated on the net area at the root of the thread. The net areas for bolts are given in Tables 1 and 2 of the Appendix.

The safe stresses for bolts are given in Table 20 of BS 449. Tables giving safe values for bolts in shear, bearing and tension can be made up and these are very useful in design. The safe values for bolts are given by the following expressions:

Single shear	= Nominal area × Permissible stress
Double shear	= 2 × Single-shear value
Enclosed bearing	= Nominal diameter × Plate thickness × Permissible stress
Bearing–single shear	= 80% of enclosed bearing value
Tension	= Net area × Permissible stress

Safe load values for close-tolerance and black bolts are given in the Appendix, and in the *Handbook on Structural Steelwork*.

The minimum pitch for bolts should not be less than $2\frac{1}{2}$ times the nominal diameter. The minimum edge distance from the centre of the hole to the edge of the plate is given in Table 21 of BS 449, part of which is given below in *Table* 2.1.

Table 2.1 (EXTRACT FROM TABLE 21, BS449)
EDGE DISTANCES OF HOLES

Diameter of hole (mm)	*Distance to sheared or hand-flame-cut edge* (mm)	*Distance to rolled, machine-flame-cut, sawn or planed edge* (mm)
26	42	36
24	38	32
22	34	30
20	30	28
18	28	26
16	26	24

2.1.2 Examples

1 A double-cover butt joint is shown in Figure 2.3. Calculate the strength of the joint by determining the strength of the cover plates, bolts and tension member.

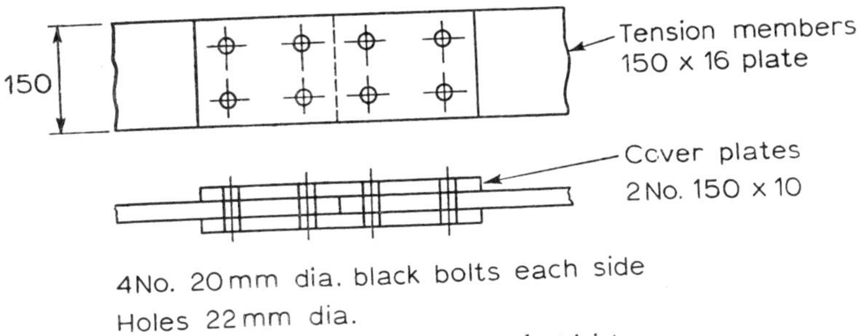

Figure 2.3 A double-cover butt joint

Permissible stresses:

Plates in tension (table 19, BS 449)	$p_t = 155\ \text{N/mm}^2$
Black bolts in shear (table 20, BS 449)	$p_t = 80\ \text{N/mm}^2$
Black bolts in bearing (enclosed) (table 20, BS 449)	$p_t = 200\ \text{N/mm}^2$

Strength of the bolts:

Double-shear value $= (\pi \times 20^2 \times 2 \times 80)/(4 \times 10^3) = 50.2$ kN

Bearing on 16 mm plate $= (20 \times 16 \times 200)/10^3 = 64$ kN

This is the same as the single bearing on 2 No. 10 mm thick cover plates.

Strength of bolts $= 4 \times 50.2 = 200.8$ kN

Strength of the tension member:

Deduct two No. 22 mm dia. holes from the cross section
Strength of member $= [150-(2\times 22)]\ 16 \times 155/10^3 = 263$ kN

Strength of cover plates $= (20/16) \times 263 = 329$ kN
The strength of the joint is 200.8 kN, the strength of the bolts in shear.

2 Figure 2.4 shows a hanger connection with black bolts in shear and tension. Determine the capacity of the connection with respect to the bolts.

Black bolts in shear—20 mm dia.:

Double-shear value = 50.2 kN (see Example 1)
Enclosed bearing on the tee stalk. The thickness of this stalk is the web thickness of $457 \times 152 \times 74$ kg/m UB, 9.9 mm
Bearing value $= (9.9 \times 20 \times 200)/10^3 = 39.6$ kN
Check single bearing on two legs of $64 \times 51 \times 6.2$ angle
Single-bearing value $= (2 \times 6.35 \times 20 \times 0.8 \times 200)/10^3 = 41.3$ kN

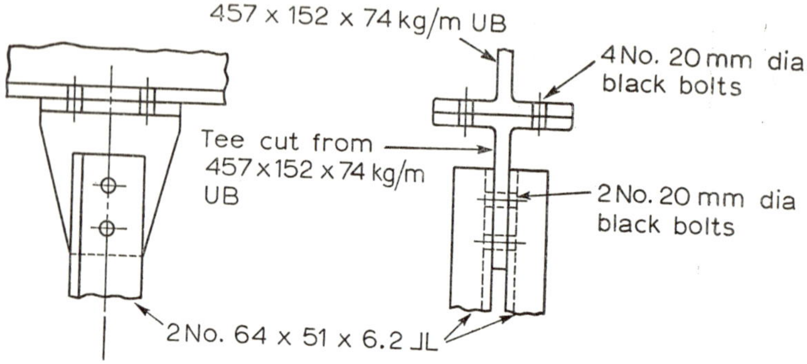

Figure 2.4 A hanger connection with black bolts in shear and tension

Minimum value of bolt = 39.6 kN
Strength of bolts in shear = 79.2 kN

Bolts in tension—20 mm dia.
Area at root of thread, i.e. Net area = 245 mm^2
Permissible stress—bolts of strength grade 4.6 = 130 N/mm^2
Strength of bolts = $(4\times245\times130)/10^3$ = 127.5 kN
The capacity of the connection is 79.2 kN.
Note: The tension value of 20 mm diameter black bolts can be read off *Table* 2 of the Appendix.

3 Determine the strength of the double splice shown in Figure 2.5. The necessary information on members and bolts is given in the figure.

Strength of bolts (two bolts in double shear, four in single shear):

Double-shear value = $(\pi\times20^2\times80\times2)/(4\times10^3)$ = 50.2 kN
Enclosed bearing = $(20\times10\times200)/10^3$ = 40 kN

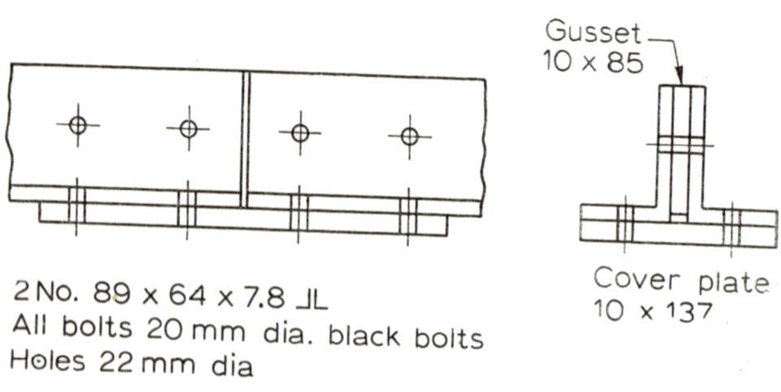

Figure 2.5 Double splice

Single bearing on two angle legs 7.93 mm thick
$= (2\times7.93\times20\times0.8\times200)/10^3 =$ 50.8 kN

Single-shear value $=$ 25.1 kN

Single bearing on 7.93 mm thick angle leg $=$ 25.4 kN

Strength of bolts $= (2\times40)+(4\times25.1)$ $=$ 180.4 kN

Note: The safe values for the bolts are given in *Table* 2 of the Appendix.

Strength of angles (gross area from steel tables $=$ 1137 mm^2 per angle):

Deduct four No. 22 mm diameter holes

Strength of angles
$= [(2\times1137)-(4\times7.93\times22)]\ 155/10^3$ $=$ 244 kN

Strength of gusset and cover plate
$= [85+137-(3\times22)]\ 10\times155/10^3$ $=$ 242 kN

Strength of joint $=$ 180.4 kN

2.1.3 Eccentric Connections

There are two types of eccentrically loaded connections:

1 Bolt group in direct shear and torsion
2 Bolt group in direct shear and tension

These connections are shown in Figure 2.6.

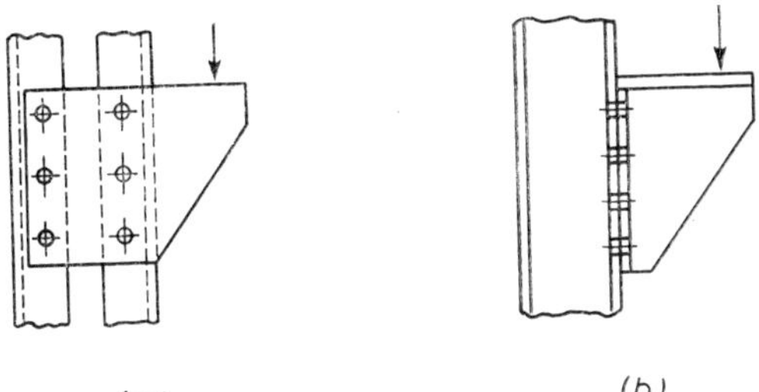

Figure 2.6 Eccentrically loaded connections. (a) Bolt group in direct shear and torsion; (b) Bolt group in direct shear and tension

In the first type of connection, the moment is applied in the plane of the connection and tends to rotate the side plates about the centre of gravity of the bolt group. A linear variation of loading due to moment is assumed with the bolt farthest from the centre of gravity of the group carrying the greatest load. The direct shear is divided equally between the bolts.

Consider the group of bolts shown in Figure 2.7. Bolts in the group A, B, etc., are at a distance r_1, r_2, etc., from the centre of gravity of the group. The coordinates of each bolt are (x_1, y_1), (x_2, y_2), etc. Let the force due to the moment on bolt A be F_T. This is the force on the

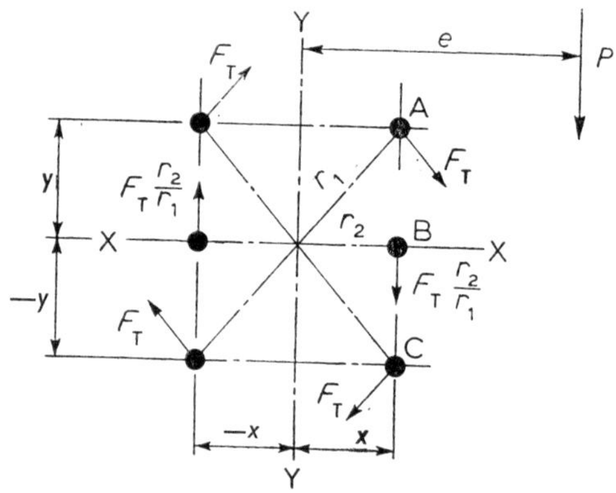

Figure 2.7

bolt farthest from the centre of rotation. Then the force on any bolt r_2 from the centre of rotation is $F_T r_2/r_1$ and so on for all the other bolts in the group. The moment of resistance of the bolt group is given by

$$M_R = F_T r_1 + F_T \frac{r_2}{r_1} r_2 + \ldots$$

$$= \frac{F_T}{r_1}(r_1^2 + r_2^2 + \ldots)$$

$$= \frac{F_T}{r_1} \Sigma r^2$$

$$= \frac{F_T}{r_1}(\Sigma x^2 + \Sigma y^2)$$

$$= \text{applied moment} = P \times e$$

The load F_T due to the moment on the maximum loaded bolt A is given by

$$F_T = \frac{Per_1}{\Sigma x^2 + \Sigma y^2}$$

The load F_S due to direct shear is given by

$$F_S = \frac{P}{\text{No. of bolts}}$$

The resultant load F_R on bolt A can be found graphically as shown in Figure 2.8(a). An algebraic formula can be derived by referring to Figure 2.8(b).

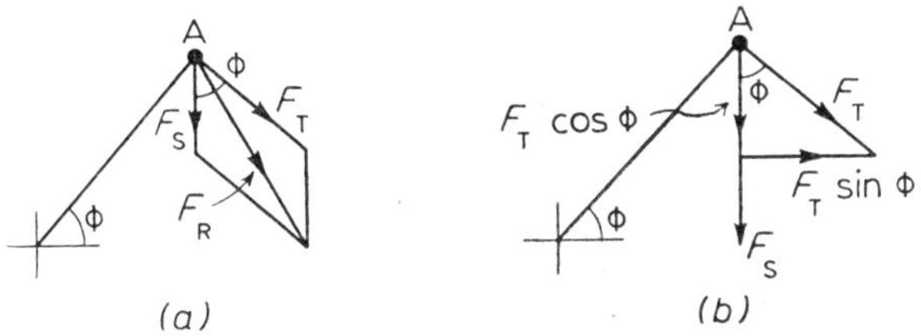

Figure 2.8

Resolve the load F_T vertically and horizontally to give:

Vertical load on bolt A $= F_S + F_T \cos\phi$

Horizontal load on bolt A $= F_T \sin\phi$

Resultant load on bolt A $F_R = [(F_T \sin\phi)^2 + (F_S + F_T \cos\phi)^2]^{1/2}$

$$F_R = (F_S^2 + F_T^2 + 2F_S F_T \cos\phi)^{1/2}$$

Using the maximum load on the bolt, the size required can be determined.

In the second type of joint the bolts are in shear and tension. A bracket subjected to a load P at an eccentricity e is shown in Figure 2.9(a). The centre of rotation is assumed to be at the bottom bolt of the group. The loads vary linearly as shown in (b).

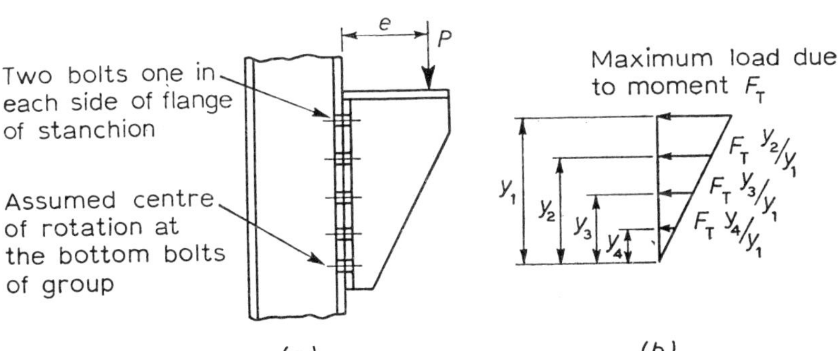

Figure 2.9 Bracket subjected to a load P at an eccentricity e

The moment of resistance M_R of the bolt group is given by

$$M_R = 2\left(F_T y_1 + F_T \frac{y_2}{y_1} + F_T \frac{y_3}{y_1} + \ldots\right)$$

$$= \frac{2F_T}{y_1} [(y_1^2 + y_2^2 + \ldots)$$

$$= \frac{2F_T \Sigma y^2}{y_1}$$

$$= \text{applied moment} = Pe$$

The load F_T due to the moment on the maximum loaded bolt is given by

$$F_T = \frac{Pey_1}{2\Sigma y^2}$$

The load F_S due to direct shear is given by

$$F_S = \frac{P}{\text{No. of bolts}}$$

Tensile stress $f_t = F_T/A$

Shear stress $f_S = F_S/A$

where A is the area of the bolt shank.

Maximum combined tensile stress $= f_t/2 + [(f_t/2)^2 + f_S^2]^{1/2}$

Maximum combined shear stress $= [(f_t/2)^2 + f_S^2]^{1/2}$

These stresses should not exceed the permissible tensile and shear stresses for the bolt material. The student should refer to theory of structures for theory of combined stresses.

The combined stress has been checked at the face of the bracket on the shank of the bolt. It may be necessary to check the bolt for tensile stress only, just under the nut on the area at the root of the thread. The shear stress here will be zero.

2.1.4 Examples

1 Check the joint shown in Figure 2.10. All data required are given on the figure.

Moment	M	$= (140 \times 430)/10^3$	$= 60$ kN m
Bolt group:	Σx^2	$= 12 \times 200^2$	$= 48 \times 10^4$
	Σy^2	$= 4(35^2 + 105^2 + 175^2)$	$= 17.1 \times 10^4$

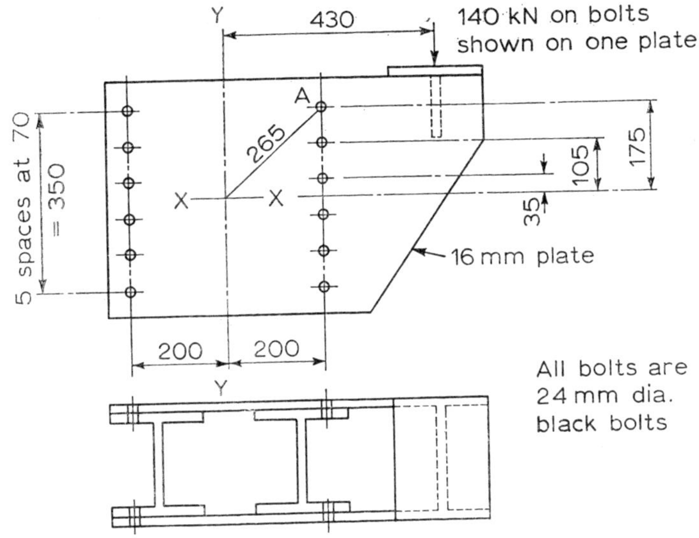

Figure 2.10

$$\Sigma x^2 + \Sigma y^2 = 65.1 \times 10^4$$
$$r_1 = (200^2 + 175^2)^{1/2} \qquad = 265$$

$\cos \phi = 200/265 = 0.755$

Bolt A is the bolt with the maximum load.

Load due to moment $= (60 \times 10^3 \times 265)/(65.1 \times 10^4) = 24.4$ kN

Load due to shear $= 140/12 \qquad = 11.7$ kN

Resultant load on bolt

$$= [24.4^2 + 11.7^2 + (2 \times 24.4 \times 11.7 \times 0.755)]^{1/2}$$
$$= 34.2 \text{ kN}$$

Single-shear value of 24 mm diameter black bolt

$$= \pi \times 24^2 \times 80/(4 \times 10^3) \quad = 36.2 \text{ kN}$$

Single bearing value on 16 mm plate

$$= 24 \times 16 \times 0.8 \times 200/10^3 \ = 61.5 \text{ kN}$$

The joint is satisfactory.

2 Determine the diameter of black bolts required for the joint shown in Figure 2.11.

Moment $\quad M = (170 \times 200)/10^3 = 34$ kN m

Bolt group $\quad \Sigma y^2 = 2(70^2 + 140^2 + 210^2 + 280^2)$

$$= 29.4 \times 10^4$$

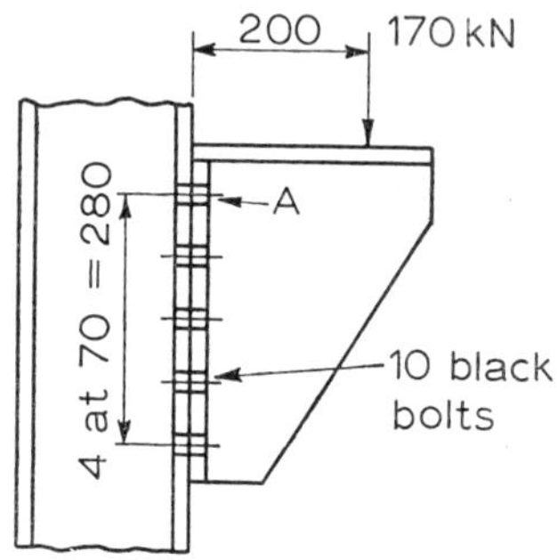

Figure 2.11

For the bolt with the maximum load at A:

Tension due to moment $F_T = (34\times10^3\times280)/(29.4\times10^4)$
$= 32.4$ kN

Shear due to load $F_S = 170/10 = 17$ kN

Check 24 mm diameter black bolts. Refer to *Table* 2 of the Appendix.

Gross area of shank $= 453$ mm^2

Net area at root of thread $= 353$ mm^2

(a) Check the combined stresses on the bolt shank:

Tensile stress $f_t = (32.4\times10^3)/453 = 71.5$ N/mm^2

Shear stress $f_S = (17\times10^3)/453 = 37.5$ N/mm^2

Maximum combined tensile stress

$= 71.5/2+[(71.5/2)^2+37.5^2]^{1/2}$

$= 35.7+41$ $= 76.7$ N/mm^2

Maximum combined shear stress $= 41$ N/mm^2

(b) Check the tensile stress at the root of the thread:

Tensile stress $f_t = (32.4\times10^3)/353$ $= 91.8$ N/mm^2

Permissible tensile stress $= 130$ N/mm^2

Permissible shear stress $= 80$ N/mm^2

The 24 mm diameter black bolts are too large and 22 mm diameter bolts could be used.

3 Check the capacity of the floor beam connection shown in Figure 2.12(a).

Bolts—Group A. 20 mm dia. black bolts.

Single-shear value $= 20^2 \times \pi \times 80/(4 \times 10^3) = 25.1$ kN

Single bearing on 7.93 mm angle leg

$= 20 \times 7.93 \times 0.8 \times 200/10^3 = 25.4$ kN

Shear capacity of bolts $= 4 \times 25.1 = 100.4$ kN

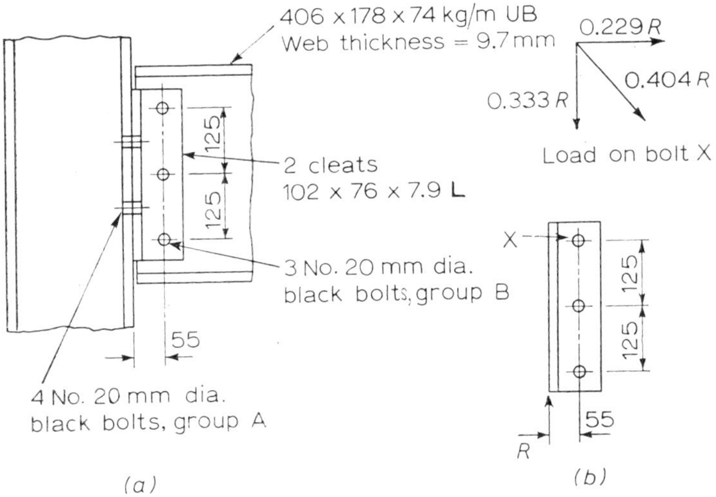

Figure 2.12 Floor beam connection

Bolts—Group B. 20 mm dia. black bolts.

Double-shear value $= 2 \times 25.1 = 50.2$ kN

Enclosed bearing on 9.7 mm thick web of UB

$= (20 \times 9.7 \times 200)/10^3 = 38.8$ kN

The reaction R is an eccentric load on this bolt group as shown in Figure 2.12(b).

Moment $M = 55R$

Bolts $\Sigma y^2 = 2 \times 125^2 = 3.12 \times 10^4$

Shear due to moment $F_T = (55R \times 125)/(3.12 \times 10^4)$

$= 0.221R$ kN

Direct shear $F_S = R/3 = 0.333R$ kN

Resultant shear on bolt $X = [(0.221R)^2 + (0.333R)^2]^{1/2}$

$= 0.4R$ kN

This is equal to the enclosed bearing value 38.8 kN

Reaction $R = 38.8/0.4 = 97$ kN

Group B bolts control the strength of the joint which is 96 kN.

2.2 HIGH-STRENGTH FRICTION-GRIP BOLTS

2.2.1 Design Considerations

High-strength friction-grip bolts consist of high-tensile steel bolts and nuts and hardened steel washers (See Figure 2.13(a)). The bolts are tightened to a predetermined shank tension so that the clamping force will transmit the force in the connected members by friction (see Figure 2.13(b)). The bolts do not act in shear or bearing as in ordinary bolted connections.

The bolts may be used to resist direct tension. The shank tension holds the plates together until it is exceeded by the applied tension.

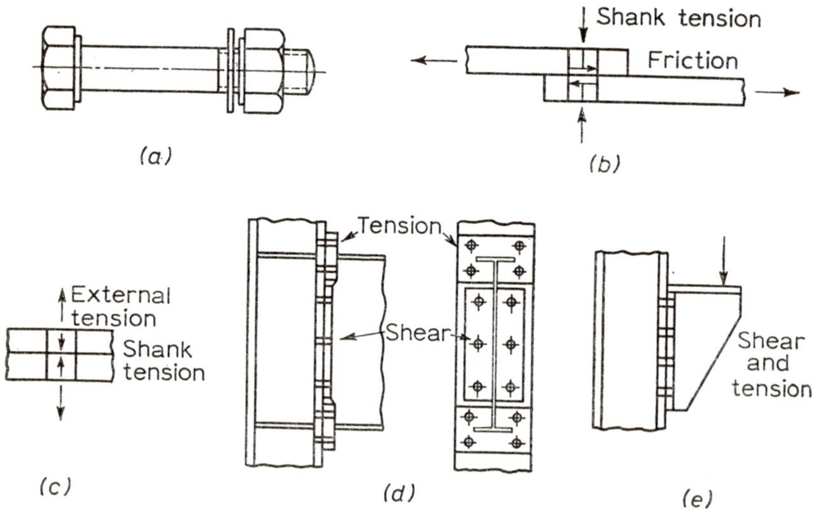

Figure 2.13

A beam-to-stanchion moment connection is shown in Figure 2.13(d), where the flange connection bolts are in tension and the web bolts are in shear. In the bracket connection in Figure 2.13(e) the top bolts are in tension and shear.

High-strength friction-grip bolts are used extensively for field connections. There is no slip or movement between the connected parts, hence this type of joint is useful where rigid connections are required.

The surfaces in contact must be free from oil, grease, scale and paint. Care must be taken to see that the bolts are tightened up to give the required tension in the shank, otherwise slip will occur and the joint will act as an ordinary bolted joint. Methods used to achieve the correct shank tension or show that it has been reached are:

1 Part turning. Here the nut is screwed up tight and then tightened a further $\frac{1}{2}$ to $\frac{3}{4}$ of a turn depending on the bolt length and diameter
2 Torque control. A calibrated power operated or hand operated torque wrench is used to deliver a stated torque to the bolt
3 Patented load-indicating bolts and washers. These have projections that flatten out when the required tension is reached

For connections subject only to shear in the plane of the friction forces, the safe load P on a joint is given by

$$P = \mu m n P_f / \phi$$

where

μ = slip factor or coefficient of friction. This is the ratio of load required to produce slip to the total shank tension in the bolt. $\mu = 0.45$ for properly prepared surfaces

m = number of contact surfaces of plates in the joint. A single-shear joint has one contact surface, a double-shear joint has two

n = number of bolts in the joint

ϕ = load factor. This is the numerical value by which the load which would cause slip in a joint is divided to give the safe working load on the joint. For structures designed to BS 449, $\phi = 1.4$ for dead and imposed loads, and $\phi = 1.2$ for wind load

P_f = proof load for the bolt given in Table 3 of BS 4395, Part 1

For example, the safe working load in single shear for a 24 mm diameter bolt, the proof load of which is 207 kN, is given by

$$\text{Safe load} = (0.45 \times 207)/1.4 = 66.5 \text{ kN}$$

Values of proof load from BS 4395, Part 1 and safe shear load, with no wind and including wind, are given in *Table* 3 of the Appendix.

For connections subject only to external tension in the direction of the bolt axis, the maximum permissible external tension on any bolt must not exceed 0.6 of the proof load. The safe values for tension only on the bolts are listed in *Table* 3 in the Appendix.

Finally the bolts in a connection may be subject to external tension in addition to shear. The externally applied tension in the direction of the bolt axis reduces the clamping action of the bolt. To allow for this effect, the safe load for the bolt is calculated using a reduced proof load obtained by subtracting from the listed proof load without tension a value of 1.7 times the externally applied tension. The clamping action of the bolt ceases when the externally applied tension reaches 0.6 of the proof load. For example, check that a 20 mm diameter bolt can safely sustain a shear of 30 kN in addition to a direct tension of 20 kN.

Proof load	= 144 kN	
Reduced proof load	= 144−(1.7×20)	= 110 kN
Safe shear	= (0.45×110)/1.4	= 35.4 kN

The bolt is satisfactory.

2.2.2 Examples

1 Design the moment and shear connection between a floor beam and stanchion in a steel frame building. The following data are given:

Floor beam	610×229×140 kg/m UB
Stanchion	254×254×132 kg/m
Moment	280 kN m
Shear	460 kN

The moment and shear are due to dead and imposed load. Set out the joint as shown in Figure 2.14.

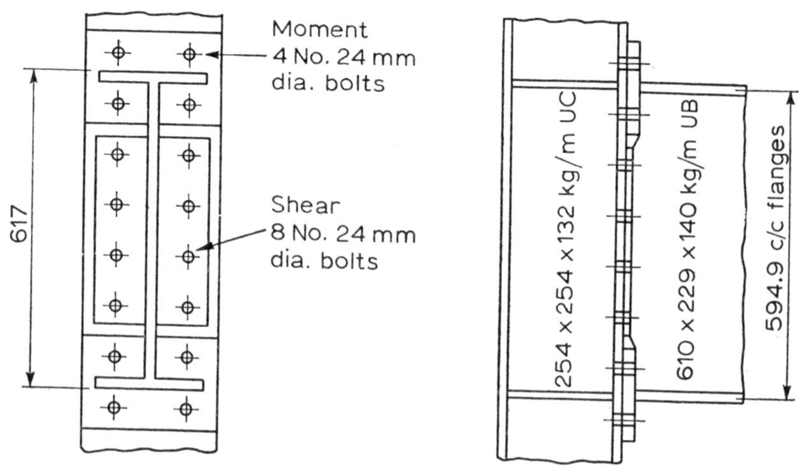

Figure 2.14

Moment—taken by flange connections in tension.

Flange force = 280/0.594 = 472 kN

For 4 No. 24 mm dia. bolts, tension value = 124×4 = 496 kN

Shear—taken by web connection.

For 8 No. 24 dia. bolts, shear value = 8×66.5 = 532 kN

The joint as set out is adequate.

2 Calculate the strength of the splice shown in Figure 2.15 with respect to shear and moment. All relevant details are given on the figure. All the bolts are high strength friction grip.

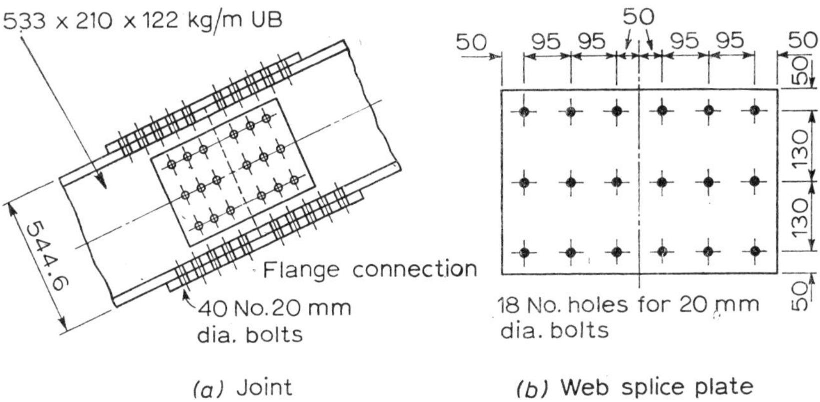

Figure 2.15

Moment connection—flange splice plates and bolts resist moment.

Shear value of one 20 mm dia. bolt = 46.2 kN

Moment of resistance = $(10\times46.2\times544.6)/10^3$ = 252 kN m

Shear connection—web splice resists all the shear, say, F kN

The joint is eccentrically loaded as shown in Figure 2.16.

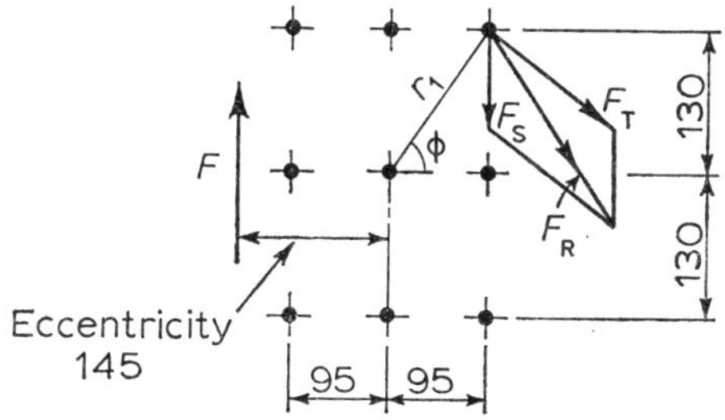

Figure 2.16

Bolt group $\Sigma x^2+\Sigma y^2$		$= (6\times 95^2)+(6\times 130^2)$	$= 15.5\times 10^4$
	r_1	$= (95^2+130^2)^{1/2}$	$= 161$
Load due to moment	F_T	$= (145F\times 161)/(15.5\times 10^4)$	$= 0.151F$ kN
Load due to shear	F_S	$= F/9$	$= 0.111F$ kN
Resultant load	F_R	$= [(0.151F)^2+(0.111F)^2+2\times 0.151\times$ $\times 0.111\times 0.59F^2]^{1/2}$	
		$= 0.234\,F$	
		$= 46.2$ kN	
Shear resistance	F	$= 197$ kN	

3 Determine the bolt size required for the bracket loaded as shown in Figure 2.17.

Bolt group	Σy^2	$= 2(100^2+200^2+300^2+400^2)$	$= 60\times 10^4$
Tension due to moment in top bolt			
	F_T	$= (200\times 300\times 400)/(60\times 10^4)$	$= 40$ kN
Direct shear	F_S	$= 200/10$	$= 20$ kN

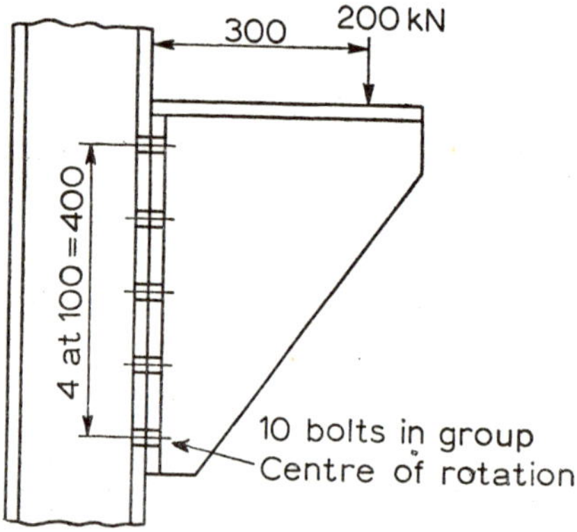

Figure 2.17

Try 20 mm diameter high-strength friction-grip bolts.

Proof load	$= 144$ kN	
Reduced proof load	$= 144-(1.7\times 40)$	
	$= 144-68$	$= 76$ kN
Shear value	$= (0.45\times 76)/1.4$	$= 24.4$ kN

This is adequate to resist the direct shear of 20 kN. 20 mm diameter bolts are satisfactory.

2.3 WELDED CONNECTIONS

2.3.1 Types of Weld and Design Data

Two types of welded joints are used, these are butt welds and fillet welds. Some of the types of butt weld are shown in Figure 2.18. The names describe the edge preparation of the plates. The strength of a butt weld is the same as the strength of the parent plate if the

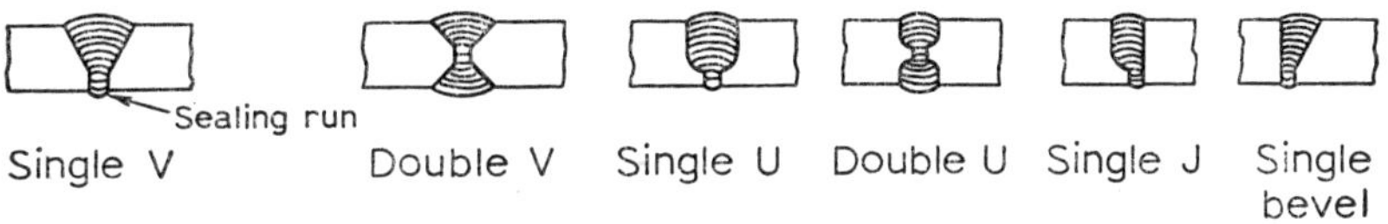

Figure 2.18 Butt welds

weld can be made from both sides or a sealing run can be placed on the side away from where the weld is laid down. The stresses are the same as for the parent metal.

Clause 54(b) of BS 449 states:

> 'Throat thickness of incomplete penetration butt welds. (BS 1856, Clause 13). Unsealed single V, U, J, or bevel butt welds, and other butt welds which are welded from one side only and are not full penetration welds, shall have a throat thickness of at least seven-eighths of the thickness of the thinner part joined. Evidence shall be produced to show that this throat thickness has been achieved. For the purpose of stress calculation and to allow for the effects of the eccentricity of the weld metal relative to the parts joined, a nominal throat thickness, not exceeding five-eighths of the thickness of the thinner part joined, shall be taken.'

All important butt welds in tension should be fully tested by radiography or ultrasonics to show that they are free from major defects. Some of the more important defects in butt welds are incomplete penetration, slag inclusions, gas pockets and cracking from various causes in the weld or heat affected zone of the adjacent metal.

Two different thicknesses of plate should not be butt welded directly together. The thicker plate should be given a taper of 1 in 5 to meet the thinner plate. The edge preparation for butt welds should be accurately made and may be machined, flame or plasma arc cut. With correct edge preparation and a properly controlled welding process butt welds are perfectly reliable.

Fillet welds are shown in Figure 2.19(a). The size of fillet weld is specified by the leg length as shown in Figure 2.19(b), while the strength of the weld is calculated on the throat thickness. For 90° fillet welds

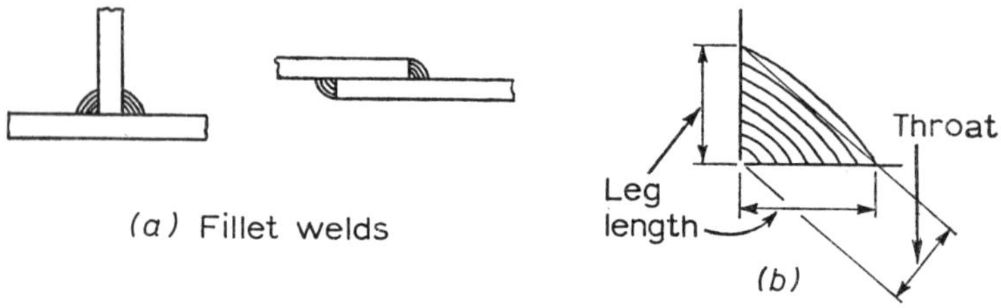

Figure 2.19

the throat length is 0.7×leg length. The allowable stress in fillet welds for grade 43 steel is 115 N/mm². Thus the strength of a 6 mm fillet weld

$$= 6 \times 0.7 \times 115 = 483 \text{ N/mm}$$

A table of the strengths of fillet welds is given in the Appendix in *Table* 4.

For intermittent fillet welds, refer to clause 54(c) of BS 449. The distance between effective lengths of consecutive welds should not exceed

(a) 16 times the thickness of thinner part for a member in compression
(b) 24 times the thickness of the thinner part or 300 mm whichever is the less for a member in tension

In modern fabrication, automatic processes are finding increasing use, and continuous fillet weld on web-to-flange welds of plate girders is standard practice. The continuous weld also has better fatigue properties and resistance to brittle fracture. The effective length of a fillet weld is taken as the actual length minus two times the weld size. This allows for craters formed at the ends of the weld where the arc is struck and broken.

Minimum leg lengths of fillet weld depend on the thickness of plate welded, and these are laid down in table 1 of BS 1856 from which the present *Table* 2.2 is made up.

Table 2.2 MINIMUM LEG LENGTH OF SINGLE RUN FILLET WELD FOR MANUAL WELDING

Thickness of thicker plate		*Minimum size of fillet weld*	
Over $\frac{3}{8}$ in up to $\frac{3}{4}$ in	9.5 mm–19 mm	$\frac{3}{16}$ in	4.8 mm
Over $\frac{3}{4}$ in up to $1\frac{1}{4}$ in	19 mm–31.8 mm	$\frac{1}{4}$ in	6.4 mm
Over $1\frac{1}{4}$ in	31.8 mm	$\frac{5}{16}$ in	7.9 mm

2.3.2 Eccentric Connections

Two types of eccentric connection are shown in Figure 2.20:

1 Torsion joint, with the load in the plane of the weld
2 Bracket connection

In both cases, the fillet welds making the joint are in shear due to direct load and moment.

In the first type of joint the weld is in direct shear and torsion. The eccentric load tends to cause rotation about the centre of gravity of the weld group. It has been customary to assume that the force in the weld due to torsion is directly proportional to the distance from the centre of gravity, thus this force is found from a torsion formula. The direct shear is taken to be uniform throughout the weld.

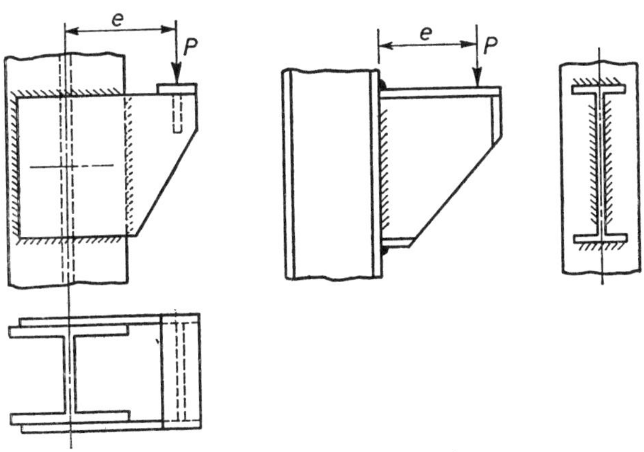

Figure 2.20 Eccentric connections

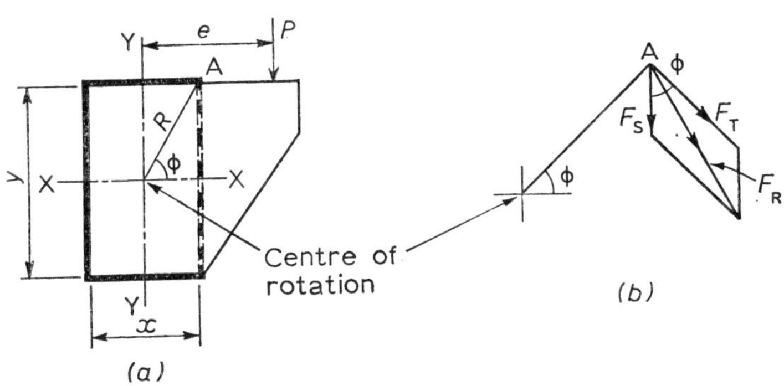

Figure 2.21

A rectangular weld group is shown in Figure 2.21(a) where the eccentric load P is taken on one plate. The weld is assumed to be of unit leg length and uniform throughout.

Direct shear $= F_S = P/\text{Length of weld} = P/2(x+y)$

Shear due to torsion $= F_T = PeR/I_p$

where I_p = polar moment of inertia $= I_{XX}+I_{YY}$

$$I_{XX} = 2y^3/12+2x(y/2)^2$$

$$I_{YY} = 2x^3/12+2y(x/2)^2$$

$$R = \tfrac{1}{2}(x^2+y^2)^{1/2}$$

The heaviest loaded length of weld is that farthest from the centre of rotation.

The resultant shear on the weld at A, the position of maximum shear, is given by

$$F_R = (F_S^2+F_T^2+2F_SF_T\cos\phi)^{1/2}$$

The weld size can be selected from *Table* 4 in the Appendix. If the weld group is not symmetrical, the centre of gravity must be found first. Then the polar moment of inertia of the group can be found and the problem solved as above.

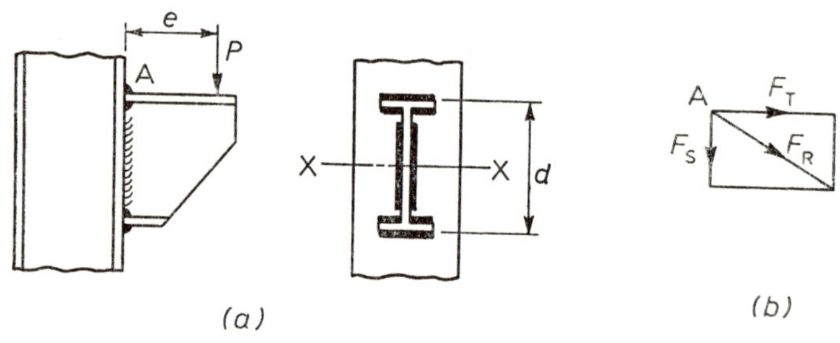

Figure 2.22 Bracket connection

A bracket connection is shown in Figure 2.22(a). The load in the weld can be found by using the beam bending formula. Assume that the weld is of unit leg length.

Direct shear force $F_S = P/\text{Length of weld}$

Load due to moment $F_T = Ped/2I_{XX}$

Resultant load $F_R = (F_T^2+F_S^2)^{1/2}$

where I_{XX} = moment of inertia of the weld group about the XX axis. Note that the resultant load on the weld is treated as a shear across the throat of the weld. The size of weld can be selected from the table of weld strengths given in the Appendix (*Table* 4). With heavily loaded brackets, full strength butt welds between the bracket flanges and stanchion flange may be necessary to carry the load. Single V or J edge preparation is required for the bracket flange in these cases.

2.3.3 Examples of Welded Connections

1 Design the weld for the direct-shear connection for the angle shown in Figure 2.23(a) where the load acts on the centroidal axis of the angle.

Use 6 mm fillet weld, strength $= 483$ N/mm

Length required $= (60\times10^3)/483 = 124$ mm

Balance the weld on each side as shown in Figure 2.23(b).

Side X, length $= (43.2/63.5)\times124 = 84.3$ effective$+12 = 98$ mm, say

Side Y, length $= 124-84.3 = 39.7$ effective$+12 = 52$ mm, say

The effective length is determined first, then twice the weld size, i.e. 12 mm is added to give the actual length.

A weld across the end of the angle may be arranged as shown in

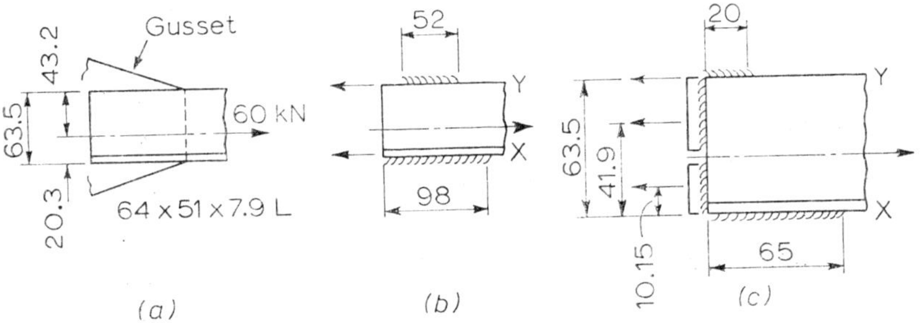

Figure 2.23

Figure 2.23(c). The length of weld on side Y, L_Y, may be found by taking moments about side X. In terms of weld lengths this gives:

$$(L_Y\times63.5)+(43.2\times41.9)+(20.3\times10.15) = (124\times20.3)$$

$$L_Y = 501/63.5 = 7.9 \text{ mm effective}+12 = 20 \text{ mm, say}$$

$$\text{Length on side X, } L_X = 124-63.5-7.9 = 52.6 \text{ mm effective}$$

$$+12 = 65 \text{ mm, say}$$

2 One side plate of an eccentrically loaded connection is shown in Figure 2.24. Determine the maximum shear force in the weld and

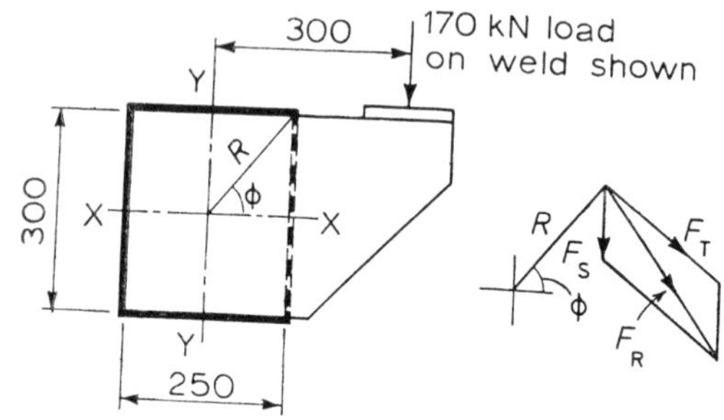

Figure 2.24

select a suitable leg length of fillet weld from the table of weld strengths given in *Table* 4 in the Appendix. The weld is of uniform leg length.

Length $L = 2(300+250) = 1100$

Inertias $I_{XX} = (2\times250\times150^2)+(2\times300^3)/12$

$= (11.25+4.5)10^6 = 15.75\times10^6$

$I_{YY} = (2\times300\times125^2)+(2\times250^3)/12$

$= (9.35+2.6)10^6 = 11.95\times10^6$

Polar $I_p = (15.75+11.95)10^6 = 27.7\times10^6$

$R = (150^2+125^2)^{1/2} = 195$

$\cos\phi = 125/195 = 0.64$

Direct shear $F_S = 170\times10^3/1100 = 154.5$ N/mm

Torsion shear $F_T = (170\times300\times195\times10^3)/(27.7\times10^6)$

$= 360$ N/mm

Resultant shear $F_S = (154.5^2+360^2+2\times154.5\times360\times0.64)^{1/2}$

$= 462$ N/mm

6 mm fillet weld, strength 483 N/mm is required.

3 The plate shown in Figure 2.25(a) is welded on three sides only. Find the maximum shear force in the weld and select a suitable fillet weld from the table of weld strengths given in the Appendix. The weld is of uniform leg length.

Find the position of the centre of gravity of the weld by taking moments about side AB. Refer to Figure 2.25(b).

$\bar{x} = 2\times200\times100/700 = 57.1$ mm

$L = 700$ mm

Eccentricity of load $= 292.9$

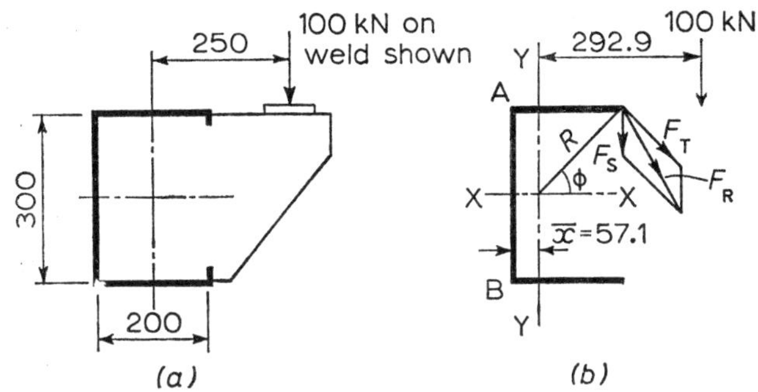

Figure 2.25

Inertia $\quad I_{XX} = (2\times200\times150^2)+300^3/12$
$\qquad = (9+2.25)10^6 \qquad = 11.25\times10^6$
$\quad I_{YY} = (300\times57.1^2)+(2\times200^3)/12+(2\times200\times42.9^2)$
$\qquad = (0.97+1.33+0.72)10^6 \qquad = 3.02\times10^6$

Polar $\quad I_p = (11.25+3.02)10^6 \qquad = 14.27\times10^6$
$\quad R = (142.9^2+150^2)^{1/2} \qquad = 207$
$\quad \cos\phi = 142.9/207 \qquad = 0.69$

Direct shear $\quad F_S = (100\times10^3)/700 \qquad = 143$ N/mm

Torsion shear $\quad F_T = (100\times10^3\times292.9\times207)/(14.27\times10^6)$
$\qquad = 426$ N/mm

Resultant: $\quad F_R = (143^2+426^2+2\times143\times428\times0.69)^{1/2}$

Shear $\qquad = 535$ N/mm.

8 mm fillet weld, strength 644 N/mm is required.

4 Determine the leg length of fillet weld required for the joint shown in Figure 2.26. The web welds are to be $\frac{1}{2}$ the leg length of the flange welds.

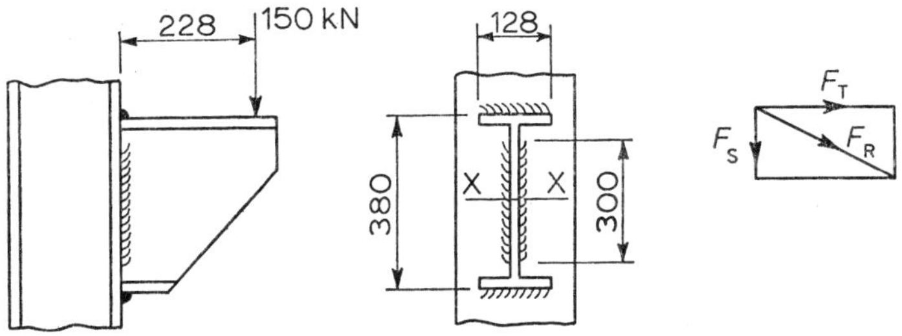

Figure 2.26

Take the flange weld to be of unit leg length.

Length $L = (2\times128)+(2\times\frac{1}{2}\times300)$

$= 256+300 \qquad = 556$

Inertia $I_{XX} = (2\times128\times190^2)+(2\times300^3)/(2\times12)$

$= (9.25+2.25)10^6 \qquad = 11.5\times10^6$

Direct shear $F_S = (150\times10^3)/556 \qquad = 270$ N/mm

Shear from moment

$F_T = (150\times10^3\times228\times190)/(11.5\times10^6)$

$= 565$ N/mm

Resultant shear $F_R = (565^2+270^2)^{1/2} \qquad = 626$ N/mm

Flange weld—8 mm fillet weld, strength $= 644$ N/mm

Web weld—4 mm fillet weld.

2.4 CONNECTIONS FOR STRUCTURAL HOLLOW SECTIONS

2.4.1 Types of Connections

Suitable joints for structural hollow sections have been developed by the British Steel Corporation, Tubes Division. An introduction to some of the more commonly used joints and their design is given here. Reference should be made to the literature available from the British Steel Corporation. Most of the joints are used in truss and lattice girder construction and further consideration will be given to this in Chapter 11.

It is important that joints for structural hollow sections should be made as neat as possible. Untidy joints ruin the pleasing effect of the tubular and box sections. The details for some typical site-bolted joints are shown in Figure 2.27. These make use of flange plates, cleats, brackets and shoes.

Shop-welded truss and lattice girder joints are shown in Figure 2.28. These show a direct butt welded joint and the connection of branch members in either circular hollow sections or rectangular hollow sections. For the joint shown in (c) on the figure, the centroidal axes of the bracing members should meet on the axis of the main chord. Where possible a weld gap of at least 12 mm should be left between the bracings.

The ends of these members, if not meeting other members at a joint, should always be closed off with an end plate to prevent internal corrosion. Unless specially protected, the minimum thickness of material

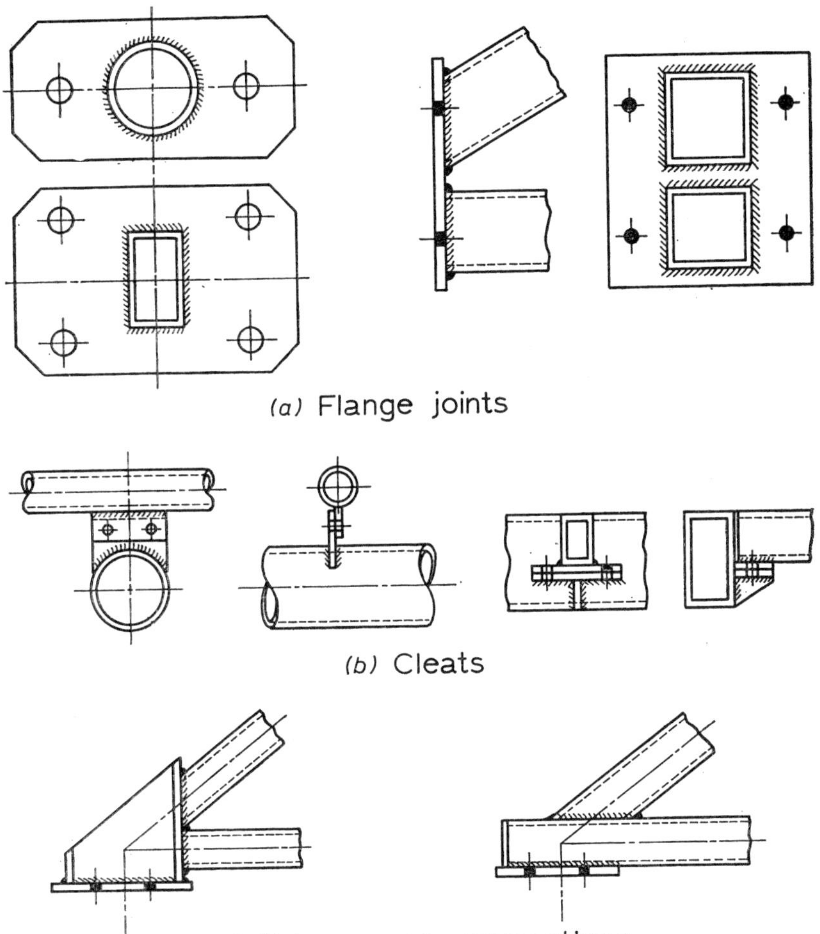

Figure 2.27 Site-bolted joints

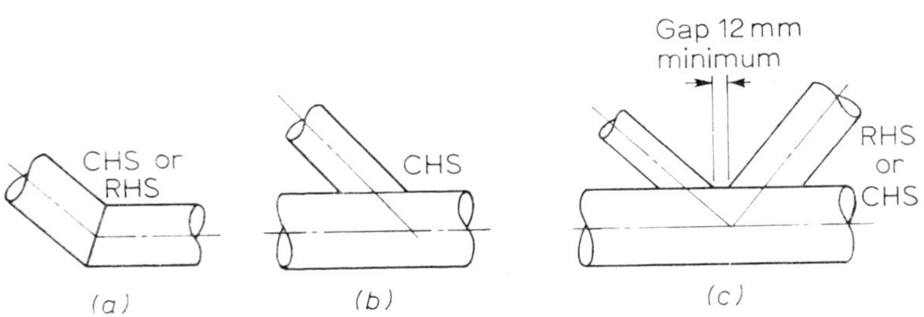

Figure 2.28 Shop-welded truss and lattice girder joints

for a sealed tube or box is 4 mm when the member is exposed to the weather and 3 mm if not so exposed. For an open section, the minimum thicknesses are 8 mm and 6 mm respectively.

2.4.2 Design of Welded Joints

Shop welded joints are the most important type used in construction in structural hollow sections. Reference should be made to BS 449, Clause 53b for welding of tubes. The main data given in this clause are summarised as follows.

1 Tubes welded end to end shall be butt welded. See Figure 2.28(a).

2 The weld connecting the end of a branch tube to the surface of the main tube (see Figure 2.28(b)) where the axes of the tubes intersect at an angle of not less than 30° shall be:

(a) a butt weld throughout, or
(b) a fillet weld throughout, or
(c) a fillet weld in one part and a butt weld in another with a continuous change from one form to the other.

Type (a) may be used whatever the diameters of the tubes joined. If type (a) is not used, then:

Type (b) shall be used when the diameter of the branch is less than $\frac{1}{3}$ of the diameter of the main tube
Type (c) shall be used where the diameter of the branch tube is equal to or greater than $\frac{1}{3}$ of the diameter of the main tube

The stress shall be calculated on an area equal to the throat thickness multiplied by the effective length of the weld.

3 The angle of intersection of tubes should generally not be less than 30°. If it is, the adequacy of the joint should be demonstrated. A large angle of intersection reduces the amount of weld required and so keeps the cost of the joint down.

Butt welds may be used on all thicknesses of sections. Edge preparation is recommended if the thickness is greater than 6 mm. End to end butt welds may be made with or without backing members. The use of a backing member ensures correct alignment of members, a sound root run and full penetration. Butt welds are shown on Figure 2.29(a). The end preparation for a mitred butt weld is shown in (b) on the figure. Backing members have to be specially made to suit the mitre.

The end of a branch is generally shaped to fit the main member and ensure that the gap between the parts to be welded does not exceed about 3 mm. See Figure 2.29(c).

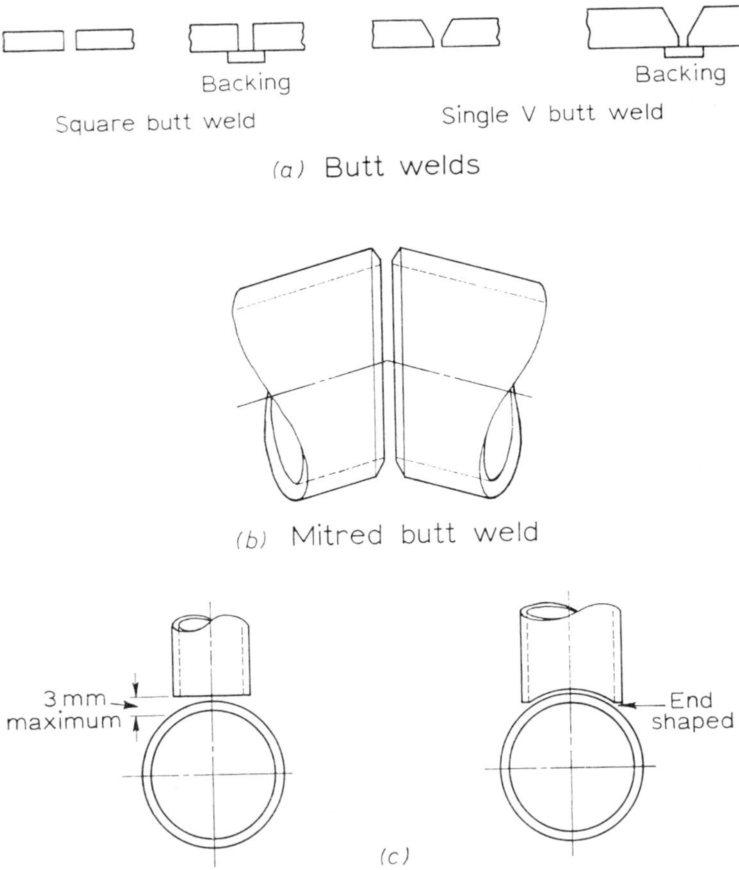

Figure 2.29

For full details on joint manufacture, see the British Steel Corporation publications. Details are given for all types of fillet and butt welds. The fit-up for the joints and types of backing members used are also given.

2.4.3 Length of Intersection

Formulae are given below for calculating the lengths of intersection for:

(a) A tube with another tube
(b) A tube with a flat plate
(c) A rectangular hollow section with another rectangular hollow section or with a flat plate

These calculations appear to be rather laborious. However, tables giving lengths of intersection for members meeting at various angles are given in the technical literature of the British Steel Corporation. A designer would make good use of these tables in practical design.

(a) *Intersection of a tube with another tube*

The determination of the length of the curve of intersection of one tube with another tube or with a flat plate is given in Appendix C of BS 449. See Figure 2.30.

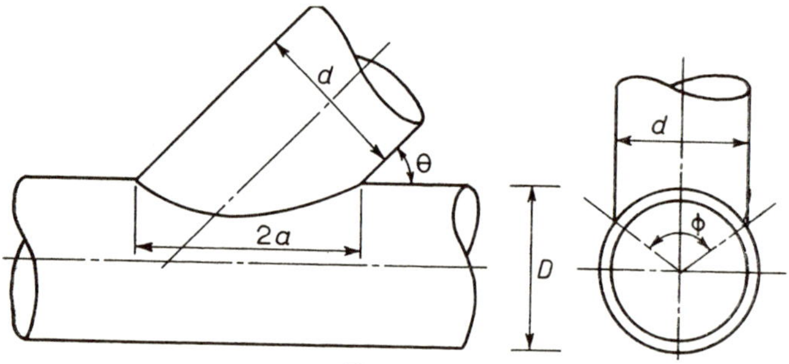

Figure 2.30

The length of the curve of intersection may be taken as:

$$P = a+b+3(a^2+b^2)^{1/2}$$

where

$$a = \tfrac{1}{2}d \operatorname{cosec} \theta$$

$$b = (d/3)\,(3-(d/D)^2)/(2-(d/D)^2) = (D/4)\phi$$

where ϕ is measured in radians and $\sin \phi/2 = d/D$. Calculate the length of the curve of intersection for the following joint.

Size of branch	$d = 60.3$ mm
Size of main member	$D = 88.9$ mm
Angle of intersection	$\theta = 45°$

$$a = \frac{60.3}{2}\times\sqrt{2} \qquad = 42.6$$

$$b = \frac{60.3}{3}\,\frac{3-(60.3/88.9)^2}{2-(60.3/88.9)^2} \qquad = 33.2$$

$$P = 42.6+33.2+3(42.6^2+33.2^2)^{1/2} \quad = 238.1 \text{ mm}$$

(b) *Intersection of a tube with a flat plate* (See Figure 2.31)
Length of curve of intersection may be taken as:

$$P = a+b+3(a^2+b^2)^{1/2}$$

where

$$a = \tfrac{1}{2}d \operatorname{cosec} \theta$$
$$b = \tfrac{1}{2}d$$

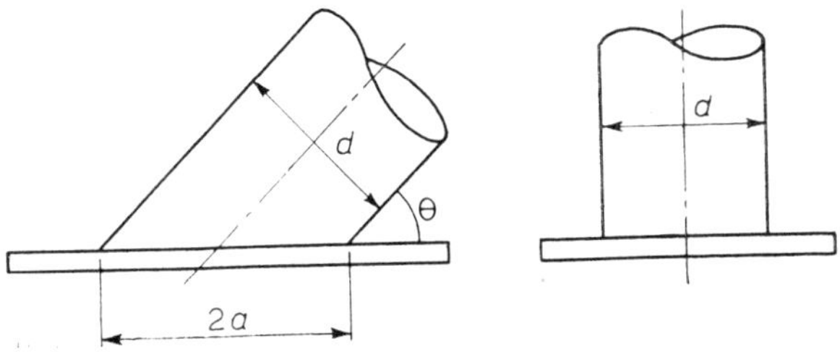

Figure 2.31

Calculate the length of the curve of intersection for the following joint:

Size of tube $d = 76.1$ mm

Angle of intersection $\theta = 45°$

$$a = (76.1/2)\sqrt{2} = 53.9$$
$$b = 76.1/2 \quad = 38.1$$
$$P = 53.9+38.1+3(53.9^2+38.1^2)^{1/2} = 290 \text{ mm}$$

(c) *Intersection of a rectangular hollow section with a flat plate or with another rectangular hollow section* (See Figure 2.32)

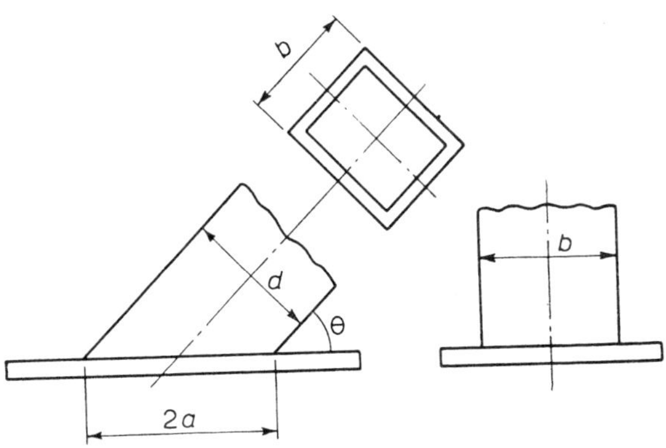

Figure 2.32

Length of intersection:

$$P = 4a+2b$$
$$a = \tfrac{1}{2}d \operatorname{cosec} \theta$$

Calculate the length of intersection for the following joint. The long side is mitred.

Rectangular hollow section—$76.2\times50.8\times3.2$
Angle of intersection $\theta = 50°$
$a = (76.2/2)\ 1.31 \qquad = 49.9$
$P = (4\times49.9)+(2\times50.8) \qquad = 301.2$ mm

2.4.4 Examples of Welded Connections

1 Determine the leg lengths of fillet weld required for the joint shown in Figure 2.33. This is the end joint A of the Warren girder shown in the figure.

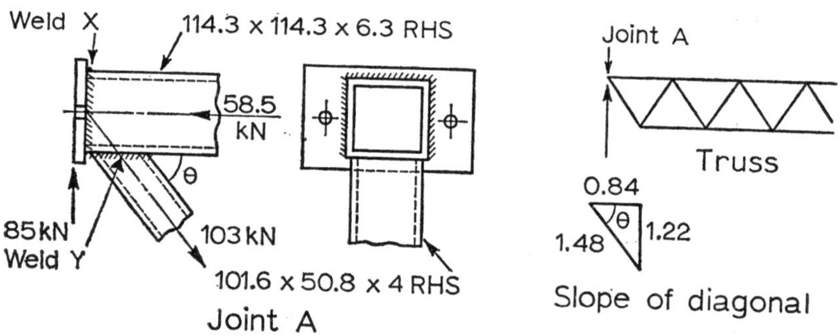

Figure 2.33

Weld X

Length $= 114.3\times4 \qquad = 457.2$ mm
Resultant load on weld $\qquad = 103$ kN
Load $= 103\times10^3/457.2 \qquad = 225$ N/mm

Use minimum weld 4 mm, strength 322 N/mm

Weld Y

Length $= (2\times50.8\times1.48/1.22)+2\times101.6 \qquad = 326.2$ mm
Load $= 103\times10^3/326.3 \qquad = 316$ N/mm

Use 4 mm fillet weld.

2 Determine the leg length of weld required for the joint shown in Figure 2.34. This is an internal joint in a tubular lattice girder.

As the diameter of the branch is greater than $\frac{1}{3}$ diameter of the main tube, a fillet-butt weld or butt weld is required.

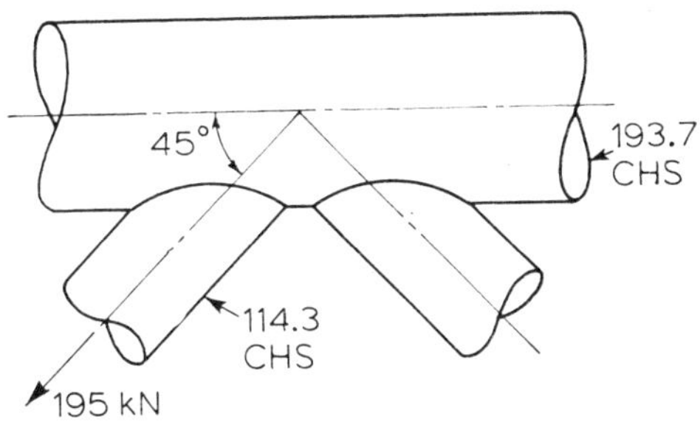

Figure 2.34

Length of curve of intersection. Refer to section 2.4.3(a):

$$a = (114.3/2)\sqrt{2} \qquad = 80.8$$

$$b = \frac{114.3}{3} \times \frac{3-(114.3/193.7)^2}{2-(114.3/193.7)^2} \qquad = 61.2$$

Length $= 80.8+61.2+3(80.8^2+61.2^2)^{1/2} \quad = 445$ mm

Load on weld $= 195\times10^3/445 \quad = 438$ N/mm

6 mm fillet-butt weld required. Strength = 483 N/mm

2.5 PROBLEMS

1 A double channel tension member carrying a load of 550 kN is to be spliced as shown in Figure 2.35.

(a) Determine the number of 20 mm diameter black bolts required to make the splice.
(b) Check the double channel member in tension.
(c) Check the splice plates in tension.

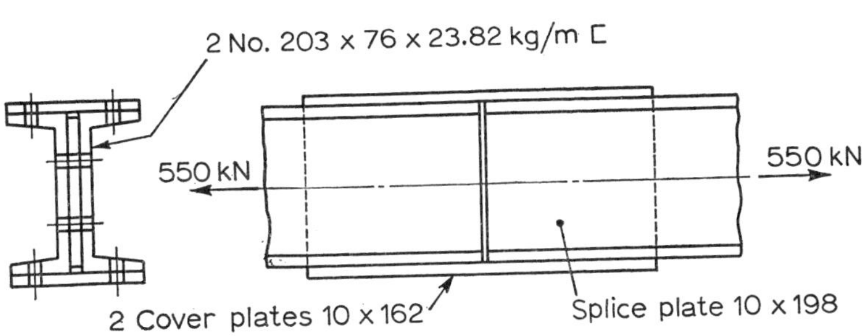

Figure 2.35

Answers (a) 6 bolts in web and 12 in flanges on each side
Strength = 602 kN
(b) Strength = 565 kN
(c) Strength = 604 kN

2 Check that 30 mm diameter close tolerance bolts are satisfactory for the bolted joint shown in Figure 2.36.

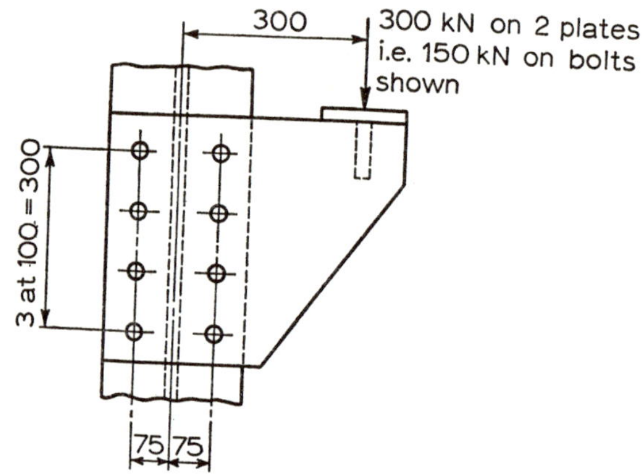

Figure 2.36

Answer Maximum resultant shear on bolt = 62.6 kN.

3 A bolted bracket is shown in Figure 2.37. Determine the stresses in the maximum loaded bolts.

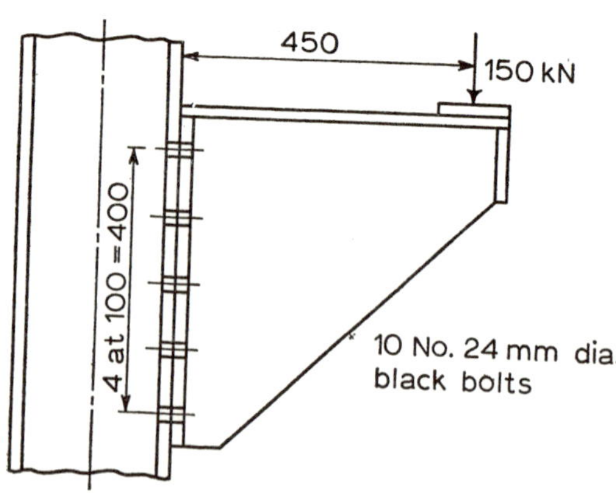

Figure 2.37

Answers (a) Shank
Maximum combined tensile stress = 109.8 N/mm²
Maximum combined shear stress = 60 N/mm²
(b) Root of thread
Tensile stress = 127.5 N/mm²

4 The joint shown in Figure 2.38 has to resist a moment of 150 kN m and a shear of 210 kN. Assuming that the top flange connection resists

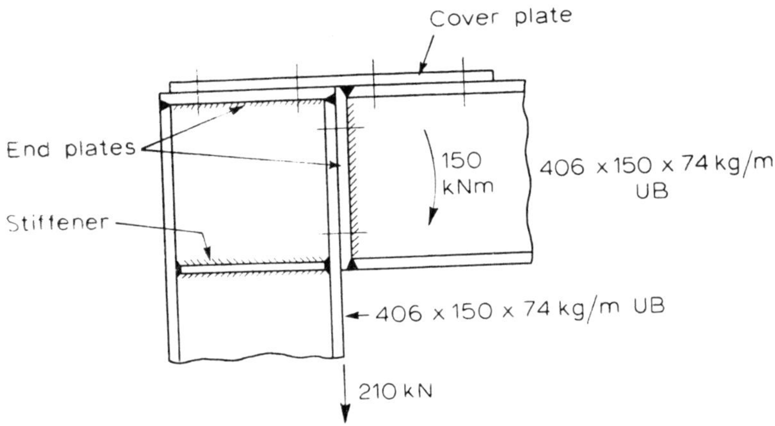

Figure 2.38

the moment and the web connection the shear, determine the number of 20 mm diameter high-strength friction-grip bolts required in the joint.

Answers Moment connection 8 bolts
Shear connection 6 bolts

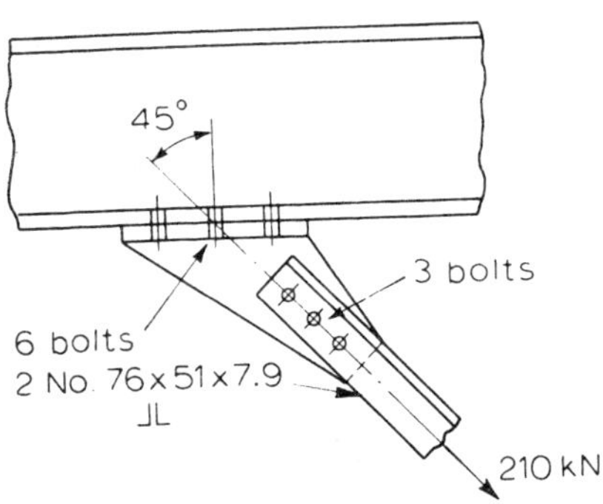

Figure 2.39

5 Check that the joint shown in Figure 2.39 is adequate to carry the load in the tie. All bolts are 20 mm diameter high-strength friction-grip bolts.

6 Determine the leg length of fillet weld required for the eccentric joint shown in Figure 2.40

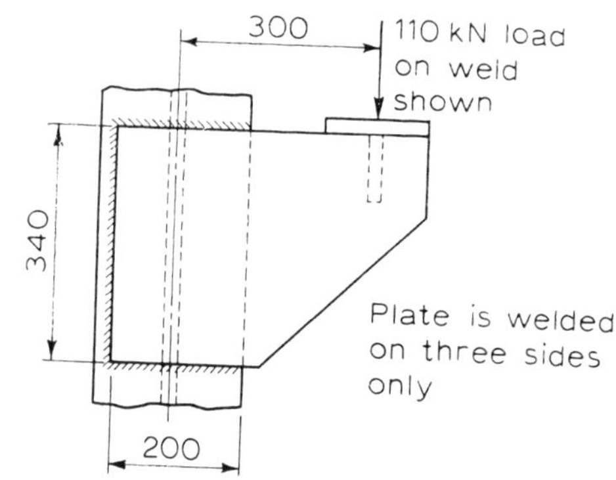

Figure 2.40

Answer 10 mm fillet weld

7 Calculate the safe loads that the brackets shown in Figure 2.41 can carry.

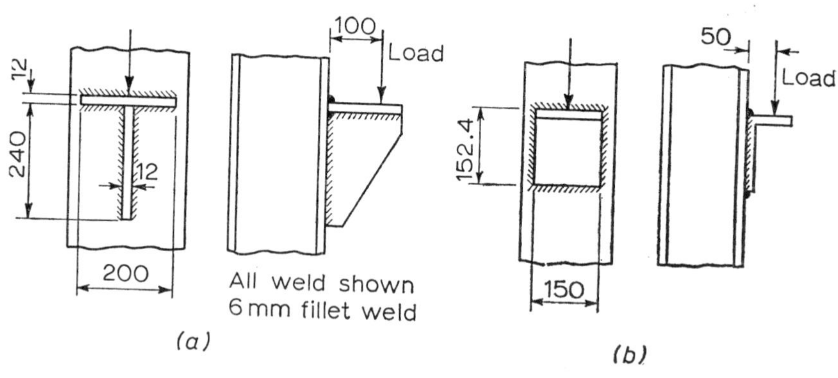

Figure 2.41

Answers (a) 281 kN

(b) 215 kN

8 Determine the leg length of weld required for the joint shown in Figure 2.42.

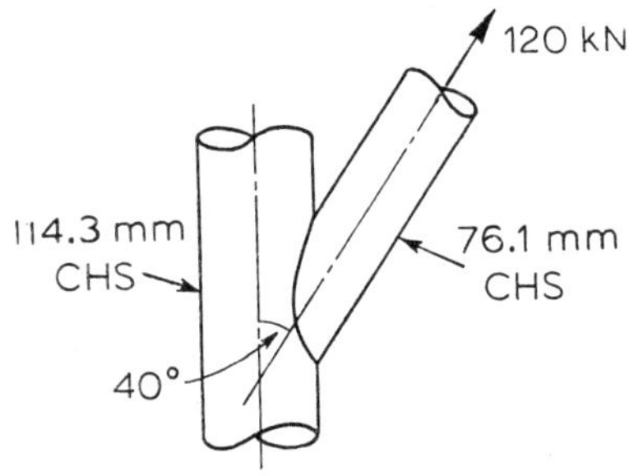

Figure 2.42

Answer 5 mm fillet-butt weld

3

Beams and compound beams

Beams are structural elements carrying lateral loads. The loads are resisted by bending and shear in the beam but local stress conditions and deflection are important considerations.

3.1 DESIGN CONSIDERATIONS

3.1.1 Types

Rolled and formed sections used for beams are shown in Figure 3.1. The compound beam consists of a universal beam and flange plates. This is used where the depth is limited; the flange plates strengthen the universal beam.

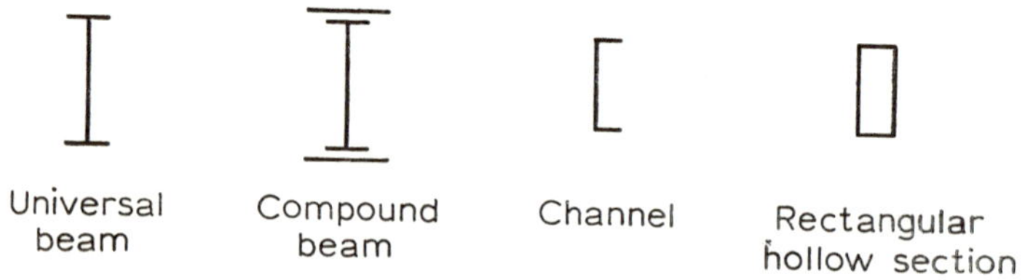

Figure 3.1 Rolled and formed sections used for beams

3.1.2 Beam Loads and Stresses

A beam and loading is shown in Figure 3.2. The loads are applied to the beam through floor slabs, floor beams or stanchions etc., carried by the beam. The loads are due to dead loads of floors, walls, etc.,

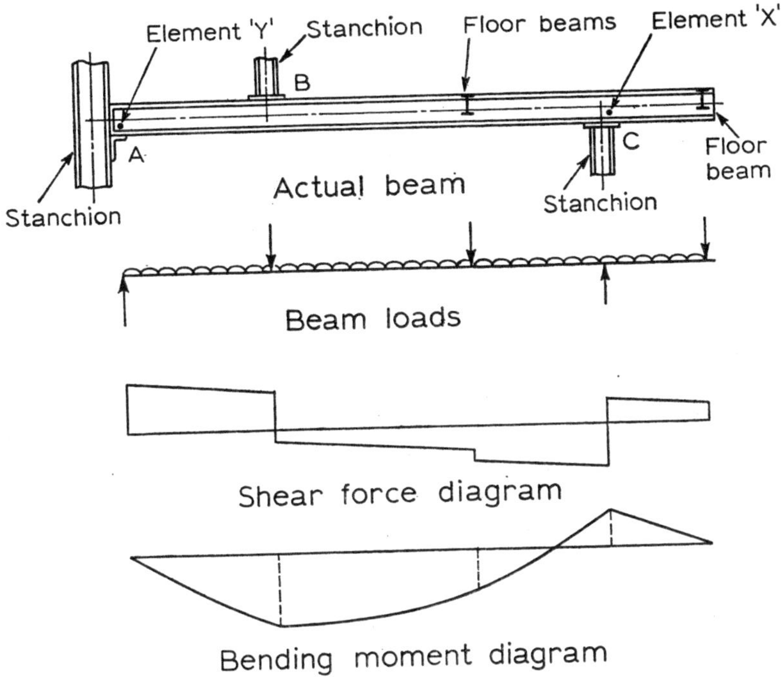

Figure 3.2

self weight and imposed floor loads. Dead and imposed loads for floors are given in CP3 1967, Part 1, Chapter V, for various types of buildings.

The primary stresses are bending and shear. The stress distributions are shown in Figure 3.3(b) and (c). The maximum bending stress generally occurs near the centre of span, for a simply supported beam, and the maximum shear stress occurs at the supports. For the cantilever support at C in Figure 3.2 and the interior supports of continuous beams and in fixed end beams, the maximum shear and bending stress occur at the same section. Note, that at the section, the maximum bending stress occurs at the extreme fibre where the shear stress is zero, while the maximum shear stress occurs at the neutral axis where the bending stress is zero.

An element such as 'X' in Figure 3.2 in the web of the beam is subjected to bending, shear and bearing stresses as shown in Figure 3.3(d). Element 'Y' is subjected to shear and bearing stresses. The stresses in these elements require investigation. Rules for calculating the equivalent stress due to combined bending, shear and bearing stresses are given in Clauses 14 (c) and (d) of BS 449. The student

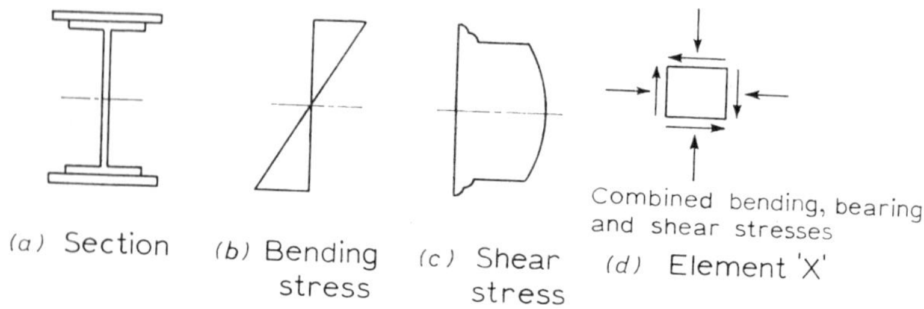

Figure 3.3

should refer to the Code and text books on strength of materials for further information on this problem.

When the top flange is left unsupported, failure can occur at a low stress as the compression flange buckles sideways. This type of failure is considered later in the chapter.

At supports A and C and at points of concentrated load as at B in Figure 3.2 high local stresses and buckling of the web can occur and these require investigation. The welds connecting the flange plates to the flanges of a universal beam are designed for horizontal shear. Finally, deflection of beams is often an important consideration.

3.2 FLANGE BENDING STRESSES

3.2.1 Permissible Bending Stresses

The permissible bending stresses for universal beams and compound beams are given in tables 2 and 3 of BS 449.

The following allowable stresses are given in table 2 of BS 449:

Tensile p_{bt} }
Compressive p_{bc} }

= 165 N/mm²—thickness of material up to and including 40 mm

= 150 N/mm²—thickness of material over 40 mm

Lower stresses are required for the thicker material as this is more liable to have laminations or other defects in quality than the thinner plate.

For universal beams, the allowable compressive stress given in table 2 applies when the compression flange is restrained against lateral buckling. The stresses given in table 3 depend on the tendency to buckling of the compression flange. This is dealt with below.

It should be noted that Clause 13 of the code permits all stresses, except for grillage beams (Clause 40), to be increased by 25 % where the increase is solely due to wind forces.

3.2.2 Curtailment of Flange Plates

In simply supported compound beams, the flange plates can be curtailed or cut off to save material where they are no longer required to resist bending moment. This is indicated in Figure 3.4 where the moment of resistance of the beam is shown plotted on the same diagram as the bending moment caused by the beam loading.

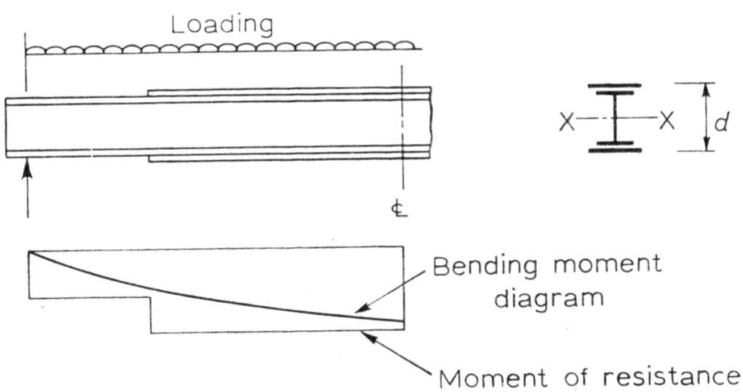

Figure 3.4

In modern practice, where automatic methods of fabrication are used, beams are often made of uniform section throughout. The simplification in fabrication and other benefits of a uniform section can offset the extra cost of material. Particular cases may require a careful estimation of costs to see if the saving is justified.

The length of the flange plate should be in accordance with Clause 27(b) of BS 449. This clause states in part:

'Each flange plate shall extend beyond its theoretical cut-off point, and the extension shall contain sufficient rivets or welds to develop in the plate the load calculated for the bending moment and girder section (taken to include the curtailed plate) at the theoretical cut-off point.'

3.2.3 Fixed End and Continuous Beams

With fixed end and continuous beams, haunches or flange plates may be added at the supports since the bending moment diagram has maximum values at these points as shown in Figure 3.5(a). A haunched and a continuous beam are shown in Figura 3.5(b) and (c) respectively. Design of these types of beams is outside the scope of this book, but the sketches are given here to show the comparison with the simply supported beam above.

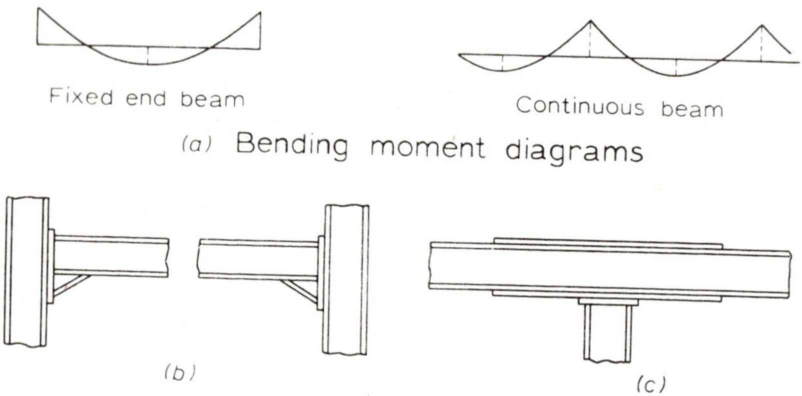

Figure 3.5

3.2.4 Bending Stresses

The commonest type of beam, an I section, has two axes of symmetry as shown in Figure 3.6(a). A crane girder section with one axis of symmetry, the YY axis, is shown in (b) in the figure. Sections with no axis of symmetry are mentioned below.

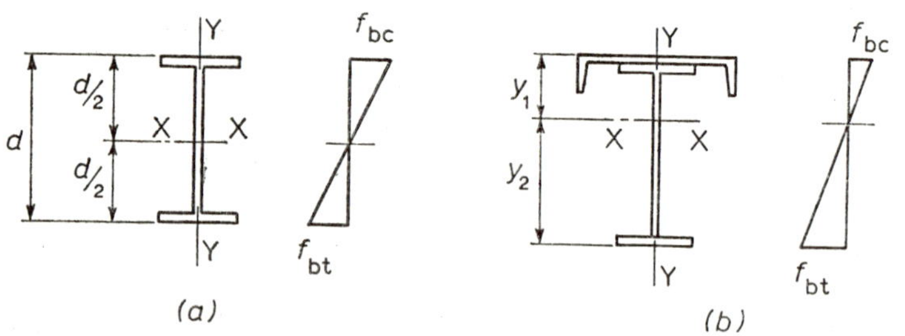

Figure 3.6

(a) Bending about one axis—XX axis. See Figure 3.6.

M = applied bending moment (N mm) at the section causing tension on the bottom of the beam

$Z_1 = I_{XX}/y_1$ mm^3—modulus of section, top flange

$Z_2 = I_{XX}/y_2$ mm^3—modulus of section, bottom flange

$Z \quad = 2I_{XX}/d$ mm³—modulus of section for a symmetrical beam

$I_{XX} =$ mm⁴—moment of inertia

For a symmetric section, the stresses in the top and bottom fibres are equal and are given by

$$f_{bc} = f_{bt} = M/Z \qquad \text{N/mm}^2$$

For an unsymmetric section, the bending stresses are:

top fibre in compression $\quad f_{bc} = M/Z_1 \quad$ N/mm²

bottom fibre in tension $\quad f_{bt} = M/Z_2 \quad$ N/mm²

Beam sections and bending stress distributions for the two cases are shown in Figure 3.6(a) and (b).

(b) Bending about two axes

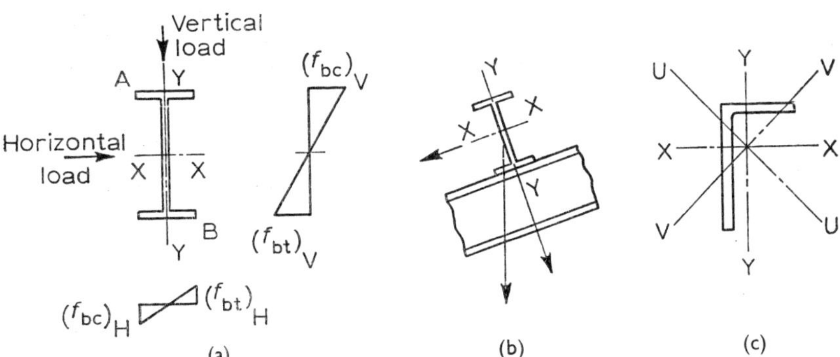

Figure 3.7

Consider first the I sections shown in Figure 3.7(a) and (b). In (a) the I section is carrying a vertical and a horizontal load. The figure in (b) shows an I section purlin on a slope carrying a vertical load. This load is resolved into a normal and a lateral component which cause bending about the XX and YY axes respectively.

Referring to Figure 3.7(a):

M_{XX} = bending moment about XX axis—N mm

M_{YY} = bending moment about YY axis—N mm

Z_{XX} = modulus of section for XX axis—mm³

Z_{YY} = modulus of section for YY axis—mm³

The maximum stresses occur at points A and B where the compressive and tensile stresses due to the horizontal and vertical bending moments are added. For the section shown the maximum stress at A in compression is given by

$$f_{bc} = (f_{bc})_V + (f_{bc})_H$$
$$= \frac{M_{XX}}{Z_{XX}} + \frac{M_{YY}}{Z_{YY}} \qquad \text{N/mm}^2$$

The maximum stress in tension at B would have the same value.

In the example where the section does not have an axis of symmetry (see the unequal angle shown in Figure 3.7(c)) bending occurs about the principal axes UU and VV. The student should refer to text books on structural mechanics for the complete theory on this example. Section properties for the four axes for the angle are given in steel tables.

(c) Checking beams

In checking a given beam, the value of the bending moment can be calculated from the external loading applied to the beam including the self weight of the beam, while the value of the section modulus is taken from the Handbook on Structural Steelwork. The stress can be calculated as set out above. The calculated stress should not exceed the allowable stress given in BS 449, table 2 or 3.

The same reasoning applies to checking a beam subjected to bending about two axes. The maximum stress equal to the sum of the stresses due to bending about the two axes should not exceed the allowable stress given in the code in table 2 or 3.

(d) Design of beams

In designing a beam for bending about one axis the value of the modulus required

$$Z = M/p_b \text{ mm}^3$$

is calculated where p_b is the allowable bending stress (table 2 or 3). A suitable beam is then selected from steel tables such that the value of modulus of section is slightly greater than that required.

In cases of bending about two axes or where the allowable compressive stress depends on the beam section (see the discussion below) the beam section cannot be determined directly. Here a beam must be selected and checked as set out above.

Beam sections for various spans and loads can be selected from safe load tables given in the *Handbook on Structural Steelwork*.

3.2.5 Lateral Instability of Compression Flange

The compression flange of a beam acts as a column and can buckle sideways under the compressive load it carries. Deep I section beams with narrow flanges buckle readily and fail at a stress lower than the allowable stress given in table 2 of BS 449. This buckling can be prevented by restraining the compression flange with lateral supports. An effective lateral support must be capable of resisting $2\frac{1}{2}\%$ of the flange force. Slabs provide lateral support. The design of the connection at the end of the beam is also important. The ends of the beam must be restrained against torsion. If the compression flange is connected to the support column or beam or is restrained against lateral bending, this greatly reduces the tendency to instability. Rules

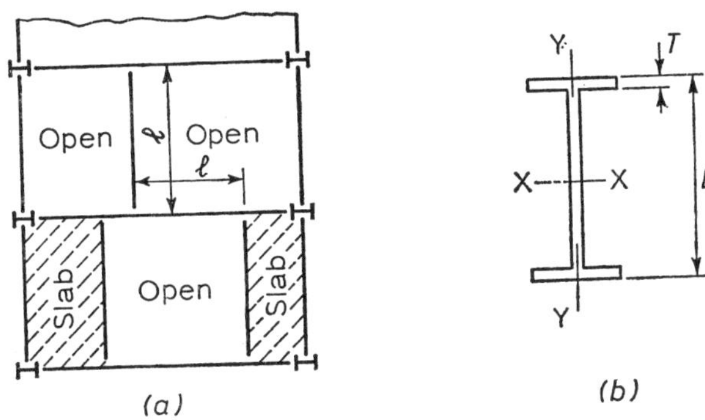

Figure 3.8

for taking this into account are given later. Laterally restrained and unrestrained beams are shown in Figure 3.8(a).

Note that for rectangular hollow sections, lateral instability does not occur if the depth of the box does not exceed four times the breadth. The rule for obtaining the allowable compressive stress will be given for universal and compound beams with equal flanges of uniform cross section throughout. Further rules are given in BS 449 for compound beams with curtailed flanges and the code should be consulted in these cases.

Referring to Figure 3.8(b):

l = effective length of compression flange between supports
D = overall depth
r_Y = radius of gyration about the YY axis
T = mean thickness of flange

The allowable bending stress in compression depends on l/r_Y and D/T and values are given in table 3 of BS 449. Rules and diagrams are given in BS 449 for determining the value of l. Some values for floor beams are shown in Figure 3.8(a). An extract from Clause 26 of BS 449 follows. The full text of the clause should be read.

(i) Simply supported beams and girders where no lateral restraint of compression flange is provided but where each end of the beam is restrained against torsion.
   1 Ends of compression flange unrestrained against lateral bending (free to rotate in plan): l = span
   2 Ends partly restrained against lateral bending: $l = 0.85 \times$ span
   3 Ends fully restrained against lateral bending: $l = 0.7 \times$ span

   Restraint against torsion is provided by flange or web cleats.

(ii) For beams provided with members giving effective lateral restraint to the compression flange at intervals along the span in addition to the end torsional restraint, the effective length of compression flange shall be taken as the distance centre to centre of restraints.

(iii) The compression flange of a beam supporting a floor slab may generally be taken as fully restrained.

3.2.6 Example

1 A 457×152×67 kg/m UB spans 6.1 m and the top flange is unrestrained laterally. The ends of the beam are restrained against torsion but unrestrained against lateral bending. Find the allowable bending stress in compression, p_{bc}.

Effective length of compression flange $l = 6.1$ m

Radius of gyration $r_Y = 31.2$ mm

$l/r_Y = 6100/31.2 = 195$

$D/T = 30$ from steel tables

$p_{bc} = 71$ N/mm² from table 3 of BS 449.

3.3 WEB STRESSES

3.3.1 Web Shear

The actual shear stress at any section of a beam is given by

$$f_q = \frac{SAY}{It}$$

where S is the shear force in the beam, A is the area outside the section where the shear stress is required, Y is the distance of the centre of gravity of A from the centroidal axis of the beam, I is the moment of inertia of the beam and t is the thickness of beam at the point where the stress is required.

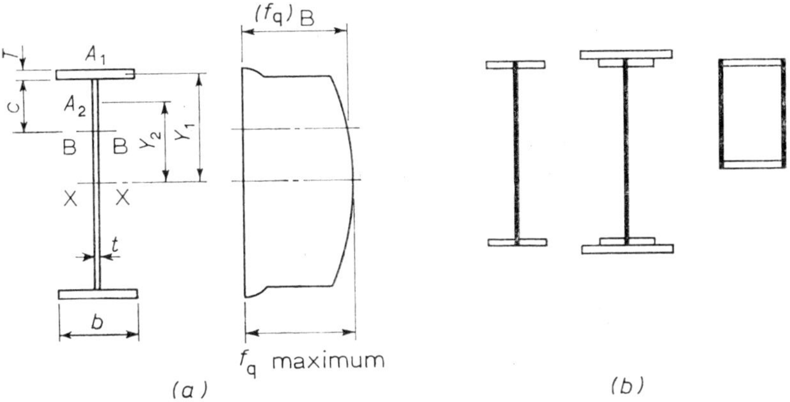

Figure 3.9

Referring to Figure 3.9(a), the shear stress at BB on the simplified beam section is

$$(f_q)_B = \frac{S(A_1y_1 + A_2y_2)}{It}$$

where

$$A_1 = bT$$

$$A_2 = ct$$

The fillets between the web and flange have been neglected. The maximum shear stress occurs at the neutral axis of the section. The stress distribution is shown on the figure. The allowable maximum shear stresses are given in table 10 of BS 449.

It can be seen from the shear stress distribution that the web carries most of the shear, so it is customary to check the average shear stress on the web of the beam. The average shear stress for universal or compound beams is

$$f'_q = \frac{\text{Shear}}{\text{Depth of universal beam} \times \text{Web thickness}}$$

For rectangular hollow sections, the average stress is

$$f'_q = \frac{\text{Shear}}{2 \times \text{Depth of section} \times \text{Web thickness}}$$

The areas of sections reckoned to carry shear are shown solid on Figure 3.9(b). The allowable average shear stresses in unstiffened webs are given in table 11 of BS 449.

3.3.2 Web Buckling

At points of concentrated load and at supports, unstiffened webs of universal beams and compound beams are likely to fail by buckling. The formula from Clause 28 of BS 449 for determining the safe load that may be applied to the beam is given below. The formula is based on spreading the load at 45°.

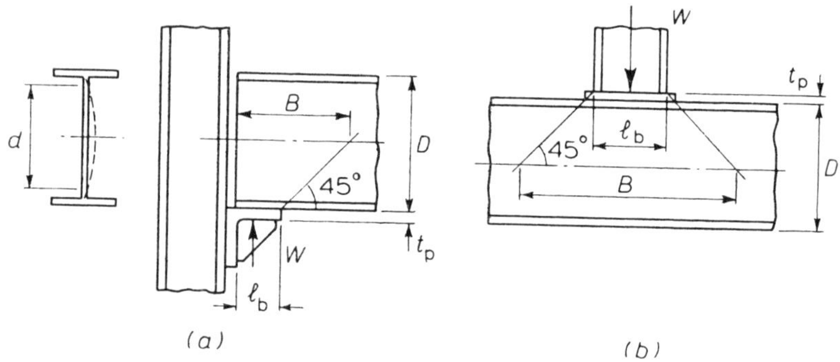

Figure 3.10

Referring to Figure 3.10, the safe concentrated load that can be carried on an unstiffened web is

$$W = p_c t B$$

where t is the web thickness, d is the clear depth of web between fillets, p_c is the allowable axial stress for struts, given in table 17 of BS 449 for a slenderness ratio of $d\sqrt{3}/t$, D is the overall depth of the beam, t_p is the thickness of bearing seat plate (if any), l_b is the length of stiff bearing (this must not be greater than $\frac{1}{2}D$ for end bearings and D for intermediate bearings) and $B = \frac{1}{2}D + t_p + l_b$ for end bearings, $B = D + t_p + l_b$ for intermediate bearings.

3.3.3 Web Crushing

The direct bearing stress at the end of the web fillet is not to exceed 190 N/mm². This stress is given in table 9 of BS 449. The load in this case is spread at 30° to the horizontal. The position of the maximum bearing stress and the lengths on which it acts for end and intermediate supports are shown on Figure 3.11.

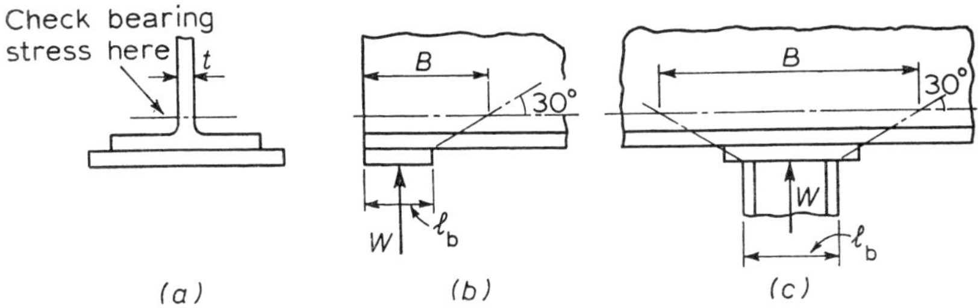

Figure 3.11

Safe load $W = 190\ Bt$

B = length where crushing strength is checked (see Figure 3.11 for determination)

t = web thickness.

Alternatively, the bearing stress may be calculated for a given load.

Safe load tables give safe values for concentrated loads on webs of universal beams for buckling and crushing, for end bearings and intermediate bearings. If the load that has to be carried exceeds the safe value for the universal beam, stiffeners must be provided to strengthen the web. Refer to the design of stiffeners for plate girders given in Chapter 4.

3.3.4 Combined Stresses

Combined stresses in beams are due to

1 Bending and shear
2 Bearing, bending and shear

The positions where these stresses occur in a beam are shown on Figure 3.2. Bearing and shear and bearing, bending and shear are checked at the end of the straight portion of the web. See Figure 3.11(a).

Formulae for calculating the effects of combined stresses are given in Clause 14 of BS 449. An extract from this clause follows.

'(c) Bending and shear. The equivalent stress f_e due to bending and shear shall not exceed the values of p_e given in Table 1.

The equivalent stress f_e is obtained from the following formulae:

$$f_e = (f_{bt}^2 + 3f_q^2)^{1/2} \quad \text{or} \quad (f_{bc}^2 + 3f_q^2)^{1/2}$$

in which f_{bc} or f_{bt} and f_q are the numerical values of the co-existent bending and shear stresses.

'(d) Combined bearing, bending and shear stresses. Where a bearing stress is combined with tensile bending and shear stresses under the most unfavourable conditions of loading, the equivalent stress f_e, obtained from the following formulae, shall not exceed the values of p_e given in Table 1.

$$f_e = (f_{bt}^2+f_b^2+f_{bt}f_b+3f_q^2)^{1/2}$$

or

$$f_e = (f_{bc}^2+f_b^2-f_{bc}f_b+3f_q^2)^{1/2}$$

in which f_{bt}, f_{bc}, f_q, f_b are the numerical values of the coexistent bending, shear and bearing stresses.'

In the above extract

f_{bt} = bending stress, tension

f_{bc} = bending stress, compression

The allowable equivalent stress p_e from Table 1 for plates, universal beams, etc., up to and including 40 mm in thickness is 230 N/mm^2 for Grade 43 steel. The application of the formulae is shown in examples later in the chapter.

3.4 WELD—COVER PLATE TO UNIVERSAL BEAM

The fillet welds between the cover plate and the flange of the universal beam are designed to resist horizontal shear. See Figure 3.12.

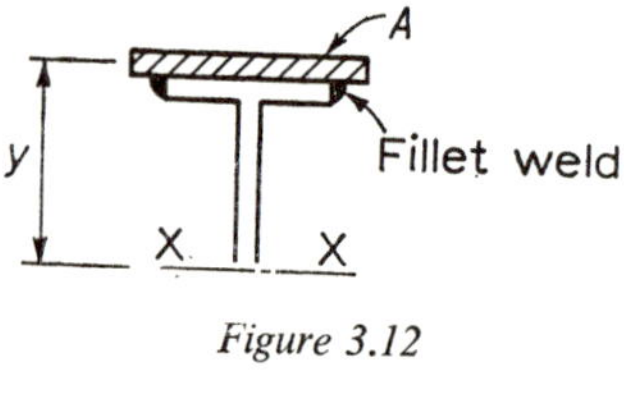

Figure 3.12

Horizontal shear in the weld = $SAy/2I_{XX}$ N/mm where S is the shear at section (N), A is the area of section connected (mm^2), y is the lever arm of A (mm) and I_{XX} is the moment of inertia of section (mm^4).

The leg length of weld can be selected from the fillet weld strengths given in the Appendix in *Table* 4. In some cases only a very small weld is required, and here the minimum size of fillet weld given in *Table* 2.2 (Chapter 2) should be used.

Intermittent welds were used extensively but with modern practice, where welds are made automatically, continuous fillet welds are used. These welds considerably reduce the likelihood of failure due to fatigue and brittle fracture.

3.5 DEFLECTION OF BEAMS

Reference is made to Clause 15 of BS 449 which states that the deflection must not be such as will impair the strength and efficiency of the structure or lead to damage to finishes or be unsightly.

Members may be cambered to offset the dead load deflection.

The maximum deflection due to loads other than dead loads shall not exceed 1/360 span.

Deflection formulae are given in engineering handbooks. Some of the most commonly used formulae are given in Figure 3.13. In any

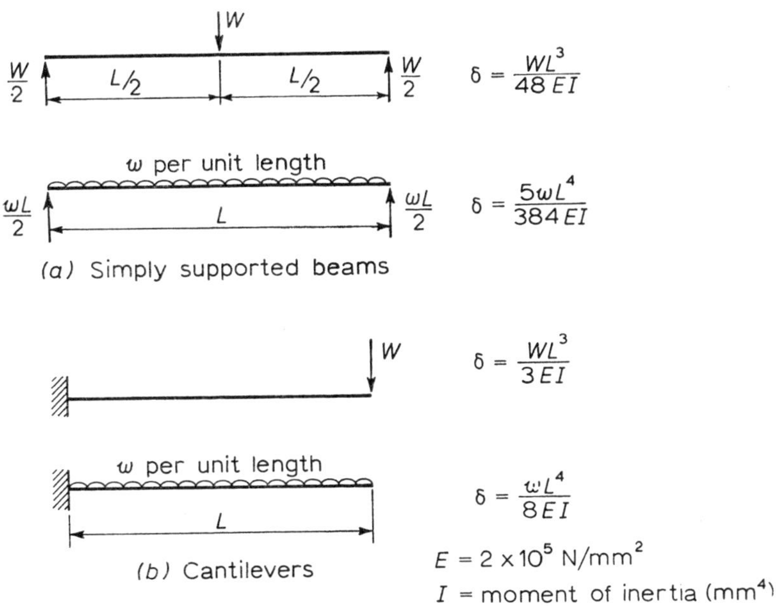

Figure 3.13 Deflection formulae

general loading case, the deflection can be calculated by the moment area method. Reference should be made to books on structural mechanics for this method.

3.6 CONSTRUCTION DETAILS

Fabrication for floor beams consists in cutting rolled stock to length, making the end connections, adding any cleats or gussets required and drilling bolt holes. Modern practice is to carry out all welding in the shop, where the process can be carefully controlled, and to make field connections by bolting. High-strength friction-grip bolts are often used in the field joints.

The member must be accurately fabricated to the correct length and the end plates must be perpendicular to the member axis. The holes must be set out and drilled as shown on the drawings. Reference should be made to Chapter V of BS 449 which deals with fabrication and erection.

The main connections required are beam to beam and beam to stanchion. Some of these connections are shown on Figure 3.14.

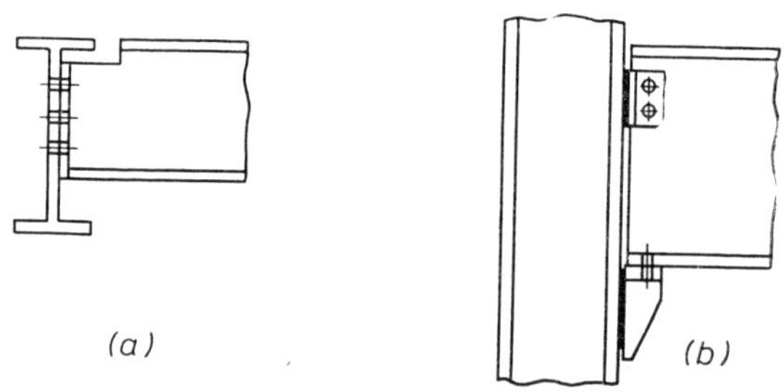

Figure 3.14

Dimensions for detailing the beam-to-beam connections are given in the *Handbook on Structural Steelwork*. Here the flange on the secondary beam is cut back to clear the flange of the main beam.

In compound beams, cover plates are welded to the rolled section. If the welding is intermittent it must be done manually. In modern practice, continuous fillet welds made by an automatic process are used. This is shown in Figure 3.15. Sometimes special precautions in welding

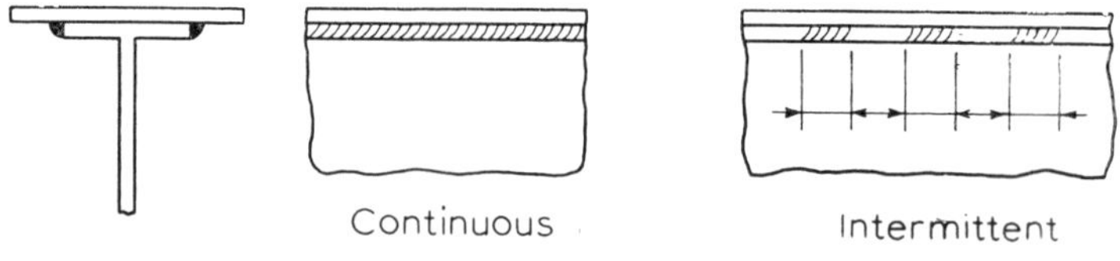

Figure 3.15

such as preheating are required to reduce the effect of distortion and prevent cracking. These requirements should be checked with a welding engineer.

After fabrication the member is checked for accuracy, given any specified protective treatment and identified for erection purposes.

Details requiring holes in the tension flange should be avoided if these holes are at or near the point of maximum moment. In determining properties of the net section for bending, holes on the tension side of the neutral axis must be taken into account. Holes on the compression side need not be considered except when the connection is made with black bolts. Holes near the neutral axis do not have any great effect on the section modulus.

3.7 EXAMPLES OF BEAM DESIGN

3.7.1 Floor Beams for an Office Building

The steel for the floor of a library with book storage is shown in Figure 3.16. The floor is a reinforced concrete slab supported on universal beams. The design loading has been estimated as:

Dead load—slab, finishes, self weight of steel, ceiling, partitions, services = 610 kg/m² = 6 kN/m²

Imposed load from Table 1 of CP3, Chap V, Part 1 = 4 kN/m²

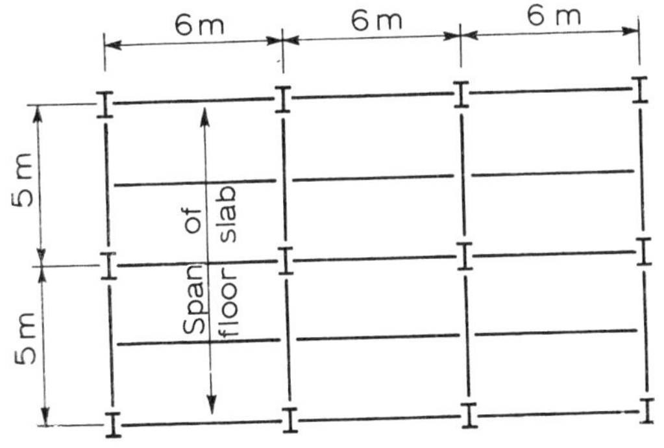

Figure 3.16 Steel for a library floor

In addition all external beams carry a loading due to the weight of a brick wall and a solid concrete casing for fire protection of 715 kg/m, that is 7 kN/m.

Determine the section required for each floor beam.

Because of symmetry, only one panel between columns need be considered. The beams in this panel are given mark numbers as shown in Figure 3.17.

On all beams the compression flange is fully restrained, so the allowable bending stress is equal to 165 N/mm².

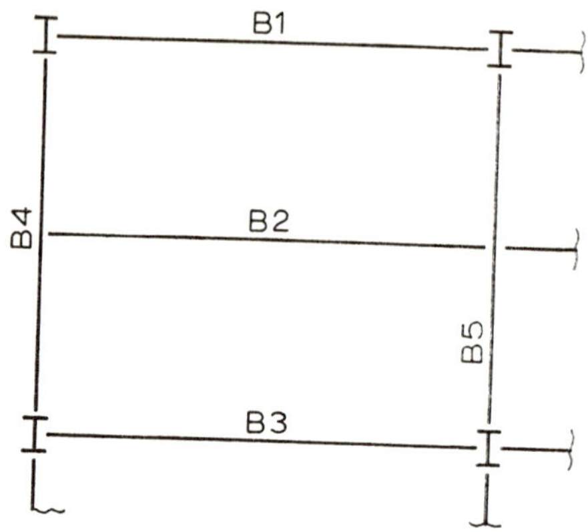

Figure 3.17

1 *External Beam BI* *Span* 6m

Loading

Dead load	$= 6\times6\times1.25$	=	45 kN
Imposed load	$= 4\times6\times1.25$	=	30 kN
Wall and casing	$= 7\times6$	=	42 kN
Total uniformly distributed load		=	115 kN
Moment M	$= (115\times6)/8$	=	86 kN m
Modulus Z	$= (86\times10^6)/165$	=	5.22×10^5 mm³

The moment of inertia to limit the deflection due to imposed load to 1/360 span is derived from

$$\frac{L}{360}=\frac{5WL^3}{384EI}$$

Inertia
$$I=\frac{5\times W\times L^2\times360}{384\times2\times10^5}=2.34\times10^{-5}\,WL^2$$

$$=2.34\times10^{-5}\times30\times10^3\times(6000)^2$$

$$=2.53\times10^7\ \text{mm}^4$$

$305\times165\times40$ kg/m UB $\quad Z_{xx}=559.6\ \text{cm}^3$

$I_{xx}=8500\ \text{cm}^4$

2 *Internal Beams B2 and B3* *Span* 6 m

Loading

Dead load	$= 6\times6\times2.5$	$= 90$ kN
Imposed load	$= 4\times6\times2.5$	$= 60$ kN
Total uniformly distributed load		$= 150$ kN
Moment M	$= (150\times6)/8$	$= 112.5$ kN m
Modulus Z	$= (112.5\times10^6)/165$	$= 6.83\times10^5$ mm^3

The moment of inertia I to limit the deflection due to the imposed load to 1/360 span is given by

$$I = 2.34\times10^{-5}\times60\times10^3\times(6000)^2$$
$$= 5.06\times10^7 \text{ mm}^4$$

356×171×45 kg/m UB $\quad Z_{XX} = 684.7$ cm^3, $\quad I_{XX} = 12\,052$ cm^4

3 *External Beam B4* *Span* 5 m

Loading

Dead Load = reaction of beam B2 = 45 kN

Imposed load = reaction of beam B2 = 30 kN

Wall and casing—uniformly distributed load
= 5×7 = 35 kN

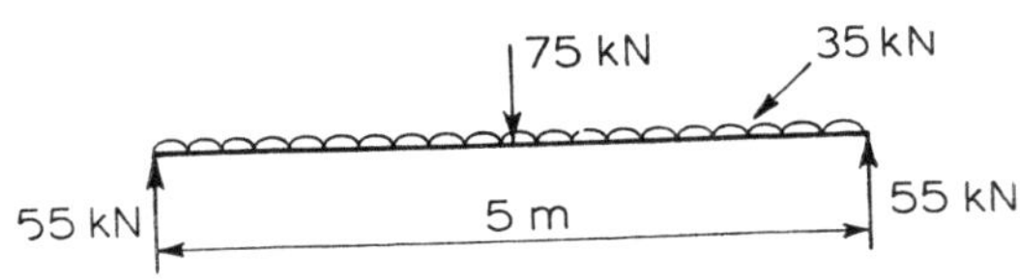

Figure 3.18

The loading on the beam is shown in Figure 3.18.

Moment M	$= (75\times5)/4+(35\times5)/8$	
	$= 94+21.9$	$= 115.9$ kN m
Modulus Z	$= (115.9\times10^6)/165$	$= 7.06\times10^5$ mm^3

406×140×46 kg/m UB $\quad Z_{XX} = 775.6$ cm^3, $\quad I_{XX} = 15\,603$ cm^4

Check deflection due to imposed load of 30 kN

$\delta = (30\times10^3\times5000^3)/(48\times2\times10^5\times15\,603\times10^4) = 2.39$ mm

δ/span $= 2.39/5000$ $= 1/2090 < 1/360$

406×140×46 kg/m UB is satisfactory

4 *Internal Beam B5* *Span* 5 m

Loading

Dead load = reaction from 2 beams B2 = 90 kN

Imposed load = reaction from 2 beams B2 = 60 kN

The loading on the beam is shown in Figure 3.19.

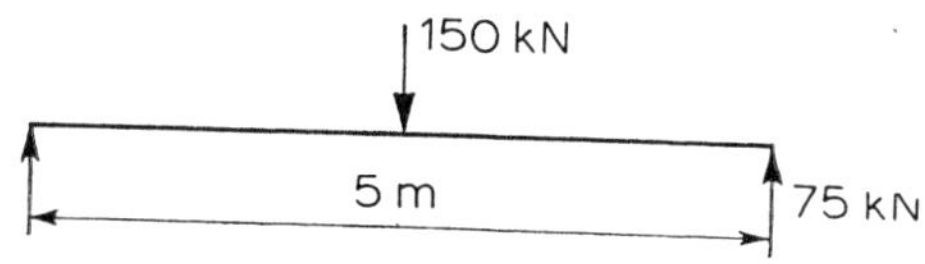

Figure 3.19

Moment M $= (150\times5)/4$ $=$ 187.5 kN m

Modulus Z $= (187.5\times10^6)/165$ $=$ 11.35×10^5 mm^3

457×152×67 kg/m UB $Z_{XX} =$ 1248 cm^3

$I_{XX} =$ 28 522 cm^4

$$\delta = \frac{60\times10^3\times5000}{48\times2\times10^5\times28\,522\times10^4} = 2.6 \text{ mm}$$

δ/span $= 2.6/5000$ $= 1/1930 < 1/360$

457×140×46 kg/m UB is satisfactory.

3.7.2 Beam with Overhanging End

A beam ABC is 10 m long and is carried on supports at A and B which are 6.5 m apart. The beam has an overhanging portion BC, 3.5 m long. The beam carries a total uniformly distributed load of 1500 kN. This includes an allowance for the self weight of the beam. The compression flange of the beam is fully restrained.

At support A, the beam is carried on a bracket connected to the flange of 305×305×198 kg/m UC. At B the beam is supported on a cap plate over 305×305×198 kg/m UC. The cap plate is 25 mm thick.

Select a suitable UB and check the web for shear, buckling, crushing and combined stresses at the supports.

The loading diagram is shown in Figure 3.20(a). The reactions, moments and shears are calculated first.

$R_B = (1500 \times 5)/6.5 = 1155$ kN

$R_A = 1500 - 1155 = 345$ kN $= S_A$

$M_B = (150 \times 3.5^2)/2 = 920$ kN m—hogging moment

$S_{BC} = 150 \times 3.5 = 525$ kN

$S_{BA} = 1155 - 525 = 630$ kN

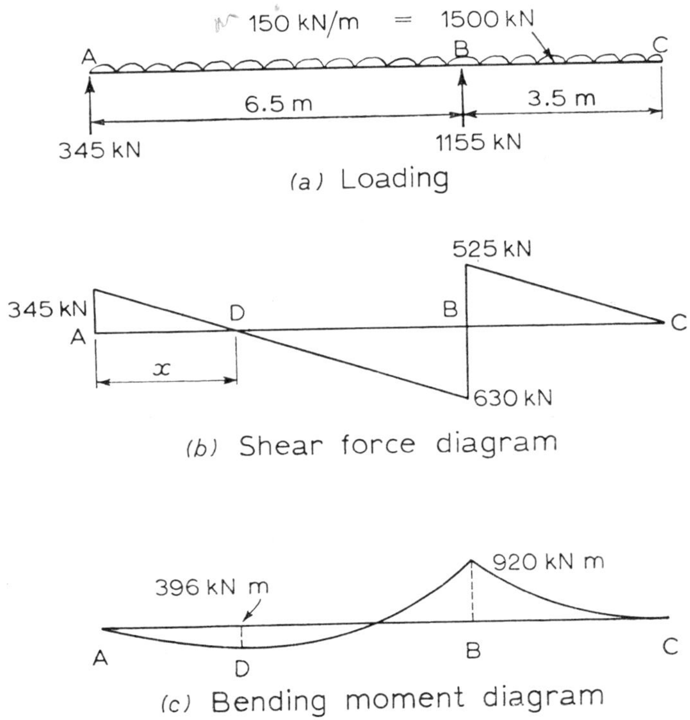

Figure 3.20

The point of maximum sagging moment is at point of zero shear.

$x = 345/150 = 2.3$ m

$M_D = (345 \times 2.3) - (\frac{1}{2} \times 150 \times 2.3^2) = 396$ kN m—sagging moment

The shear force and bending moment diagrams areshown in Figure 3.20(b) and (c).

Modulus Z $= (920\times10^6)/165 = 5.57\times10^6$ mm³

838×292×176 kg/m UB $Z = 5879$ cm³

Shear stress $= (630\times10^3)/(834.9\times14) = 54$ N/mm²—safe

The web thickness is 14 mm.

A check for deflection will not be carried out as part of this problem. The deflection could be calculated using the moment area method from structural mechanics.

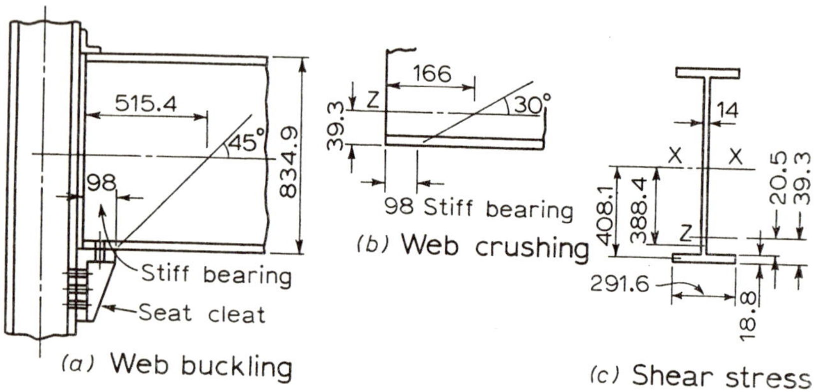

Figure 3.21

Support A A typical arrangement for this joint is shown in Figure 3.21. A stiff bearing length of 98 mm has been assumed.

1 Web buckling

Slenderness ratio $= (756.4\times\sqrt{3})/14 = 93.5$

Permissible axial stress $p_c = 86.5$ N/mm²

Actual axial stress $f_c = (345\times10^3)/(515.4\times14) = 48$ N/mm²

2 Web crushing

Bearing stress $f_b = (345\times10^3)/(166\times14) = 149$ N/mm²

Permissible bearing stress $= 190$ N/mm²

3 Combined stress

This is checked at the end of the straight portion of the web at Z on Figure 3.21(c). Dimensions for calculating the actual shear stress at this point are shown on the figure.

Moment of inertia = 245 412 cm⁴

Shear stress—refer to section 3.3.1 above.

$$f_q = \frac{345\times10^3[(291.6\times18.8\times408.1)+(14\times20.5\times388.4)]}{245\,412\times10^4\times14}$$

$$= 23.8 \text{ N/mm}^2$$

Combined bearing and shear—refer to section 3.3.4 above.
Equivalent stress

$$f_e = (149^2+3\times23.8^2)^{1/2} = 154 \text{ N/mm}^2$$

Allowable equivalent stress $p_e = 230 \text{ N/mm}^2$

Support B A typical arrangement for this joint is shown in Figure 3.22. The stiff bearing length has been taken as the full depth of the universal column.

(i) Web buckling

Actual stress $f_c = (1155\times10^3)/(1224.7\times14) = 67.7 \text{ N/mm}^2$

(ii) Web crushing

Bearing stress $f_b = (1155\times10^3)/(561.3\times14) = 147 \text{ N/mm}^2$

(iii) Combined stress

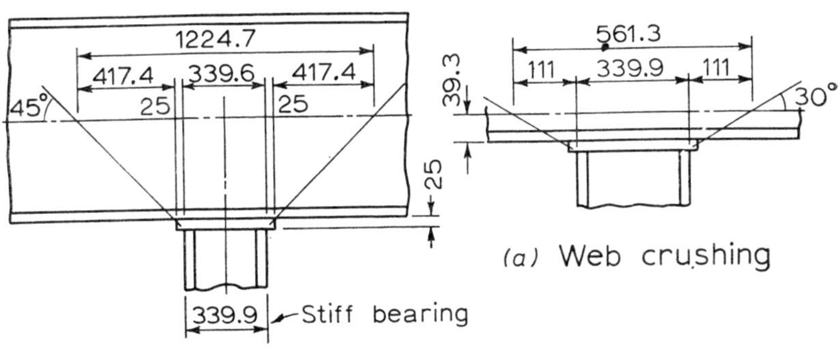

(a) Web buckling

(a) Web crushing

Figure 3.22

In this case coexistent compression due to bending, shear and bearing stresses occur. The combined stress is checked at the end of the straight portion of the web.

Bending stress

$$f_{bc} = (920\times10^6\times378.2)/(5879\times417.45\times10^3) = 142 \text{ N/mm}^2$$

Shear stress. Refer to the calculation for support A above, the stress at support B is found from this by proportion.

$$f_q = (630/345)\times23.8 = 43.5 \text{ N/mm}^2$$

Bearing stress $f_b = 147 \text{ N/mm}^2$

Equivalent stress $f_e = (142^2+147^2-142\times147+3\times43.5^2)^{1/2}$

$= 163 \text{ N/mm}^2$

Allowable equivalent stress $p_e = 230 \text{ N/mm}^2$

3.7.3 Compound Girder

Design a compound girder to carry a uniformly distributed load of 1000 kN. The girder is simply supported at each end and has a span of 11 m. Allow 23 kN for the weight of the girder. The overall depth of the girder must not exceed 700 mm. The length of stiff bearing at the supports is 150 mm. Check the web at the supports for shear, buckling, crushing and combined stresses. Determine the theoretical cut-off points for the flange plates and design the cover plate to universal beam weld. Adequate lateral support is provided for the compression flange.

(a) Girder Section

The total load carried by the beam is 1023 kN, that is 93 kN/m.

$$\text{Moment} = (1023\times 11)/8 = 1408 \text{ kN m}$$

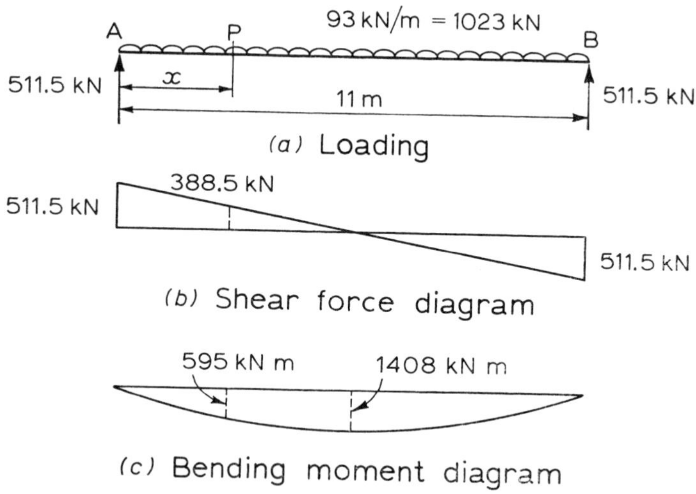

Figure 3.23

The loading, shear force and bending moment diagrams are shown in Figure 3.23.

$$\text{Modulus} \quad Z = (1408\times 10^6)/165 = 8.54\times 10^6 \text{ mm}^3$$

Assume that the flange plates are less than 40 mm thick, so that the permissible bending stress $p_{bc} = 165$ N/mm^2. Try 610×229×140 kg/m UB and two No. 300×30 flange plates. The girder section is shown in Figure 3.24.

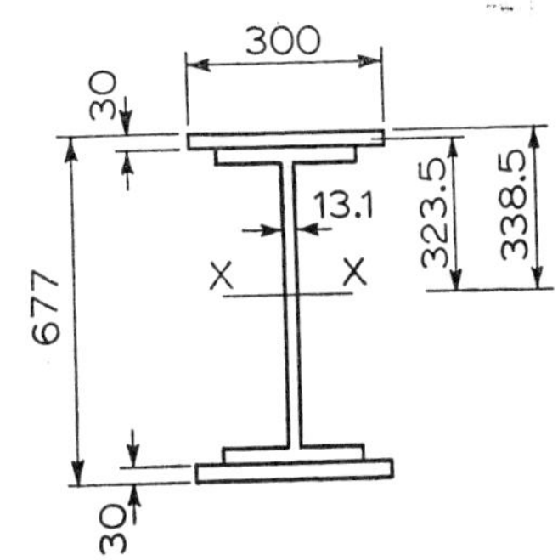

Figure 3.24

Properties for the universal beam taken from the *Handbook on Structural Steelwork* are:

$$A = 178.2 \text{ cm}^2$$
$$I_{XX} = 111\,673 \text{ cm}^4$$
$$Z_{XX} = 3620 \text{ cm}^3$$

The overall depth of the compound girder is 677 mm and this is satisfactory.

Properties for the compound girder are calculated:

$$I_{XX} = 11.17\times10^8+60\times30\times323.5^2$$
$$= 29.92\times10^8 \text{ mm}^4$$
$$Z_{XX} = (29.92\times10^8)/338.5 = 9.15\times10^6 \text{ mm}^3$$

Thus the girder is adequate for bending.

(b) Theoretical Cut-off Points for Flange Plates

The cut-off points for the flange plates are found by determining the position along the beam where the bending moment is equal to the moment of resistance of the universal beam.

Moment of resistance of the universal beam

$$= (3620\times10^3\times165)/10^6 = 595 \text{ kN m}$$

Referring to Figure 3.23(a), determine the position of P, where the bending moment in the beam is 595 kN m, from the following equation:

$$511.5x-\tfrac{1}{2}\times93x^2 = 595$$
$$46.5x^2-511.5x+595 = 0$$
$$x^2-11x+12.8 = 0$$
$$x = \frac{11\pm(121-51.2)^{1/2}}{2} = \frac{11\pm8.35}{2} = 9.68 \text{ m or } 1.32 \text{ m}$$

Calculate the percentage saving in material due to curtailment of flange plates.

Area of UB $= 1.782 \times 10^4$ mm²

Area of flange plates $= 1.8 \times 10^4$ mm²

Volume of compound beam

$= (1.782 \times 10^4 \times 11 \times 10^3) + (1.8 \times 8.36 \times 10^7) = 34.6 \times 10^7$ mm³

Volume of material saved $= 2.64 \times 1.8 \times 10^7 = 4.75 \times 10^7$ mm³

Theoretical saving in material $= 13.7\%$

Flange plates must extend beyond this point as stated in section 3.2.2 above.

(c) Flange Plate to UB welds

Shear at cut-off point $= 511.5 - (1.32 \times 93) = 388.5$ kN

Horizontal shear on two fillet welds

$= (388.5 \times 10^3 \times 300 \times 30 \times 323.5)/(29.92 \times 10^8) = 378$ N/mm

Shear per weld $= 189$ N/mm

Use the minimum leg length fillet weld, 6 mm, strength 483 N/mm. Continuous fillet weld is recommended. If intermittent fillet welding is required, try:

150 mm of weld and 150 mm spacing

Effective length of weld $= 150 - 2 \times$ weld leg length

$= 138$ mm

Horizontal shear on 300 mm length

$= (300 \times 189)/10^3 = 56.7$ kN

Weld strength $= (138 \times 483)/10^3 = 66.5$ kN

This arrangement would be suitable.

(d) Web

Average shear stress

$$f'_q = (511.5 \times 10^3)/(617 \times 13.1) = 63.2 \text{ N/mm}^2$$

Web buckling Refer to Figure 3.25(b) where all data for the calculations are shown.

Slenderness ratio $= (543.1 \times \sqrt{3})/13.1 = 71.8$

Permissible axial stress $p_c = 113$ N/mm²

Actual axial stress $f_c = (511.5 \times 10^3)/(458.5 \times 13.1)$

$= 85.5$ N/mm²

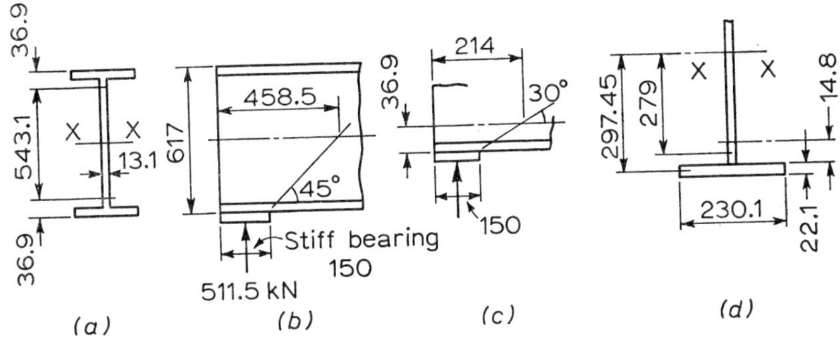

Figure 3.25

Web crushing Refer to Figure 3.25(c).

Bearing stress $f_b = (511.5\times10^3)/(214\times13.1) = 182\ \text{N/mm}^2$

Combined stress

Actual shear stress. Refer to Figure 3.25(d).

$$f_q = \frac{511.5\times10^3\,[(230.1\times22.1\times297.45)+(14.8\times13.1\times279)]}{111\,673\times10^4\times13.1}$$

$$= 54.6\ \text{N/mm}^2$$

Equivalent stress

$$f_e = (182^2+3\times54.6^2)^{1/2}$$
$$= 206\ \text{N/mm}^2$$

Allowable equivalent stress $= 230\ \text{N/mm}^2$

3.7.4 Beam Subjected to Bending about two Axes

A beam of span 5 m with simply supported ends restrained against torsion and with its major principal axis inclined at 30° to the horizontal is shown in Figure 3.26. This beam is supported at its ends on

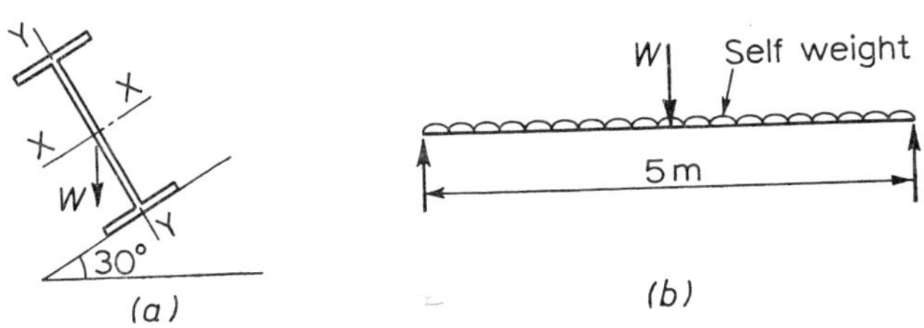

Figure 3.26

sloping roof girders. The unrestrained length of compression flange is 5 m and the beam section is 381×152×52 kg/m UB. Find the maximum concentrated load that can be supported at the centre of the beam.

Let the centre load on the beam = W kN

Moments $M = (W\times 5)/4+(52\times 9.81\times 5^2)/(10^3\times 8)$

$= (1.25W+1.59)$ kN m

$M_{XX} = M \cos 30°$

$= (1.08W+1.38)$ kN m

$M_{YY} = M \sin 30°$

$= (0.625W+0.795)$ kN m

Section properties of 381×152×52 kg/m UB are:

$Z_{XX} = 842.3$ cm^3

$Z_{YY} = 89.96$ cm^3

$r_{YY} = 3.21$ cm

$D/T = 30.7$

Permissible stress for bending about XX axis:
Effective length of compression flange = 5 m

$l/r_{YY} = 5000/32.1 = 156$

$p_{bc} = 98$ N/mm^2—Table 3 of BS 449

Stresses:

$(f_{bc})_{XX} = (1.08W\times 10^6)/(842.3\times 10^3)+(1.38\times 10^6)/(842.3\times 10^3)$

$= (1.28W+1.64)$ N/mm^2

$(f_{bc})_{YY} = (0.625W\times 10^6)/(89.96\times 10^3)+(0.795\times 10^6)/(89.96\times 10^3)$

$= (6.95W+8.85)$ N/mm^2

Combined stress must not exceed 98 N/mm^2

$1.28W+1.64+6.95W+8.85 = 98$

$8.23W = 87.51$

Safe load $W = 10.6$ kN

3.8 PROBLEMS

1 The floor plan for an office building is shown in Figure 3.27. The floor is precast concrete slabs supported on steel beams. The design loading has been estimated as follows:

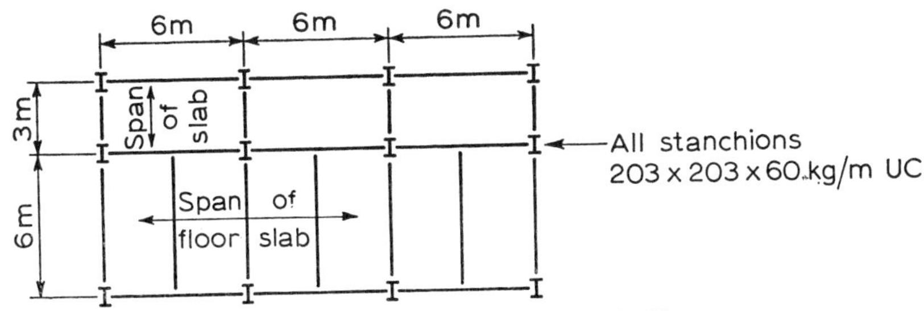

Figure 3.27 Floor plan for an office building

Dead load including self weight of steel, concrete slabs, finishes, partitions etc. = 633 kg/m^2 = 6.2 kN/m^2

Imposed load = 2.5 kN/m^2

In addition all external beams carry loading from brick walls and solid concrete casing of 745 kg/m, that is 7.3 kN/m.

(a) Give the beams mark numbers, calculate the loading and show the reactions at the ends of each beam on a sketch of the floor.
(b) Design the floor beams.
(c) Sketch connections at the ends of the beams.

2 The beam loaded as shown in Figure 3.28 is supported on a stanchion at A and by a web connection to the flange of the stanchion at B. The point loads on the beam are from secondary beams that frame

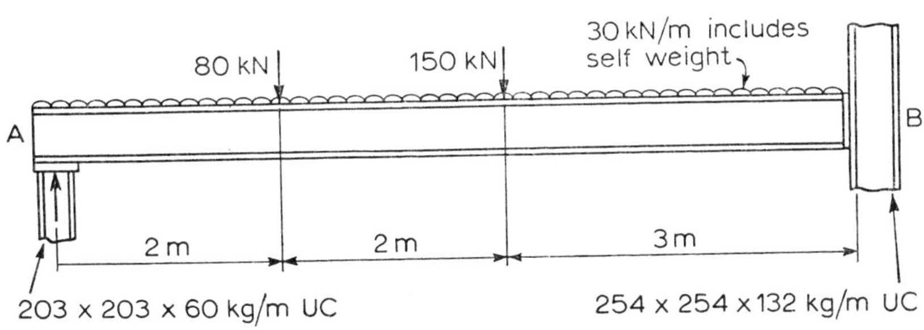

Figure 3.28

into web of beam AB. The top flange of the beam is restrained against lateral movement.

(a) Select a suitable universal beam.
(b) Check web shear, buckling, crushing and the combined stresses at support A.

Answer 686×254×140 kg/m UB.

3 A simply supported girder is required to span 7.2 m. The total load including an allowance for the weight of the girder has been estimated at 630 kN uniformly distributed. The overall depth of the girder must not exceed 500 mm and it is proposed to use a compound girder consisting of a universal beam and two flange plates. The girder will be supported on brackets at each end with a stiff bearing length of 80 mm. The flange plates are not to be curtailed. The compression flange is adequately restrained laterally.

(a) Check that a section consisting of 457×191×98 kg/m UB and two No. 15×250 flange plates is satisfactory.
(b) Design the flange-plate-to-universal-beam weld.
(c) Check web shear, buckling, crushing and combined stresses at the supports.

4 The end framing for a steel-frame building is shown in Figure 3.29(a). The beam AB forming a lintel for the doorway carries vertical

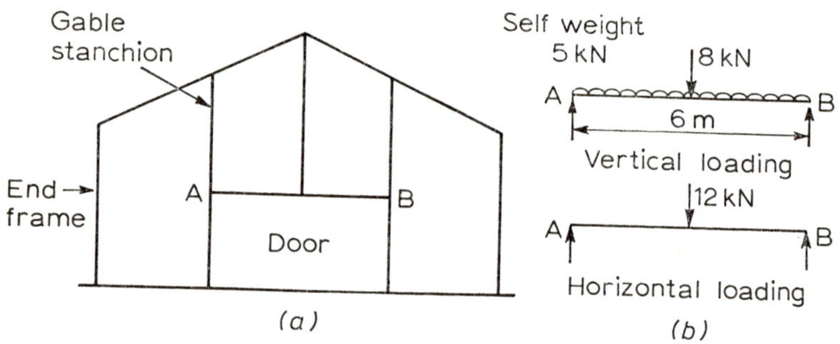

Figure 3.29 End framing for a steel-frame building

loading from wall framing and self weight and horizontal loading from wind on the end of the building. These loads on the beam are shown in (b) on the figure.

(a) Determine the size of universal beam required for beam AB.
(b) Calculate the horizontal and vertical deflections due to loading on the beam.

Note that the allowable stresses given in Tables 2 and 3 of BS 449 may be increased by 25% when the stresses are due to wind loads. Refer to Clause 13 of BS 449.

5 A sketch of a steel stairway is shown in Figure 3.30. The loading on the stairs is as follows:

Dead load—self weight = 61 kg/m^2 = 0.6 kN/m^2

Imposed load = 5.0 kN/m^2

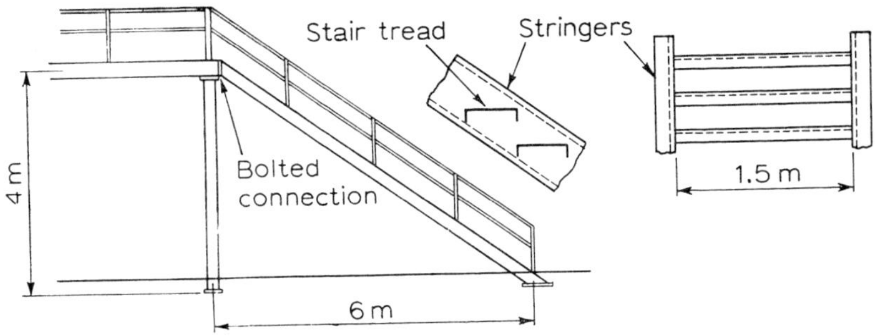

Figure 3.30 A steel stairway

Determine the size of channels required for the stringers to carry the stairs.

Note that the loads are vertical so the bending moments can be determined on a simply supported beam of 6 m span.

4

Plate girders

Plate girders are used to carry larger loads over longer spans than are possible with universal or compound beams. Plate girders are used in buildings for long span floor girders, for crane girders and in bridges.

4.1 DESIGN CONSIDERATIONS

4.1.1 Types

Plate girders are constructed by welding steel plates together to form I sections. The term box girder is used for closed box sections. Typical sections for plate and box girders are shown in Figure 4.1.

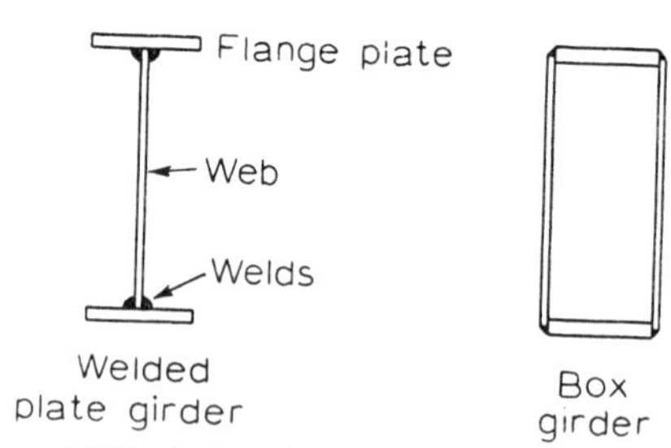

Figure 4.1 Typical sections for plate and box girders

4.1.2 Loads and Stresses

Loads are applied to plate girders through floor slabs, floor beams and stanchions carried by the girder.

The stresses calculated are primarily those of bending and shear. Nearly all of the bending moment is carried by the flange plates while the web plate resists the shear force.

Plate girders are constructed from thin plates and precautions must be taken to see that the girder section acts effectively, and the individual plates do not fail at low loads by buckling. Rules are set out in BS 449 to govern the width/thickness ratios of plates and ensure that buckling does not occur. Rules are also given for the provision of stiffeners at load points and intermediate positions along the girder. These rules have been derived from theoretical and practical considerations.

4.1.3 Depth of Plate Girder

In some cases, the depth of a plate girder may be fixed by headroom requirements. Often the depth can be selected by the designer. The depth is usually made about $\frac{1}{10}$ of the span for average loading. For lightly loaded girders of long span, carrying roof loads for example, the depth can be made $\frac{1}{15}-\frac{1}{20}$ of the span. The breadth of flange plate is usually made about $\frac{1}{3}$ of the depth.

The deeper the girder is made, the smaller the flange plates required. However, the deeper the web the thicker the web plate needed or, alternatively, more stiffeners are required with thinner plate to prevent web buckling. In some cases, designs at various depths should be made to establish the minimum weight girder. Shallow girders can be very much heavier than the minimum weight.

4.1.4 Deflection of Plate Girders

Deflection requirements are covered by Clause 15 of BS 449. This clause was discussed in Chapter 3 and the student should refer back to this chapter.

4.2 FLANGE BENDING STRESSES

4.2.1 Permissible Stresses

Permissible bending stresses are given in Tables 2, 5, 6, 7 and 8 of BS 449. Tensile stresses are given in Table 2 and lower stresses must be used when the plate thickness exceeds 40 mm.

All of the tables deal with allowable compressive stresses. Table 2 gives the stresses when the compression flange is fully restrained. Tables 5, 6, 7 and 8 are concerned with cases of I sections where the compression flange is likely to buckle sideways.

4.2.2 Variation in Girder Section

Flange plates can be curtailed or reduced in thickness where the reduction in the bending moment permits. This is shown in Figure 4.2(a) and (b). In (b) the depth of girder is kept constant throughout. For simply supported girders where the bending moment is a maxi-

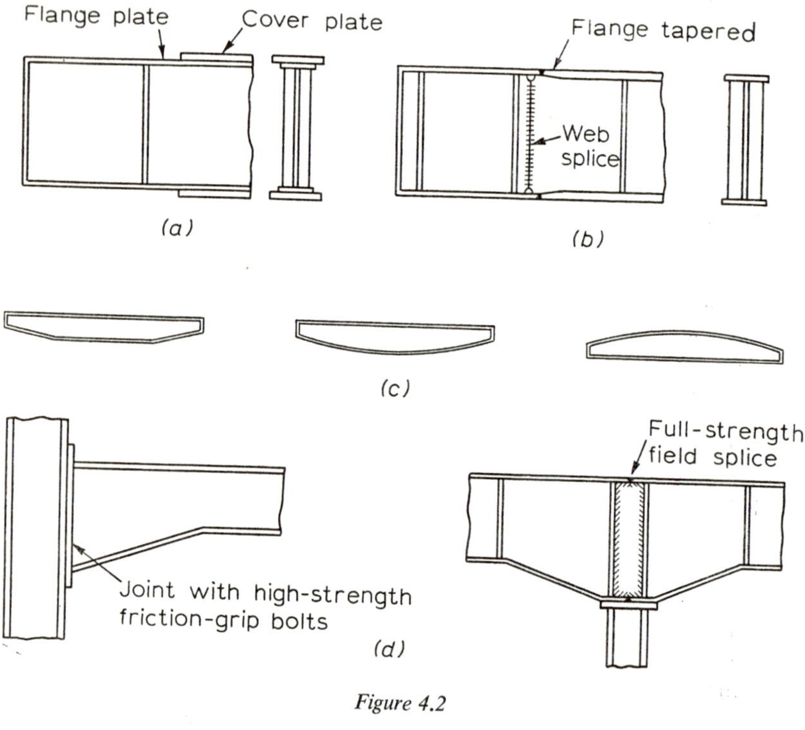

Figure 4.2

mum at the centre of the girder the section depth may be varied as shown in Figure 4.2(c). In the past, hog-back and fish-belly girders were commonly used. However, with automatic methods of fabrication it is often advantageous to make girders of uniform depth and section throughout.

In rigid-frame construction and in continuous girders the maximum bending moments occur at the supports where the girders are haunched to resist these moments as shown in Figure 4.2(d). The treatment of these girders involving rigid-frame analysis is outside the scope of this book.

4.2.3 Bending Stresses

The methods of calculating bending stresses for beams, dealt with in Chapter 3, apply also for plate girders. The section properties for plate girders, such as area, moment of inertia, modulus of section, radii of gyration, etc., must be calculated from first principles for each trial section. Moments of inertia for various sizes of flange plate at various depths apart and for web plates are given in the Handbook on Structural Steelwork. These can be used to obtain the properties of a given girder section.

Rules governing sectional areas for girders are set out in Clauses 17, 27 and 32 of BS 449. These provisions must be satisfied to ensure that the girder section is fully effective in bending. Plate width-to-thickness ratios and holes in the flange plates are dealt with.

With regard to plate thickness ratios, the provisions of the code are summarised as follows:

1 The maximum outstand for flange plates with unstiffened edges is given in Table 14. The following values are for Grade 43 steel:

compression flange	$16t$
tension flange	$20t$

where t is the thickness of the flange plate.

2 For box girders the maximum width of a plate in compression should not exceed $90t$ and only $50t$ should be reckoned as resisting compression.

These requirements are set out in Figure 4.3.

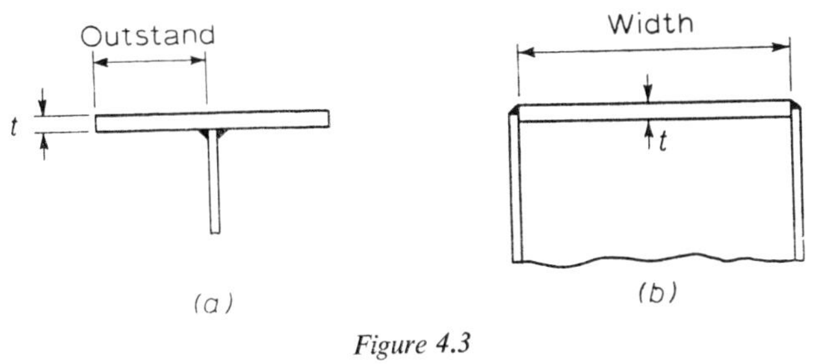

Figure 4.3

The rules regarding holes are:

1 Holes drilled in the tension flange must be deducted.
2 For the compression flange only open holes and holes for pins and black bolts need be deducted from the area.

Details requiring holes to be drilled in the flange plates at points of maximum moment should be avoided. Provisions regarding web plates are dealt with later in the chapter.

4.2.4 Lateral Instability of Compression Flange

The following information is extracted from Clause 20 of BS 449. Only the case where the flanges have equal moments of inertia about the YY axis will be considered here. For nonuniform girders, the student is referred to the full clause in BS 449. A girder section is shown in Figure 4.4(a) where l is the effective length of compression flange between supports, r_Y is the radius of gyration about the YY axis, D is the overall depth of the girder, T is the thickness of flange plate if the girder is of uniform section, t is the web thickness and d is the clear depth of web plate.

Values of the term A depend on l/r_Y and D/T and are given in Table 7 of BS 449. For the case under consideration, the critical stress C_s is given by

$$C_s = A \qquad \text{N/mm}^2$$

except that the value of C_s shall be increased by 20% when T/t is not greater than 2 and d/t is not greater than 85 for Grade 43 steel.

The allowable stress in bending in compression p_{bc} is given in Table 8 of BS 449 for values of C_s.

For values of l, the effective length of compression flange, refer to the discussion in Chapter 3.

4.2.5 Example

Refer to the girder section shown in Figure 4.4(b). Determine the allowable compressive stress in the top flange if the effective length of this flange between lateral restraints is 15 m.

Girder properties:

$$A = (2\times450\times40)+(1700\times10) = 53\times10^3 \text{ mm}^2$$

$$I_{YY} = (2\times40\times450^3)/12+(\text{Neglect web}) = 6.08\times10^8 \text{ mm}^4$$

$$r_Y = [(6.08\times10^8)/(53\times10^3)]^{1/2} = 107 \text{ mm}$$

$$l/r_Y = (15\times10^3)/107 = 140$$

$$D/T = 1780/40 = 44.5$$

From Table 7 $A = 176 = C_s$

From Table 8 $p_{bc} = 72$ N/mm^2

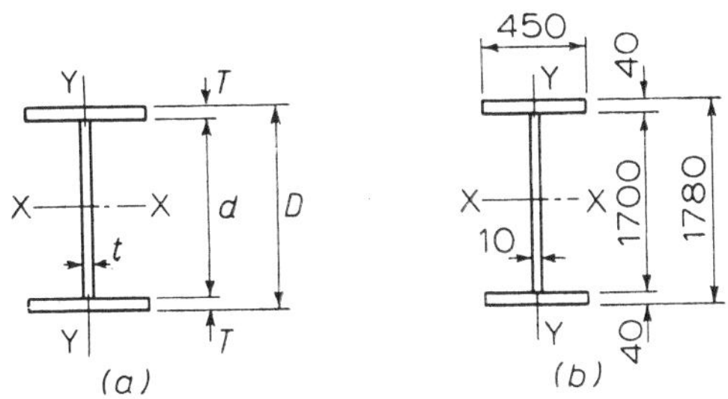

Figure 4.4

4.3 WEB SHEAR STRESSES

The allowable web shear stresses are given in Table 12 of BS 449. The average web shear stress for a plate girder is

$$f'_q = \frac{\text{Shear}}{\text{Depth of web plate} \times \text{Web plate thickness}} \quad \text{N/mm}^2$$

Compare this formula with that for a universal beam given in Chapter 3.

Referring to Table 12 of BS 449, the allowable shear stress depends on values of d/t and the stiffener spacing where d is the clear distance between flange plates and t is the thickness of web plate.

If the ratio d/t is greater than 85 for Grade 43 steel, vertical intermediate stiffeners must be provided at a distance apart not exceeding $1.5d$.

The thickness of web plate must not be less than $\frac{1}{180}$ of the smaller panel dimension for vertically stiffened webs, where the panel dimension is either the depth of web plate or the distance between stiffeners. (Refer to section 4.4.1.)

The above discussion has been restricted to girders with vertical stiffeners only. Further rules are given in Clause 27 of BS 449 for girders with vertical and horizontal stiffeners. Stiffener arrangements for plate girders are shown in Figure 4.5.

4.4 STIFFENERS IN PLATE GIRDERS

The web plates of plate girders are designed to resist shear for which thin plates only are generally required and these may require stiffening to prevent buckling. Stiffening is also required at all load points to prevent local buckling.

4.4.1 Intermediate Stiffeners

Intermediate stiffeners divide the web into panels and prevent buckling due to shear and compressive bending stresses.

Arrangements of stiffeners are shown in Figure 4.5.

Flats are the most common stiffener section used, though angles and tees are used when heavy stiffening is required. Stiffeners can be single

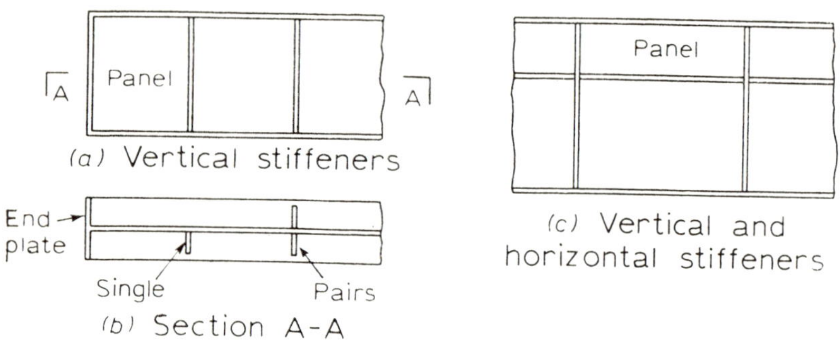

Figure 4.5 Arrangements of stiffeners

or in pairs as shown in Figure 4.5(b). Stiffeners should extend from flange to flange but should not have their ends either connected to, or fitted against the flange plates. Stiffeners are usually welded to web plates with continuous or intermittent fillet welds. On deep girders, horizontal as well as vertical stiffeners are used as shown in Figure 4.5(c). Treatment of this type of girder is outside the scope of this book. Requirements for intermediate stiffeners are set out in Clause 28(b) of BS 449. This states that the section for a vertical stiffener should be such that

$$I \text{ is not less than } (1.5\times10^{-4}\times d_1^3\times t^3)/s^2$$

where I is the moment of inertia in cm^4 of the complete stiffeners about the centre of the web, s is the maximum permitted clear distance in mm between stiffeners for thickness t, t is the minimum required thickness of web in mm and d_1 is the clear depth of web in mm.

Generally the designer chooses the size of stiffener to be used from his own experience. He can check this size by the formula given above. Flats are generally used for intermediate stiffeners and the maximum outstand must not exceed $12t$, where t is the thickness of the flat. The code also gives rules for horizontal stiffeners.

4.4.2 Load-Bearing Stiffeners

Load-bearing stiffeners must be provided at all load points and at supports to prevent a failure of the web there. Requirements are set out in Clause 28(a) of BS 449.

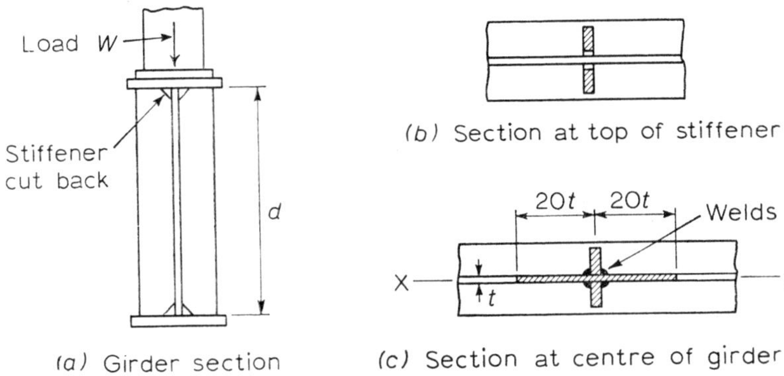

Figure 4.6

Consider a load applied to the top of the girder as shown in Figure 4.6. The load is first transmitted by bearing into the top of the stiffener. It is assumed that the net area of stiffener carries all of the load in bearing.

Safe load $\quad W = 190A_b/10^3 \quad$ kN

where A_b is the net area at the top of stiffener (mm^2). (190 N/mm^2 = permissible bearing stress—Table 9, BS 449)

Note that the stiffener at the top is cut back to clear the web to flange welds. Practical details for plate girders will be discussed in section 4.6.

The load is now transmitted from the stiffeners into the web plate by the stiffener-to-web-plate welds. These are designed to take the load W and may be intermittent or continuous fillet welds.

Finally, the complete stiffener section is designed as a strut with an effective length of 0.7 d. Twenty times the web thickness on each side of the centre line of the stiffener is counted in as resisting compression. The maximum outstand of the stiffener should not exceed $12t$ for flats and $16t$ for sections unless the edge is continuously stiffened. Here t is the thickness of the stiffener. See Figure 4.6(c). A is the area of the strut (= area of stiffener$+40t^2$ mm^2), I_{XX} is the moment of inertia of the stiffener about the XX axis (mm^4), r_{XX} is the radius of gyration of the stiffener about the XX axis (= $(I_{XX}/A)^{1/2}$ mm),

l/r_{XX} is the slenderness ratio of stiffener ($= 0.7d/r_{XX}$), p_c is the safe axial stress (N/mm^2) in compression, from Table 17 of BS 449, for the calculated value of slenderness ratio and

$$\text{Safe load} \quad W = p_c A/10^3 \quad \text{kN}$$

4.5 FLANGE-TO-WEB WELD

The flange-to-web welds are usually fillet welds for plate girders but full-strength welds can be used on heavy box girders. See Figure 4.7.

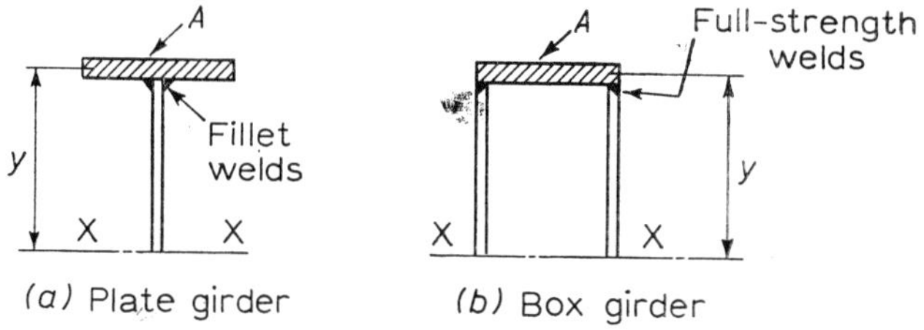

Figure 4.7

The welds are designed for horizontal shear given by

$$\text{Horizontal shear per weld} = SAy/2I_{XX} \quad \text{N/mm}$$

Refer to Chapter 3 for the design of the weld between the cover plate and the flange in a compound beam and for the definition of terms in the equation.

The fillet weld can be intermittent or continuous but continuous welds are preferred.

4.6 CONSTRUCTION DETAILS

4.6.1 Girder Fabrication

Plates are cut to size by flame cutting or plasma-arc cutting. An edge strip, say, 2 mm wide, is then machined off to give a good finish on metal unaffected by the cutting process. This is particularly important with the high-strength steels since these methods of cutting leave a hardened edge zone where failure from fatigue or brittle fracture can begin. Flame-cut edges may be acceptable on plates under 25 mm thick. Weld edge preparation can be made by either flame-cutting or machining.

The plates are assembled in a jig or manipulator and tack welded to hold them in position. The web-to-flange welds are made by an automatic welding machine running on ways beside the girder. Submerged-arc welding or gas-shielded welding is used. The girder is rotated in the jig, so the various welds are made in the downhand position. Welding is continuous with this method of fabrication. Intermittent welding is carried out manually. Special precautions are needed to minimise distortion. These involve preheating thick plates, setting out the welding sequences, etc. The expert advice of a welding engineer is generally required.

4.6.2 Stiffeners

Load-bearing stiffeners are usually welded to the load-bearing flange but they may also be machined for a bearing fit. Fillet weld or full-strength weld is used. Intermediate stiffeners should not be welded to the flanges.

Welding across the tension flange of a girder should be avoided. This causes a high stress concentration or notch at the weld and this can initiate a failure of the girder due to fatigue or brittle fracture.

Stiffeners are usually welded to the web of the girder with continuous fillet weld. This is done manually. Intermittent fillet welding is also used. The inner corners of stiffeners are cut back to clear the flange-to-web welds. This is shown in Figure 4.6.

4.6.3 Connections

Typical connections of beam-to-plate girders are shown in Figure 4.8(a). End supports for plate girders are shown in (b).

4.6.4 Splices

Splices may be necessary in long plate girders. They may be welded or bolted using high-strength friction-grip bolts or a combination of welding and bolting. Splices are shown in Figure 4.9.

In the welded splice shown in Figure 4.9(a), both flange and web are connected by full-strength welds. Thick flange plates require edge preparation for welding from both sides. The weld on the tension flange should be checked for soundness by radiography. The web plates are usually cut back to avoid a stress concentration which would occur where the web splice weld would meet the flange-to-web welds. The web and flange splice welds are staggered as shown. If the flange plates to be spliced are of different thickness, the thicker plate must be tapered to the thickness of the thinner plate. (Refer to Figure 4.9(c).) This reduces the stress concentration here.

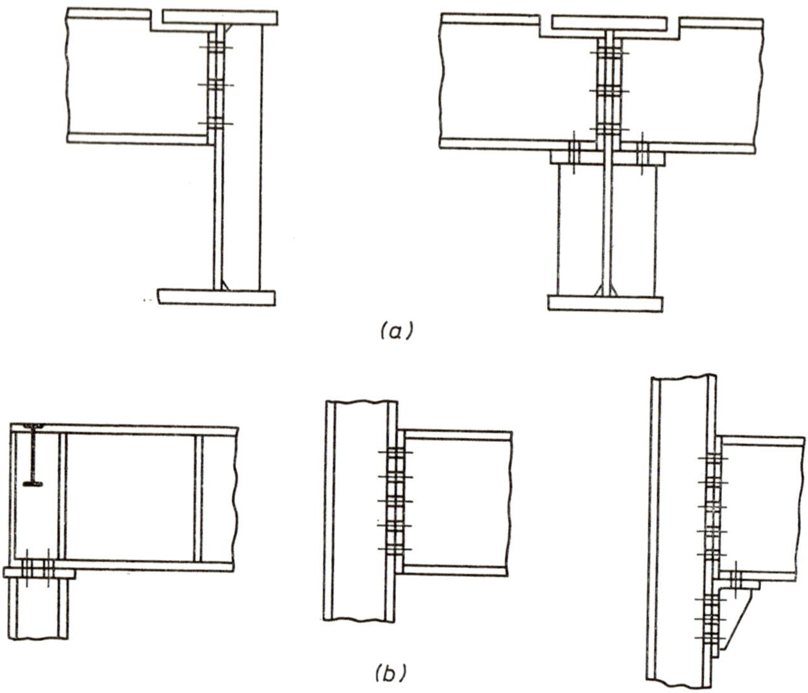

Figure 4.8 Connections for beam-to-plate girders

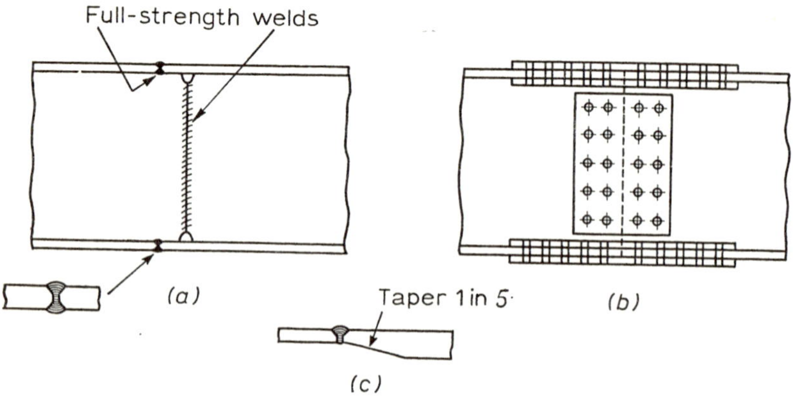

Figure 4.9

4.6.5 Inspection

The girder should be checked for accuracy of fabrication as follows.

1 The length must be correct. Clause 57 of BS 449 states that 'The erection clearance for cleated ends of members connecting steel to steel shall not be greater than 2 mm. The erection clearance at ends of beams without web cleats shall not be greater than 3 mm. . .'
2 The flange and web plates must be true and square and the girder free from twists and distortion.
3 Holes must be set out and drilled correctly to the specified diameter.
4 All stiffeners, seatings, end plates, etc., must be true in position and line.
5 Full-strength welds may require testing. Radiography and ultrasonics are used for this.

4.6.6 Protective Treatment

It is general practice to blast-clean completed fabrications and then to apply the protective treatment. However, the component plates of large built-up members can be blast cleaned prior to fabrication to remove all loose scale and rust and then given a coat of prefabrication primer. After fabrication the steel is thoroughly cleaned to remove dirt, oil, welding slag, spatter, etc., and coated with one coat of red-lead primer; it is delivered to site in this condition. After erection, rust is removed from areas where the primary coat has been damaged; these areas are made good and a second priming coat may be applied. This may be followed by the specified undercoat and finishing coat of micaceous iron oxide paint. The contact surfaces for joints made with high-strength friction-grip bolts should be masked prior to painting.

A general painting specification has been described. There are many variations of this and special paints are available for special conditions. Expert advice should always be sought, particularly if the steelwork is to be erected in a corrosive environment. The student should refer to CP 2008: *Protection of Iron and Steel Structures from Corrosion.*

4.7 DESIGN OF A WELDED PLATE GIRDER

A simply supported welded plate girder has a span of 12 m and carries two concentrated loads of 800 kN each, one load at 4 m from each end of the girder. In addition the girder carries a uniformly distributed load of 30 kN/m which includes an allowance for the self weight of the girder. The two concentrated loads are applied to the top flange of the girder and the compression flange is adequately restrained laterally. The girder is supported on brackets at each end. Design the girder.

(a) Loading, Moments and Shears

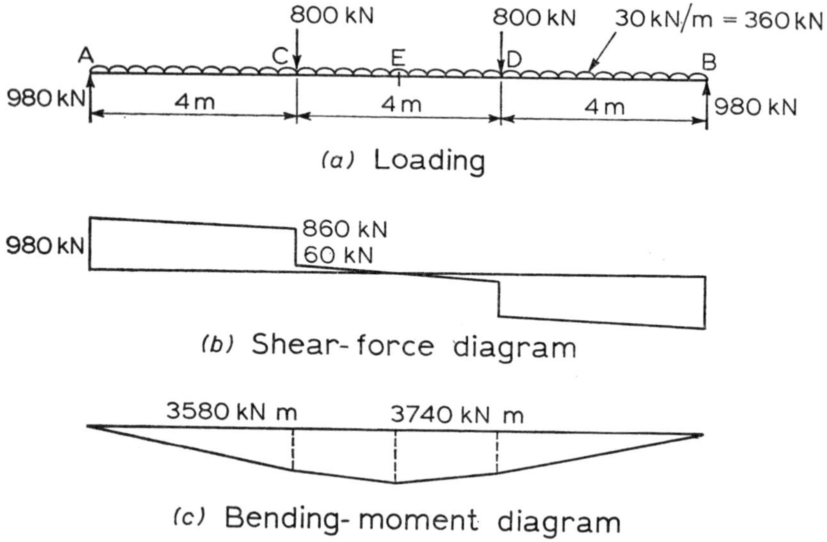

Figure 4.10

Referring to Figure 4.10(a) the loading and reaction for the girder are shown and the shear force diagram can be drawn from this as shown in (b). The moments are:

$$M_C = (980\times4)-(30\times4\times2) = 3580 \text{ kN m}$$

$$M_E = (980\times6)-(800\times2)-(30\times6\times3) = 3740 \text{ kN m}$$

The bending moment diagram is shown in Figure 4.10(c).

(b) Girder Section

Take the overall depth of girder as Span/10, that is 1200 mm. Assume that the flange plates are over 40 mm thick. Then, from Table 2 the allowable stress in bending $p_b = 140$ N/mm². As a first approximation assume that the flange plates resist all the bending moment. That is, the moment is resisted by a couple with a lever arm of, say, 1160 mm as shown in Figure 4.11(a). Then the approximate flange area can be found from:

$$\text{Flange area} = (3740\times10^6)/(1160\times140) = 2.3\times10^4 \text{ mm}^2$$

Try flange plates 450 mm$\times$50 mm.

Refer to Table 12 from which it will be assumed that the allowable shear stress is 100 N/mm² for an initial determination of the thickness of web plate.

The depth of web plate is 1100 mm and the shear is 980 kN at the girder supports.

Thickness of web plate $= (980\times 10^3)/(1100\times 100) = 8.99$ mm

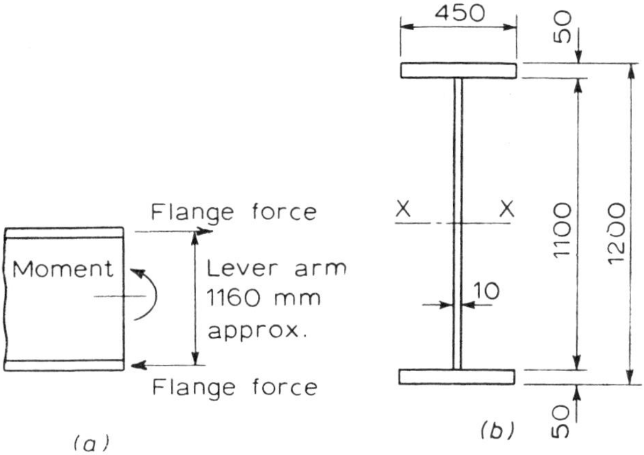

Figure 4.11

Adopt a thickness of 10 mm. The trial section for the girder is shown in Figure 4.11(b). The section properties are calculated and the bending stress checked:

$$I_{XX} = (2\times 50\times 450\times 575^2)+(10\times 1100^3)/12$$
$$= 15.96\times 10^9 \text{ mm}^4$$
$$f_{bc} = (3740\times 10^6\times 600)/(15.96\times 10^9) = 139 \text{ N/mm}^2\text{—safe}$$

For the web plate the ratio $d/t = 110$ so that intermediate stiffeners must be provided. Load-bearing stiffeners are of course required at the supports and under the concentrated loads.

The spacing of the intermediate stiffeners is not to exceed

the lesser of: (i) $1.5d = 1650$ mm

(ii) $180t = 1800$ mm

Adopt the following spacing:

Between A and C—spacing = 1000 mm = $0.91d$

Between C and D—spacing = 1333 mm

Then from Table 12 the allowable shear stress $p'_q = 100$ N/mm². This is the value assumed above. Thus the girder section is satisfactory.

(c) Stiffener under 800 kN Load

Try two stiffeners each 160 mm×20 mm, arranged as shown in Figure 4.12.

The bearing area at the top of the stiffener is shown in Figure 4.12(b).

$$\text{Bearing stress} = (800\times10^3)/(2\times135\times20) = 148 \text{ N/mm}^2\text{—safe}$$

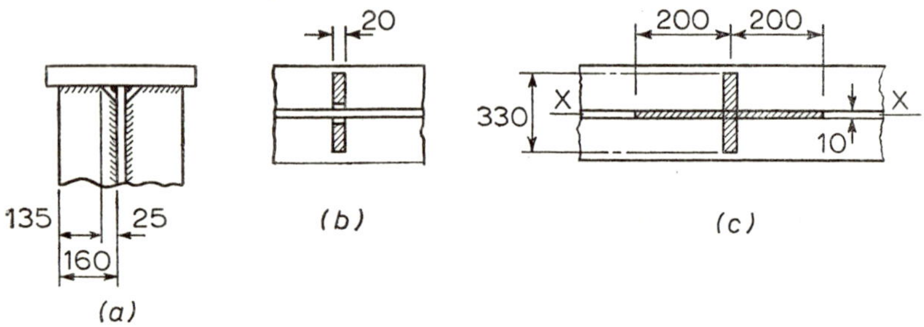

Figure 4.12

The area at the centreline of the web acting as a strut is shown in (c). The following properties are calculated from the dimensions shown on the figure.

$$I_{XX} = (20\times330^3)/12 = 6\times10^7 \text{ mm}^4$$

$$A = (400\times10)+(160\times20\times2) = 10.4\times10^3 \text{ mm}^2$$

$$r_{XX} = [(6\times10^7)/(10.4\times10^3)]^{1/2} = 75.8 \text{ mm}$$

$$l/r_{XX} = (0.7\times1100)/75.8 = 10.2$$

$$p_c = 151 \text{ N/mm}^2 \quad \text{from Table 17}$$

$$f_c = (800\times10^3)/(10.4\times10^3) = 77 \text{ N/mm}^2$$

A smaller size of stiffener could be used but it is customary to design conservatively. The weld length required using 5 mm fillet weld:

Weld strength = 402 N/mm

Length = $800\times10^3/402 = 2000$ mm

There are four welds. Use continuous fillet weld.

(d) End Plate

Design of the end plate is on the same lines as that for the load-bearing stiffener above. The arrangement of the stiffener is shown in Figure 4.13(a). An end plate 20 mm thick has been assumed. The bearing area at the bottom of the stiffener is shown in Figure 4.13(b).

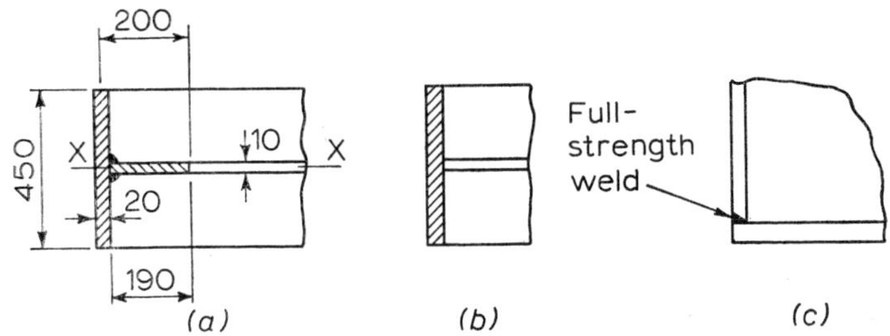

Figure 4.13

Check the bearing stress assuming that the end plate takes all the load

$$\text{Bearing stress} = (980\times 10^3)/(450\times 20) = 108.9\ \text{N/mm}^2 - \text{safe}$$

Maximum outstand of stiffeners is 220 mm, that is $11t$ where t the thickness of stiffener is 20 mm.

This is satisfactory as the maximum outstand of a member in compression should not exceed $12t$ for a flat unless the oustanding leg is stiffened.

Check the section acting as a strut:

$$A = (450\times 20)+(190\times 10) = 10.9\times 10^3\ \text{mm}^2$$

$$I_{XX} = (20\times 450^3)/12 = 15.2\times 10^7\ \text{mm}^4$$

$$r_{XX} = [(15.2\times 10^7)/(10.9\times 10^3)]^{1/2} = 118\ \text{mm}$$

$$l/r_{XX} = (0.7\times 1100)/118 = 6.52$$

$$p_c = 152.5\ \text{N/mm}^2 \quad \text{from Table 17}$$

$$f_c = (980\times 10^3)/(10.9\times 10^3) = 90\ \text{N/mm}^2$$

Weld length required using 6 mm fillet weld:

$$\text{Weld strength} = 482\ \text{N/mm}$$

$$\text{Length} = (980\times 10^3)/482 = 2040\ \text{mm}$$

Two No. 6 mm continuous fillet welds are required between the end plate and the web. A full-strength weld is required between the end plate and top and bottom flange plates as shown in Figure 4.13(c).

(e) Intermediate Stiffeners

Try stiffeners consisting of pairs of flats each 100 mm$\times$10 mm thick. The maximum outstand is $10\,t$ and this is satisfactory. A section through the stiffener is shown in Figure 4.14.

The moment of inertia I of the complete stiffener about the centre-line of the web is given by

$$I = (10\times 210^3)/12 = 7.73\times 10^5\ \text{mm}^4$$

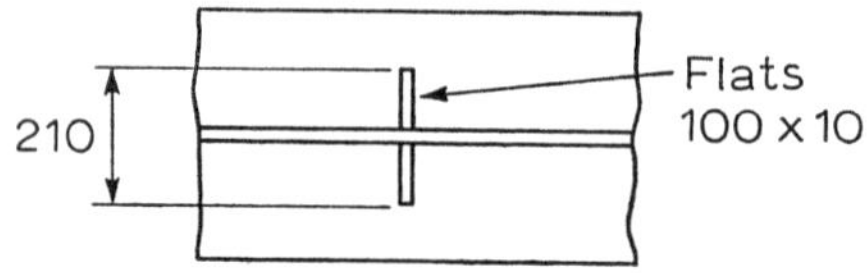

Figure 4.14

Refer to section 4.4.1 where the formula for the minimum moment of inertia of the stiffener is given.

s = maximum permitted clear distance between stiffeners
= 1800 mm—see (b) above.

t = minimum required thickness of web
= 1333/180 = 7.35 mm

or t = 8.99 mm for shear strength—see (a) above. Then I should not be less than

$$(1.5\times1100^3\times8.99^3)/(10^4\times1800^2) = 4.5\times10^5 \text{ mm}$$

The stiffener section chosen is adequate.

(f) Web-to-Flange Welds

Horizontal shear per weld

$= (980\times10^3\times450\times50\times575)/(15.96\times10^9\times2)$

$= 396$ N/mm

Use minimum weld size, that is, 8 mm fillet weld. Use continuous welding since the girder would be manufactured by an automatic process.

(g) A sketch of the girder is shown in Figure 4.15.

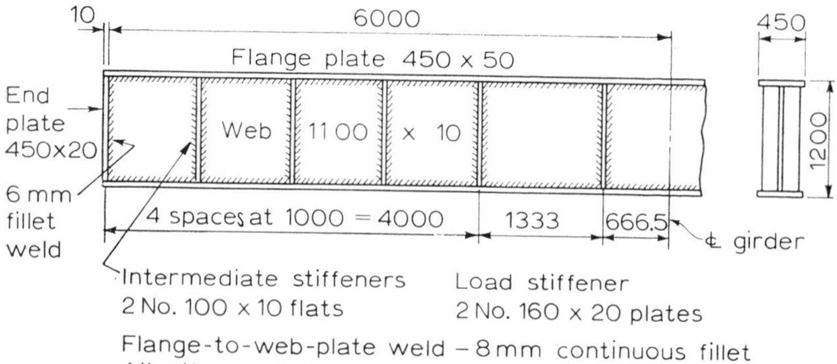

Figure 4.15

4.8 PROBLEMS

1 A welded mild steel plate girder carries a concentrated load of 1440 kN at the centreline. This load is transmitted to the girder from a 254×254×107 kg/m UC that is carried on the top flange. The girder is 15 m span and is simply supported at its ends. The compression flange has adequate lateral supports. Assume the weight of the girder is 60 kN and that the girder is supported on brackets at each end.

(a) Design a section for the girder.
(b) Design the load-bearing stiffeners and web-to-flange weld.
(c) Sketch all details and the arrangement of the girder.

2 The framing plans for a four-storey building are shown in Figure 4.16. The front elevation is to have a plate girder at first-floor level to

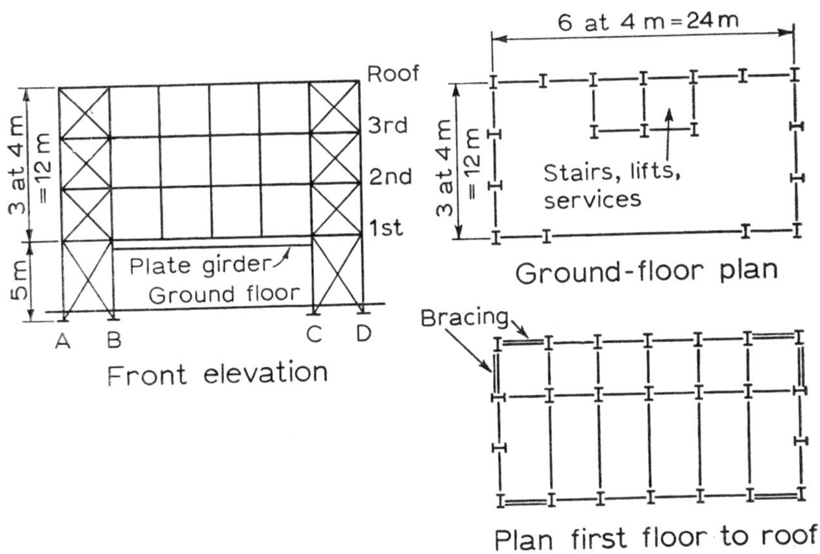

Figure 4.16 Framing plans for a three-storey building

carry wall and floors and give clear access between columns B and C. The plate girder is simply supported with a shear connection between the girder end plates and the column flanges. Columns B and C are 305×305×158 kg/m UC. The loading from floors, roof and wall is as follows.

Dead loads:

Front wall between B and C includes glazing and stanchions	= 71 kg/m^2 = 0.7 kN/mm^2
Floors including slab, steel, screed, finish, ceilings, partitions, etc.	= 610 kg/m^2 = 6.0 kN/m^2
Roof including slab, steel, screed, finish, ceiling etc.	= 408 kg/m^2 = 4.0 kN/m^2

Imposed loads:

Roof	= 1.5 kN/m^2
Floors	= 2.5 kN/m^2

(a) Calculate the loads on the plate girder.

(b) Design the plate girder and show all design information on a sketch.

5

Crane Girders

Crane girders carry hand operated or electric overhead cranes in industrial buildings such as factories, workshops, steelworks, etc. The girders are subjected to vertical and horizontal loads from the weight of the crane, hook load and dynamic loads.

5.1 DESIGN CONSIDERATIONS

5.1.1 Types and Uses

Types of girder used are shown in Figure 5.1. These girders are subjected to horizontal as well as vertical loading, hence a horizontal

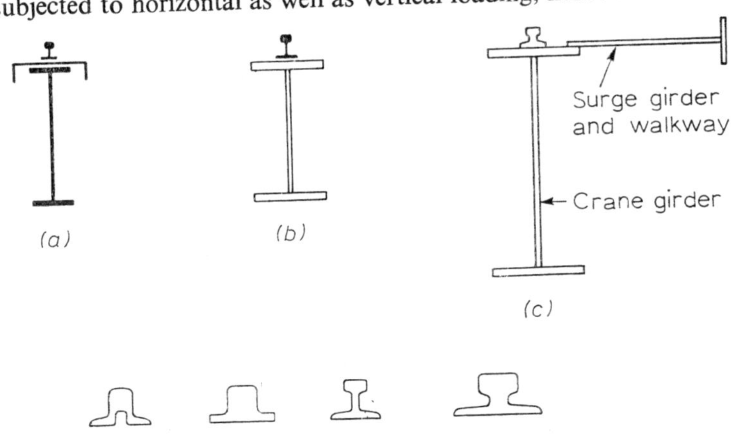

Figure 5.1

beam or larger flange is provided at the top of the girder. Light girders consist of a universal beam and channel as shown in (a). A built-up section is shown in (b). Heavy crane girders consist of a plate girder with a surge girder acting as a walkway as shown in (c). Box girders are also used for heavy cranes. Only the girder type shown in (a) will be considered here.

Various bridge rail sections are shown in the figure in (d). These are not taken into account in calculating the strength of the section. Details of crane rail sections are given in literature of The British Steel Corporation. The size of rail used depends on the capacity of the crane. The fixing of crane rails will be dealt with under practical details later in this chapter. Only simply supported crane girders will be considered here.

5.1.2 Crane Data

Electric overhead cranes or hand operated cranes are considered here. There are other types such as jib cranes and monorail hoists.

The crane wheel loads and dimensions should be obtained from the manufacturer. Dimensions and wheel loads are given in crane manufacturers' handbooks. Crane data required are:

Crane capacity	kN
Span	m
Weight of crane	kN
Weight of crab	kN
End carriage wheel centres	m
Minimum hook approach	m
Maximum static wheel load	kN
End clearance	mm

The data are shown in Figure 5.2.

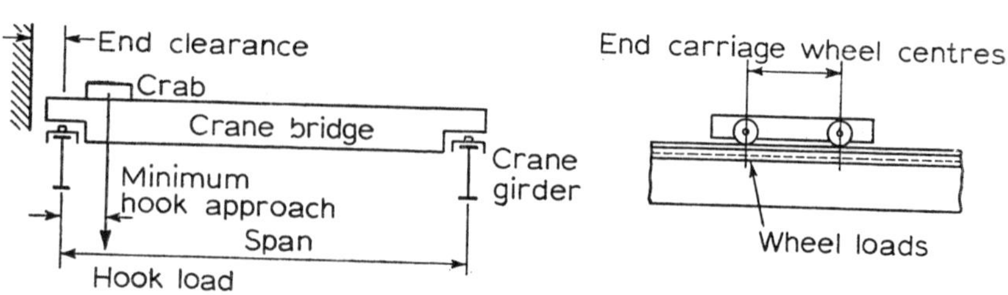

Figure 5.2

Only data required for the design of a crane girder have been listed above. In order to make the arrangement and detail drawings for the building, complete manufacturers' drawings for the crane are required.

5.2 MAXIMUM DESIGN DATA

5.2.1 Loads on Crane Girders

Crane girders are subjected to vertical loads from the weight of the crane, hook load and impact and horizontal loads from crane surge. The dynamic loading given below is taken from BS 449, Chapter 3.

'The following allowances shall be deemed to cover all forces set up by vibration, shock from slipping of slings, kinetic action of acceleration and retardation and impact of wheel loads:

'1 For loads acting vertically, the maximum static wheel loads shall be increased by the following percentages:

for electric overhead cranes	25%
for hand operated cranes	10%

'2 The horizontal force acting transverse to the rails shall be taken as a percentage of the combined weight of the crab and the load lifted as follows:

for electric overhead cranes	10%
for hand-operated cranes	5%

This force shall be taken into account when considering the lateral rigidity of the rails and their fastenings.

'3 Horizontal forces acting along the rails shall be taken as a percentage of the static wheel loads which can occur on the rails as follows:

for overhead cranes, either electric or hand operated 5%

The forces specified in either 2 or 3 above shall be considered as acting at rail level.

Gantry girders and their vertical supports shall be designed on the assumption that either of the horizontal forces 2 or 3 may act at the same time as the vertical load. An increase of 10% on the permissible stresses specified shall be allowed for combinations of loadings 1 and 2 above for design of girders and supporting structures.'

The application of these clauses from the code will be shown later in an example.

5.2.2 Maximum Moments and Shears

The wheel loads are rolling loads and these must be placed in position to give the maximum moment and shear. For two equal wheel loads: The maximum shear occurs when one load is nearly over a support. The maximum moment occurs when the centre of gravity of the loads

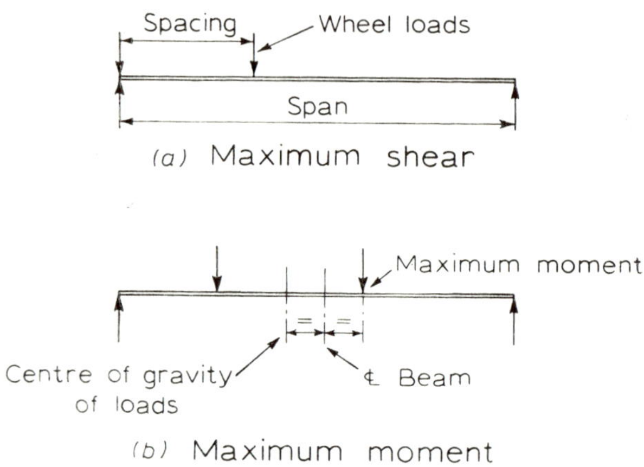

Figure 5.3

and one load are placed equidistant about the centre line of the girder. The maximum moment occurs under the wheel load nearest the centre of the girder.

The theory for these cases is given in structural mechanics textbooks. These load cases are shown in Figure 5.3. If the spacing between loads is greater than 0.586 of the span of the girder for two equal loads the maximum moment will be given by placing one wheel load at the centre of the girder.

5.3 STRESSES

The sections resisting vertical and horizontal bending moment are shown in Figure 5.4 in (a) and (b) respectively. The horizontal bending is resisted by the top flange and channel only.

M_{XX} = vertical bending moment (N mm)

M_{yy} = horizontal bending moment (N mm)

For the vertical bending moment, the bending stresses are given by:

top flange $(f_{bc})_V = M_{XX}/Z_1$ (N/mm^2)—compression

bottom flange $(f_{bt})_V = M_{XX}/Z_2$ (N/mm^2)—tension

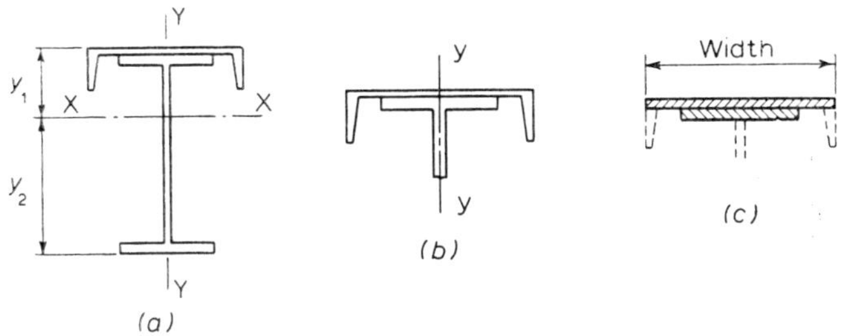

Figure 5.4

where

$Z_1 = I_{XX}/y_1$ = modulus of section for top flange (mm^3)

$Z_2 = I_{XX}/y_2$ = modulus of section for bottom flange (mm^3)

The allowable tensile stress is given in Table 2 of BS 449. The allowable compressive stress for the top flange must be determined by taking lateral instability into account.

Referring to Clause 19(a) of BS 449, the allowable compressive stress depends on values of D/T and l/r_Y and is given in Table 3 in the code where D is the overall depth of the girder, T is the mean thickness of the flange (= area of horizontal portion of the flange divided by the width. See Figure 5.4(c)), l is the effective length of the compression flange, that is the length of the girder and r_Y is the radius of gyration of the beam section about the YY axis. See Figure 5.4(a). These properties may be calculated from first principles. The Handbook on Structural Steelwork lists the properties for crane girder sections composed of universal beams and channels.

The horizontal bending stress is given by

$$(f_{bc})_H \quad \text{or} \quad (f_{bt})_H = M_{yy}/Z_{yy}$$

where Z_{yy} is the modulus of section for the top flange only of the girder. Refer to (b) in the figure. Values of Z_{yy} are given in the *Handbook on Structural Steelwork* for various crane girder sections.

The maximum compressive stress in the top flange is then:

$$(f_{bc})_V + (f_{bc})_H$$

This should not exceed the allowable compressive stress determined above, increased by 10%, where the vertical and horizontal stresses have been added.

Web shear stress, buckling, crushing and combined stresses should be checked as for beams. The channel-to-universal-beam welds are designed for horizontal shear as set out in Chapter 3.

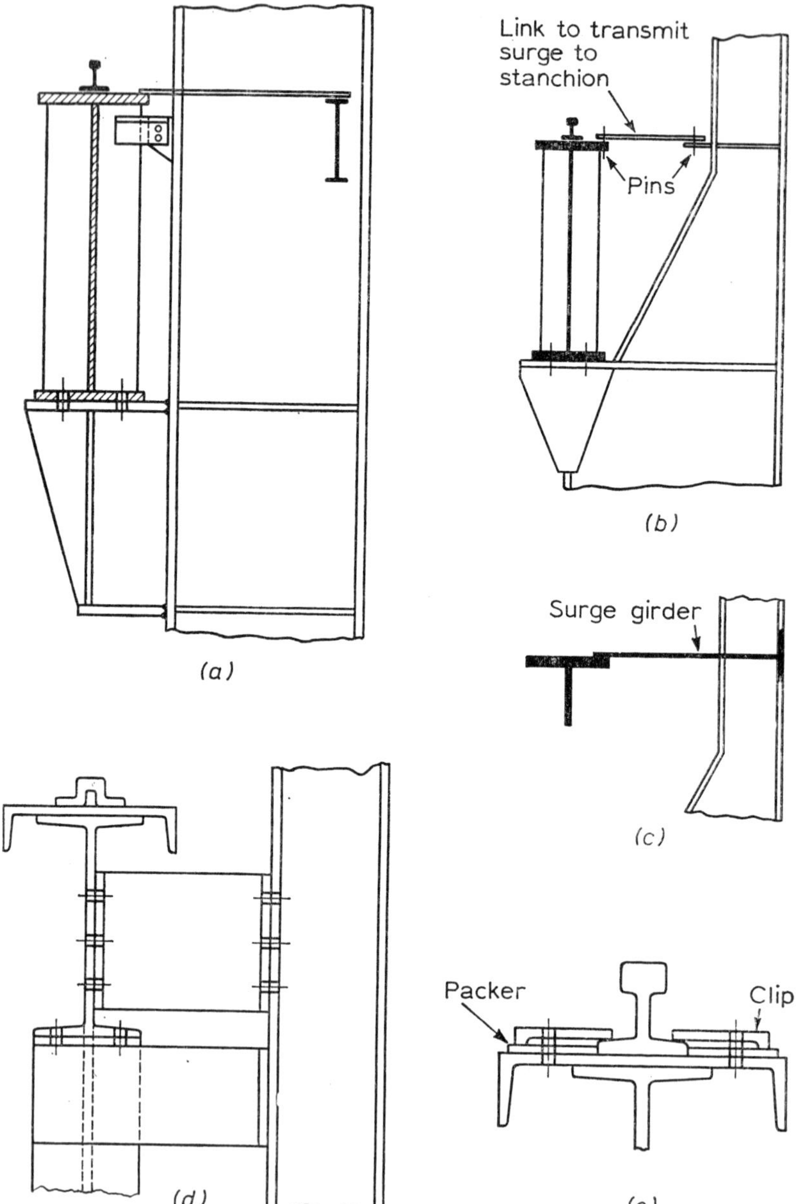
Link to transmit surge to stanchion
Pins
(a)
(b)
Surge girder
(c)
Packer
Clip
(d)
(e)

Figure 5.5

Brackets or links have to be designed to transmit the horizontal load from the top flange to the crane stanchion. Detail design of these parts will not be given here but the arrangement is shown in Figure 5.5.

5.4 CONSTRUCTION DETAILS

Typical details are given in Figure 5.5 for crane rail fixing and the connection of the crane girder to the bracket and stanchion.

Crane rails are sometimes supported on a continuous rubber pad to reduce impact stresses.

5.5 DESIGN OF A COMPOUND CRANE GIRDER

Design a crane girder for an electric overhead crane. The design data are as follows:

Crane capacity	= 300	kN
Span of crane between rails	= 18.3	m
Weight of crane	= 250	kN
Weight of crab	= 50	kN
Minimum hook approach	= 1.06	m
End carriage wheel centres	= 3.64	m
Span of crane girder	= 7.6	m
Self weight of crane girder—estimate	= 16	kN

(a) Maximum Wheel Loads

The crane loads are shown in Figure 5.6.

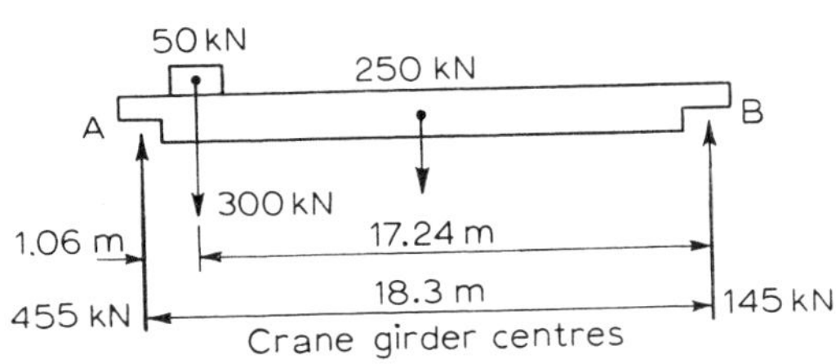

Figure 5.6

The maximum static wheel load at A $= 250/2+(350\times 17.24/18.3)$
$= 455$ kN

This load is applied through two wheels in the end carriage. The vertical wheel load including impact is given by:

Maximum static wheel load	= 227.5 kN
Impact allowance 25%	= 56.9 kN
Total wheel load	= 284.4 kN
The horizontal surge load	= 10% of hook load and weight of crab
Weight of crab	= 10% of (300+50)
	= 35 kN

The surge load is assumed to be transmitted by friction through the four crane wheels to the crane girders on each side. Hence the surge load per wheel is 8.75 kN. Some designers assume that the surge load is applied by the flanges on the wheels and so apply the whole surge load to one girder.

(b) Maximum Bending Moments in Girders

The vertical and horizontal loading causing maximum moments in the crane girder are shown in Figure 5.7(a) and (b) and the positions of the loads to cause maximum shear are shown in (c).
Maximum vertical bending moment—Refer to Figure 5.7(a)

$$R_B = 8+(284.4/7.6)(1.07+4.71) = 225 \text{ kN}$$
$$R_A = 16+(2\times 284.4)-225 = 359.8 \text{ kN}$$
$$M_C = (225\times 2.89)-16\times(2.89/7.6)\times(2.89/2) = 641.3 \text{ kN m}$$

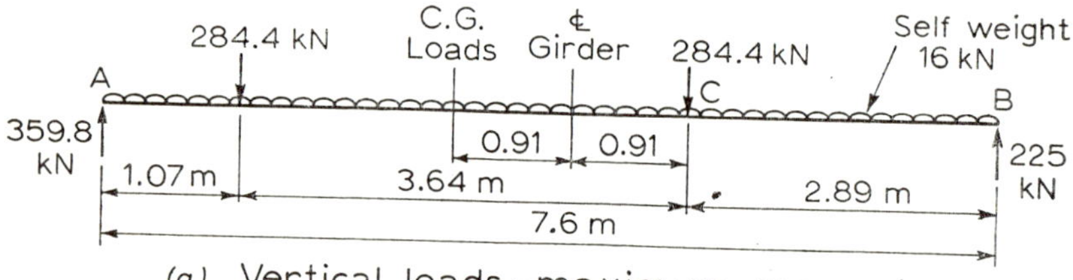

(a) Vertical loads - maximum moment

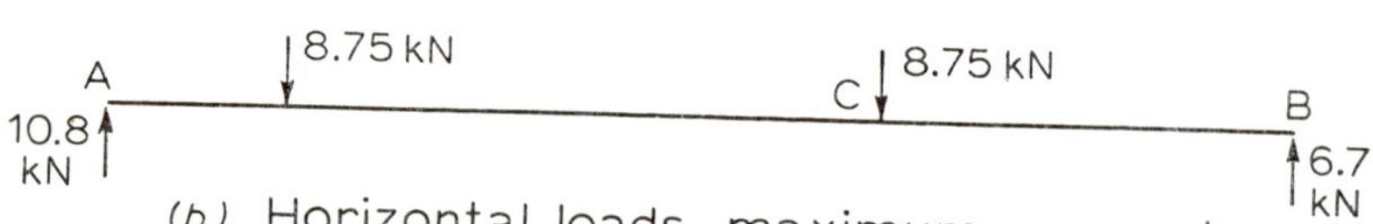

(b) Horizontal loads - maximum moment

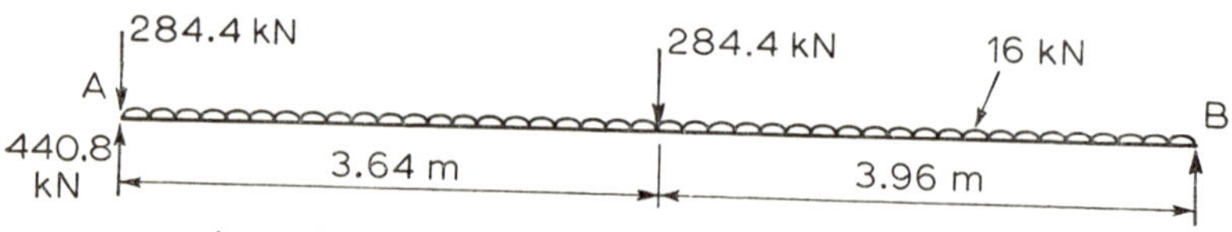

(c) Loads causing maximum vertical shear

Figure 5.7

Maximum horizontal bending moment—Refer to Figure 5.7(b)

$R_B = (8.75/7.6)(1.07+4.71) \quad = \quad 6.7 \text{ kN}$

$R_A = (2\times8.75)-6.7 \quad = \quad 10.8 \text{ kN}$

$M_C = 6.7\times2.89 \quad = \quad 19.4 \text{ kN m}$

Maximum vertical shear—Refer to Figure 5.7(c)

$= 284.4+8+284.4\times(3.96/7.6) = 440.8 \text{ kN}$

(c) Trial Section Properties and Allowable Stresses

Try 610×229×140 kg/m universal beam and 305×89×42 kg/m channel

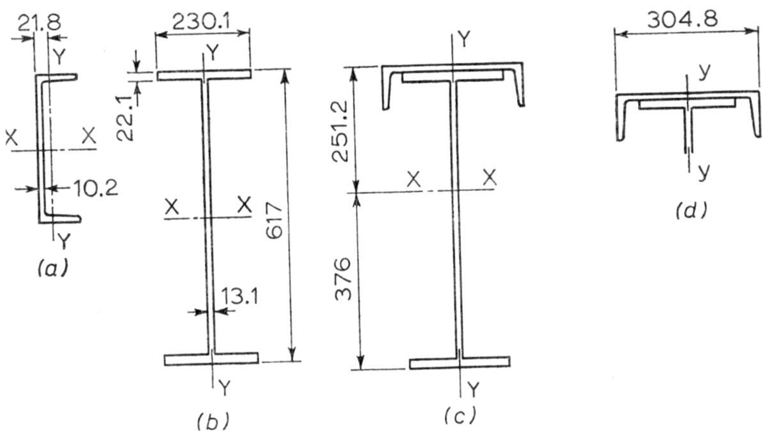

Figure 5.8

The properties from the *Handbook on Structural Steelwork* are:

610×229×140 kg/m UB

$I_{XX} = 111\,673 \text{ cm}^4$

$I_{YY} = 4253 \text{ cm}^4$

$A = 178.2 \text{ cm}^2$

305×89×42 kg/m channel

$I_{XX} = 7\,061 \text{ cm}^4$

$I_{YY} = 325.4 \text{ cm}^4$

$A = 53.1 \text{ cm}^2$

Crane girder—Refer to Figure 5.8(c)

$A = (1.782+0.531)10^4 \qquad = 2.313\times 10^4 \text{ mm}^2$

$\bar{y} = [(1.782\times 10^4\times 308.5)+(531\times 10^4\times 605.4)]/(2.313\times 10^4)$

$= 376 \text{ mm}$

$I_{XX} = 1.117\times 10^9+(1.782\times 10^4\times 67.5^2)$

$+(0.003\times 10^9)+(0.531\times 10^4\times 229.4^2) \qquad = 1.479\times 10^9 \text{ mm}^4$

top flange $Z_1 = (1.479\times 10^9)/251.2 \qquad = 5.9 \times 10^6 \text{ mm}^3$

bottom flange $Z_2 = (1.479\times 10^9)/376 \qquad = 3.94\times 10^6 \text{ mm}^3$

$I_{YY} = (4.253+7.061)10^7 \qquad = 11.314\times 10^7 \text{ mm}^4$

$r_{YY} = [(11.314\times 10^7)/(2.313\times 10^4)]^{1/2} = 70 \text{ mm}$

Mean thickness of top flange

$T = (1/304.8)[(230.1\times 22.1)+(304.8\times 10.2)] = 27 \text{ mm}$

$D/T = 627.2/27 \qquad = 23.2$

Top Section—Refer to Figure 5.8(d)

$I_{yy} = 7.061\times 10^7+(22.1\times 230.1^3)/12 \qquad = 9.311\times 10^7 \text{ mm}^4$

$Z_{yy} = (9.311\times 10^7)/152.4 \qquad = 6.11\times 10^5 \text{ mm}^3$

Allowable compressive stress for top flange

$l/r_{YY} = 7600/70 = 108.5$

$D/T = 23.2$

$p_{bc} = 153 \text{ N/mm}^2$ from Table 3 of BS 449

Allowing for a combination of vertical and surge loads:

$p_{bc} = 153+10\% = 168 \text{ N/mm}^2$

Allowable tensile stress for bottom flange:

$p_{bt} = 165 \text{ N/mm}^2$ from Table 2 of BS 449

Allowable shear stress for web:

$p_q' = 100 \text{ N/mm}^2$

(d) Stresses in Girder

Compressive stress in top flange:

$f_{bc} = (641.3\times 10^6)/(5.9\times 10^6)+(19.4\times 10^6)/(6.11\times 10^5)$

$= 109+31.8 \qquad = 140.8 \text{ N/mm}^2$

Tensile stress in bottom flange:

$f_{bt} = (641.3\times 10^6)/(3.94\times 10^6) \qquad = 163 \text{ N/mm}^2$

Web shear stress:

$f'_q \quad = (440.8 \times 10^3)/(617 \times 13.1) \qquad = \quad 54.5\ N/mm^2$

The girder section is satisfactory.

5.6 PROBLEMS

1 A simply supported crane girder for a 200 kN capacity electric overhead crane spans 7 m. The maximum static wheel loads from the

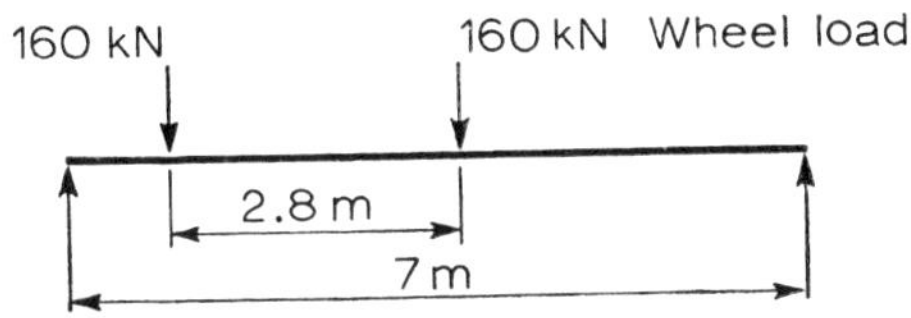

Figure 5.9

end carriage are shown in Figure 5.9. It is proposed to use a crane girder consisting of

533×210×122 kg/m UB and
305× 89× 42 kg/m Channel

The weight of the crab is 40 kN and the self weight of the girder may be taken as 15 kN. Check that this girder is adequate to carry the loads.

2 The following data for an electric overhead travelling crane have been supplied to the design office:

Hook load	=	150 kN
Span of crane	=	15 m
Weight of crane bridge	=	180 kN
Weight of crab	=	40 kN
No. of wheels in end carriage	=	2
Centres of wheels in end carriage	=	3 m
Minimum hook approach	=	1 m

The factory building will have crane stanchions at 8 m centres and the crane girders are to be simply supported. Design the crane girder.

6

Axially loaded stanchions

Stanchions, columns and struts primarily resist axial load. Stanchions are vertical members supporting floors and roofs in buildings.

6.1 DESIGN CONSIDERATIONS

6.1.1 Types and Loading

Some of the sections used for stanchions are shown in Figure 6.1. Rolled, compound and built-up sections are used.

Axial loading carried by stanchions is due to loads from floors,

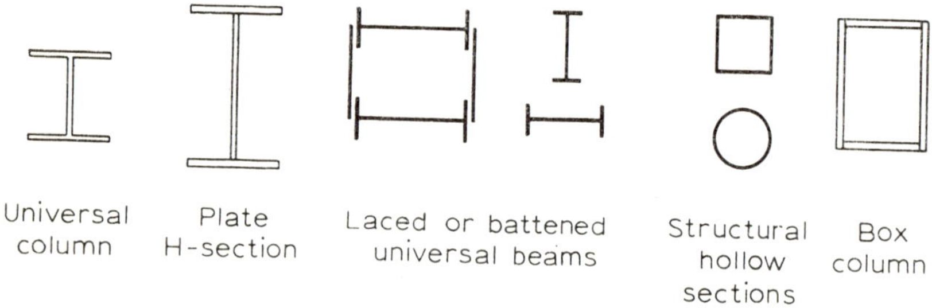

Figure 6.1 Stanchion sections

roofs, walls and self weight. Other loading from cranes, wind, etc., causes bending moments as well as axial loads. Rigid joints in frames cause moments to be transmitted from beams into the stanchions. Eccentrically loaded stanchions are treated in the following chapters.

6.1.2 Strut Action

Stanchions may be classified by length. A short stanchion, post or pedestal fails by crushing of the material as shown in Figure 6.2(a). Long stanchions fail by buckling as shown in (b). Practical stanchions are of this second type.

The safe load that a stanchion can carry depends on

1 member dimensions—area, shape and length
2 end conditions

The shape is important and the tubular stanchion shown in Figure 6.2(c) will carry a much higher load than the round-bar stanchion of the same cross-sectional area, length and end conditions. This is

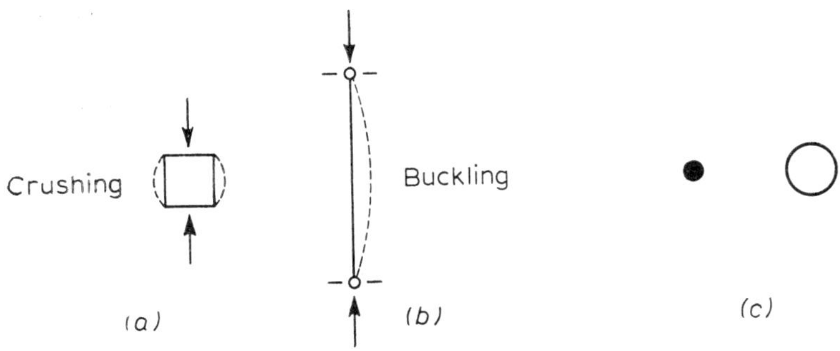

Figure 6.2

because the tube has a higher resistance to buckling. The ideal stanchion section to resist axial load is circular with equal strength in all directions. The practical sections shown in Figure 6.1 approximate to this ideal.

6.2 STRUT THEORY AND CODE REQUIREMENTS

6.2.1 Safe Load on a Strut

The safe axial load on a stanchion depends on:

A = area of cross section
r = least radius of gyration of the section. This measures the effectiveness of the shape of the section
l = effective length which depends on the actual length L and the end conditions of the strut

8*

The safe stress depends on:

$$\text{Slenderness ratio} = \frac{\text{Effective length}}{\text{Least radius of gyration}} = \frac{l}{r}$$

$$\text{where the least radius of gyration} = \left(\frac{\text{Least moment of inertia}}{\text{Area of section}}\right)^{1/2}$$

The safe stresses are given in Table 17 of BS 449. The safe stresses are derived from the Perry Robertson formula. Text books on structural mechanics should be consulted for a full treatment of strut theory.

6.2.2 Effective Length

In Clause 31 of BS 449 rules are given for determining the effective length of a strut. These depend on the end conditions. From this clause the effective length l shall be taken as follows.

1 Effectively held in position and restrained in direction at both ends — $l = 0.7L$
2 Effectively held in position at both ends and restrained in direction at one end — $l = 0.85L$
3 Effectively held in position at both ends, but not restrained in direction — $l = L$
4 Effectively held in position and restrained in direction at one end and at the other partially restrained in direction but not held in position — $l = 1.5L$
5 Effectively held in position and restrained in direction at one end, but not held in position or restrained in direction at the other end — $l = 2.0L$

where L is the length of strut from centre to centre of intersections with supporting members.

These rules are summarised in Figure 6.3 where the deflected shapes of the struts are shown.

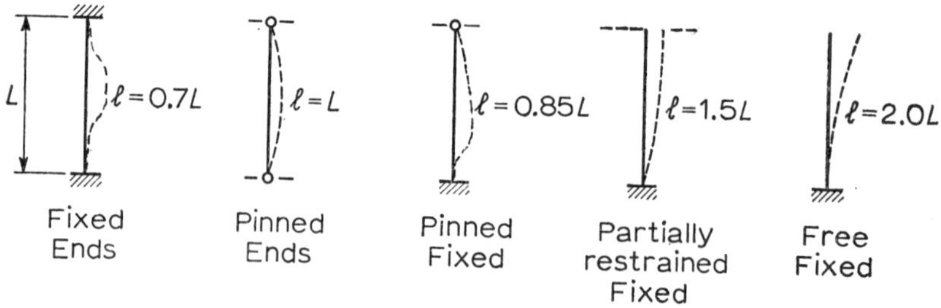

Figure 6.3 Effective lengths of struts

Diagrams are given in Appendix D of BS 449 showing the effective lengths of stanchions for various conditions at the base and at floor beam connections. Some cases for multi-storey building stanchions are shown in Figure 6.4. These have been adapted from the appropriate figures in BS 449. The code should be consulted and read in conjunction with this section.

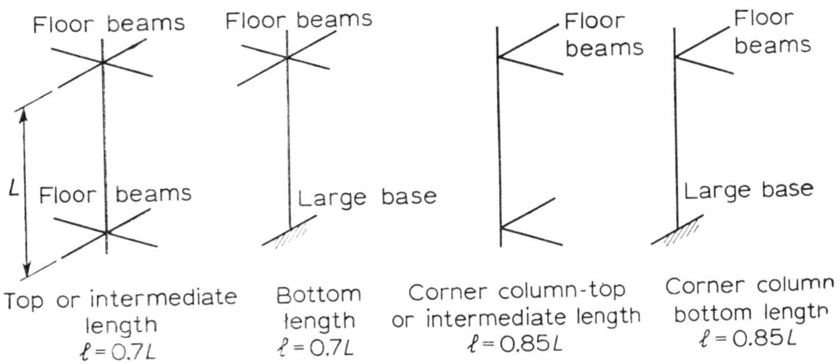

Figure 6.4

6.2.3 Maximum Slenderness Ratio

The rules governing the maximum slenderness ratios for struts are taken from Clause 33 of BS 449. These are:

1 For a member carrying dead or imposed loads, the maximum slenderness ratio is 180

2 For a member carrying wind forces only, provided that the deformation of such a member does not cause an increase in stress in any part of the structure beyond the permissible stress, the maximum slenderness ratio is 250

6.3 EXAMPLES—AXIALLY LOADED STANCHIONS

6.3.1 Comparison of End Conditions

To show the effect of different end conditions on the carrying capacity of a strut, consider the following example. A 203×203×52 kg/m UC has an actual length of 3.6 m. Calculate the safe axial load for

(a) fixed ends
(b) pinned ends

Referring to Figure 6.5, the section properties are $A = 66.4\ \text{cm}^2$,

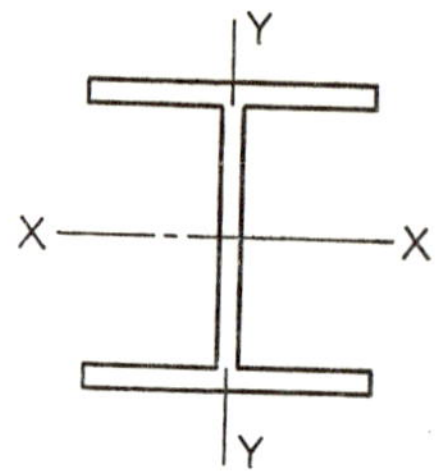

Figure 6.5

$r_{YY} = 5.16$ cm.

(a) Fixed ends

$l/r_{YY} = (3600\times 0.7)/51.6 = 49$

Allowable axial stress in compression from Table 17 of BS 449:

$p_c = 134$ N/mm^2

Safe load $= (66.4\times 10^2\times 134)/10^3 = 890$ kN

(b) Pinned ends

$l/r_{YY} = 3600/51.6 = 70$

$p_c = 115$ N/mm^2

Safe load $= (66.4\times 10^2\times 115)/10^3 = 764$ kN

6.3.2 Internal Stanchion Stack in a Building

A part plan of an office floor and the elevation of centre stanchion stack A are shown in Figure 6.6. The roof and floor loads are as follows.

Roof	Dead load (total) = 510 kg/m^2	= 5 kN/m^2
	Imposed load	= 1.5 kN/m^2
Floors	Dead load (total) = 715 kg/m^2	= 7 kN/m^2
	Imposed load	= 3.0 kN/m^2

Design stanchion A for axial load only. The self weight of the stanchion including fire protection may be taken as 102 kg/m, that is 1 kN/m. The roof and floor steel have the same layout.

When calculating the loads on the column lengths, the imposed loads may be reduced in accordance with Table 2 of CP3, Chapter V, Part 1, 1967. This allows for the fact that it is unlikely that all floors will be fully loaded simultaneously. An extract from this table is given in *Table* 6.1.

Table 6.1

Number of floors carried by member	*Reduction in Imposed Load (%)*
1	0
2	10
3	20

The roof is regarded as a floor for load calculation. The reductions at the critical points are shown on the sketch of the stanchion in Figure 6.8.

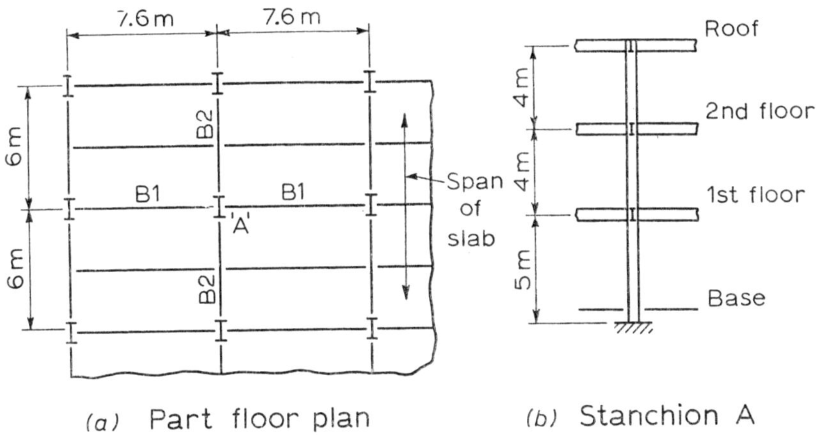

Figure 6.6

The slabs for the floor and roof are one-way slabs spanning as shown on the figure. The dead and imposed loads will be calculated separately.

(a) Loading

Four floor beams are supported at stanchion A. These are designated as B1 and B2 on Figure 6.6(a). The reactions from these beams in terms of a uniformly distributed load are shown in Figure 6.7.

Load on beam B1 $= 7.6 \times 3 \times \omega = 22.8\omega$ kN

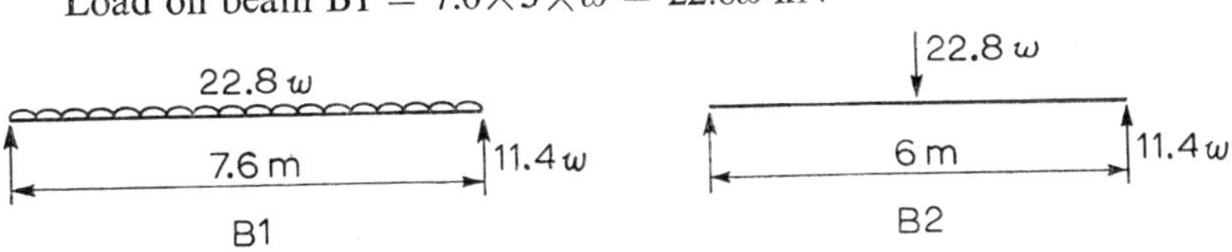

Figure 6.7

where ω is the uniformly distributed load. This is the total dead load including floor steel or the imposed load. Note this treatment of self weight of beam B2 is a simplifying approximation.

The loading on the stanchion may now be set out on a chart as shown in Figure 6.8. Loads for design are required just above first floor, second floor and base.

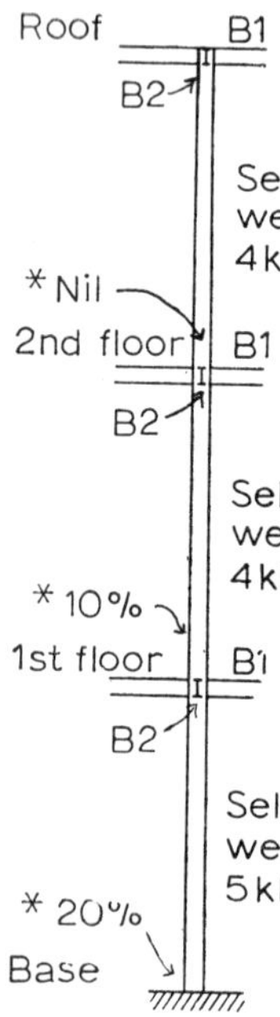

Dead load (kN)	Super load (kN) Reduction 0	10%	20%	Total design load
$\omega = 5\,kN/m^2$	$\omega = 1.5\,kN/m^2$			
2 No. B1 = 114	34.2	30.8	27.4	
2 No. B2 = 114	34.2	30.8	27.4	
Self weight = 4				
Total 232	68.4			300.4
$\omega = 7\,kN/m^2$	$\omega = 3\,kN/m^2$			
2 No. B1 = 160		61.6	54.7	
2 No. B2 = 160		61.6	54.7	
Self weight = 4				
Total 556		184.8		740.8
$\omega = 7\,kN/m^2$				
2 No. B1 = 160			54.7	
2 No. B2 = 160			54.7	
Self weight = 5				
Total 881			273.6	1154.6

Note * Permitted super load reduction

Figure 6.8

(b) Stanchion Design

Top Length—Roof to second floor
Try 152×152×23 kg/m UC.

$A = 29.8\ cm^2$

$r_{YY} = 3.68\ cm$

$l/r_{YY} = (0.7\times4000)/36.8 = 76$

$p_c = 108\ N/mm^2$ from Table 17 (BS 449)

$f_c = (300.4\times10^3)/(29.8\times10^2) = 101\ N/mm^2$

Section is satisfactory.

Intermediate length—Second floor to first floor
Try 203×203×46 kg/m UC.

$A = 58.8\ \text{cm}^2$
$r_{YY} = 5.11\ \text{cm}$
$l/r_{YY} = (0.7\times4000)/51.1 = 54.8$
$p_c = 130\ \text{N/mm}^2$ from Table 17 (BS 449)
$f_c = (740\times10^3)/(58.8\times10^2) = 126\ \text{N/mm}^2$

Section is satisfactory.

Bottom Length—Second floor to base
Try 203×203×86 kg/m UC.

$A = 110.1\ \text{cm}^2$
$r_{YY} = 5.32\ \text{cm}$
$l/r_{YY} = (0.7\times5000)/53.2 = 66$
$p_c = 120\ \text{N/mm}^2$ from Table 17
$f_c = (1154.6\times10^3)/(110.1\times10^2) = 105\ \text{N/mm}^2$

Section is satisfactory.

Several trials may be necessary to find the most economical section. Safe load tables in the Handbook on Structural Steelwork may also be used. These give the safe loads that universal columns can carry for various effective lengths.

6.3.3 Example—Comparison of Stanchion Sections

The purpose of this example is to compare the following stanchion sections:

1 Universal column
2 Circular hollow section
3 Rectangular hollow section

Design a stanchion in the three sections mentioned above to carry an axial load of 750 kN. In all cases the stanchion has an effective length of 4.2 m.

(a) Universal column
Try 203×203×60 kg/m UC.

$A = 75.8\ \text{cm}^2$
$r_{YY} = 5.19\ \text{cm}$
$l/r_{YY} = 4200/51.9 = 81$
$p_c = 102\ \text{N/mm}^2$
$f_c = (750\times10^3)/(75.8\times10^2) = 99\ \text{N/mm}$

(b) Circular hollow section

Try 219 O.D.×9.5×49.2 kg/m CHS

$$A = 62.6 \text{ cm}^2$$
$$r = 7.4 \text{ cm}$$
$$l/r = 4200/74 = 56.8$$
$$p_c = 128 \text{ N/mm}^2$$
$$f_c = (750\times 10^3)/(62.6\times 10^2) = 120 \text{ N/mm}^2$$

(c) Rectangular hollow section

Try 178×178×9.5×50 kg/m RHS

$$A = 63.60 \text{ cm}^2$$
$$r = 6.85 \text{ cm}$$
$$l/r = 4200/68.5 = 61.4$$
$$p_c = 124.5 \text{ N/mm}^2$$
$$f_c = (750\times 10^3)/(63.6\times 10^2) = 118 \text{ N/mm}^2$$

Thus a saving in weight of about 16% is made if the structural hollow sections are used in this case. This is due to the larger radius of gyration of the closed section giving a lower slenderness ratio and higher permissible stress. The Tubes Division of the British Steel Coporation give safe load tables for structural hollow sections.

6.4 BUILT-UP STANCHIONS

6.4.1 Code Requirements

The two main types of stanchions built up from steel plates are the H section and the box stanchions shown in Figure 6.9.

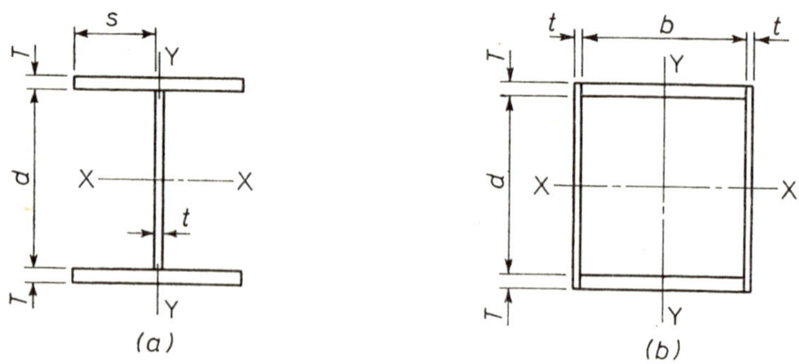

Figure 6.9

The design of these types of stanchion should be controlled by Clause 32 of BS 449. The requirements of this clause are:

'1 The thickness of an outstanding leg of any member in compression, unless the leg is stiffened, shall not be less than one-sixteenth of the outstand for grade 43 steel.'
Referring to Figure 6.9(a), s must not be greater than $16t$.

'2 Unless effectively stiffened, the unsupported width of a plate forming any part of a member primarily in compression, measured between adjacent lines of rivets, bolts or welds connecting the plates to other parts of the section shall not exceed the following:

$$90t \text{ for grade 43 steel}$$

where t is the thickness of a single plate, or the total thickness of two or more plates effectively tacked together'

The clause continues:

'However, in computing the effective area and radius of gyration the unsupported width of plate shall be reckoned as not more than $50t$ for grade 43 steel.... In computing other section properties, the full area of the plate shall be taken.'

The application of the code requirements is shown in the following example.

6.4.2 Example

Determine the safe axial load that the stanchion section shown in Figure 6.10(a) can carry. The effective length of the stanchion is 8 m.

Flange outstand $= 445 = 14.8T$

where T is the flange thickness, 30 mm. This is satisfactory. The clear width of web plate is $84t$, where t is the thickness of web. This does not

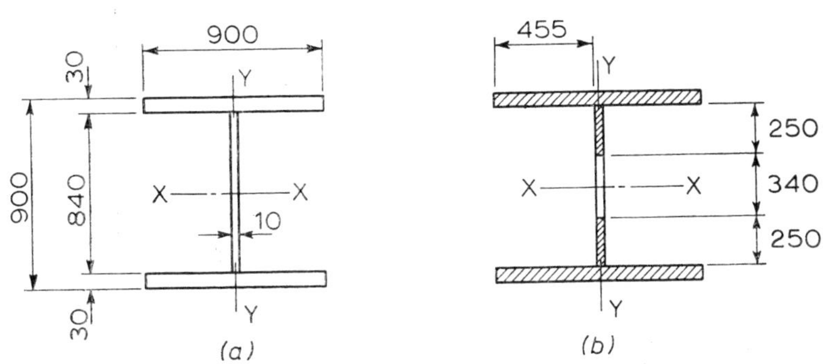

Figure 6.10

exceed $90t$, but only $50t$—that is 500 mm of the width—is used in calculating the effective area and radius of gyration. The effective section is shown in Figure 6.10(b). The properties of this section are:

$$A = (900\times60)+(500\times10) = 5.9\times10^4\ \text{mm}^2$$
$$I_{YY} = (60\times900^3)/12+(\text{neglect web}) = 3.65\times10^9\ \text{mm}^4$$
$$r_{YY} = [(3.65\times10^9)/(5.9\times10^4)]^{1/2} = 248\ \text{mm}$$
$$l/r_{YY} = 8000/248 = 32.1$$
$$p_c = 142\ \text{N/mm}^2 \quad \text{from Table 17 (BS 449)}$$
$$\text{Safe load} = (142\times5.9\times10^4)/10^3 = 8360\ \text{kN}$$

6.5 LACED AND BATTENED STANCHION

It is not the intention here to present detailed designs for these stanchions. Only some of the broad aspects will be examined. This is because modern practice tends towards the use of built-up H and box sections for main stanchions. These sections are better able to be protected against corrosion. Laced sections are often used for secondary struts.

The purpose of lacing two members together is that each member supports the other through the lacing or battening against buckling about the weaker axis.

6.5.1 Laced Stanchion, Code Requirements

The arrangement of lacing bars for one of these stanchions is shown in Figure 6.11(a). These are single lacing systems. Double lacing systems are also used.

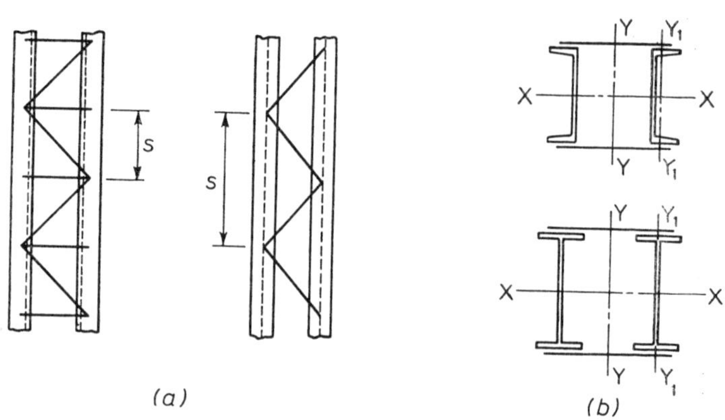

Figure 6.11

For the complete stanchion section shown in Figure 6.11(b), r_{XX} and r_{YY} are calculated. Then the slenderness ratio for buckling about these axes may be determined depending on the end conditions for the stanchion. Refer to section 6.2 above and Appendix D of BS 449. For the individual member, the effective length for buckling about the $Y_1 Y_1$ axis is taken as the spacing of the lacing bars.

Rules for the design of laced struts are given in Clause 35 of BS 449. The following information is summarised from this clause for welded construction:

(a) Referring to Figure 6.11(b), the spacing of the two members should be such that the radius of gyration about the YY axis is not less than that about the XX axis.
(b) The lacing system should be uniform throughout the length of the strut.
(c) The lacing of compression members shall be proportioned to resist a transverse shear force of $2\frac{1}{2}\%$ of the axial force in the member. This shear force shall be considered to be divided equally between all lacing systems in parallel planes. That is if the member is laced on each side, half the shear would be taken by the lacing on each side. If the member resists moment as well as axial force, then the shear due to moment must be added to that above.
(d) The section of lacing bars for compression members shall be such that the l/r ratio shall not exceed 140. For single lacing systems the effective length in welded construction is taken as the distance between the inner ends of the weld connecting the bars to the members.
(e) The thickness of flat lacing bars shall be not less than 1/40 of the length between the inner ends of welds for single lacing systems. Rolled sections can be used.
(f) The angle of inclination shall be not less than 40° nor more than 70° to the axis of the member.
(g) The maximum spacing of lacing bars shall be such that the maximum slenderness ratio l/r of the components of the strut between consecutive connections is not greater than 50 or 0.7 times the most unfavourable slenderness ratio of the strut as a whole, whichever is the less. Here l is the distance between the centres of the connections of the lacing bars to each component.
(h) In welded construction, lacing bars should overlap the main members by four times the thickness of the bar or mean thickness of flange of member connected. Welding must be provided along each side of bar for the full lap.
(i) Tie plates must be provided at the ends of members. The length of tie plates shall be not less than the perpendicular distance between the welds connecting the plates to the main members.

The main parts of the clause have been set out in (a)–(i). These will be used to make a simple design of a laced strut. The complete clause in BS 449 should be studied and other references on structural design in steel should be consulted if the design of large laced compression members is to be undertaken.

6.5.2 Example—Laced Stanchion

It is proposed to use a laced member consisting of two channels, as shown in Figure 6.12(a), to carry an axial compressive load of 650 kN.

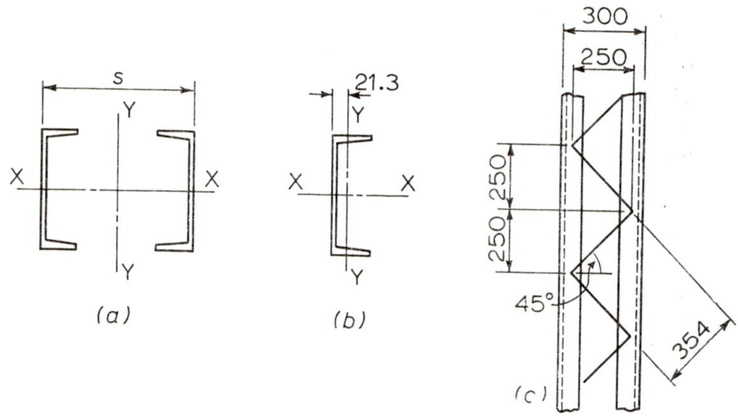

Figure 6.12

The actual length L of the member is 8 m and the effective lengths for buckling about the two axes are:

XX axis—effective length $\quad l_{XX} = 0.7L$

YY axis—effective length $\quad l_{YY} = 1.0L$

Design the strut and determine the section and arrangement of the lacing bars.

Try two No. 203×76×23.82 kg/m channels. Refer to Figure 6.12(b). The properties for a channel section are:

$$A = 30.34\ \text{cm}^2$$
$$r_{XX} = 8.02\ \text{cm}$$
$$r_{YY} = 2.23\ \text{cm}$$
$$I_{YY} = 151.4\ \text{cm}^4$$

Make the spacing of the channels such that buckling about the XX axis controls the strength of the strut. Then

$$l_{XX}/r_{XX} = (0.7\times 8000)/80.2 = 70$$

$$p_c = 115 \text{ N/mm}^2 \qquad \text{from Table 17 (BS 449)}$$

$$\text{Safe load} = (115\times 2\times 30.34\times 10^2)/10^3 = 696 \text{ kN}$$

The strut can safely carry the required load.

It is now necessary to determine the spacing of the channels such that the slenderness ratio for buckling about the YY axis is 70.

Required $\qquad r_{YY} = 8000/70 = 114.3$ mm

Try a spacing s of 300 mm, then:

$$I_{YY} = 2\times 151.4\times 10^4 + 2\times 30.34\times 10^2\times 128.7^2$$

$$= 103.6\times 10^6 \text{ mm}^4$$

actual $\qquad r_{YY} = [(103.6\times 10^6)/(2\times 30.34\times 10^2)]^{1/2} = 130$mm

The spacing is satisfactory.

Adopt flat lacing bars at 45° to the axis of the member. The arrangement of the bars and approximate lengths for preliminary calculations are shown in Figure 6.12(c). This arrangement will be checked and the size of lacing bars determined using these dimensions.

Slenderness ratio of the component of strut between lacing bars $l/r = 500/22.3 = 22.4$

This value should not exceed

(i) 50

(ii) $0.7\times 70 = 49$

The inclination of the bars is 45°, that is between the limits 40° and 70°.

Thickness of bar = 354/40 = 10 mm, say

Try lacing bar 40 mm × 10 mm

Minimum radius of gyration $r = 10/\sqrt{12} = 2.89$ mm

Slenderness ratio $l/r = 354/2.89 = 123$

This is less than 140 so it is satisfactory.

Allowable compressive stress $p_c = 57$ N/mm² from Table 17 (BS 449).

Force in one lacing bar $= (0.025\times 650\times 354)/(2\times 250) = 11.5$ kN

Stress $\qquad f_c = (11.5\times 10^3)/(10\times 40) = 28.9$ N/mm²

Section is satisfactory.

A detail could now be drawn for the lacing bars and the final arrangement could be checked against the code requirements. Finally the welding required can be determined.

6.5.3 Battened Stanchion, Code Requirements

The arrangement of batten plates is shown in Figure 6.13. Battened compression members are dealt with in Clause 36 of BS 449. The following information is summarised from this clause.

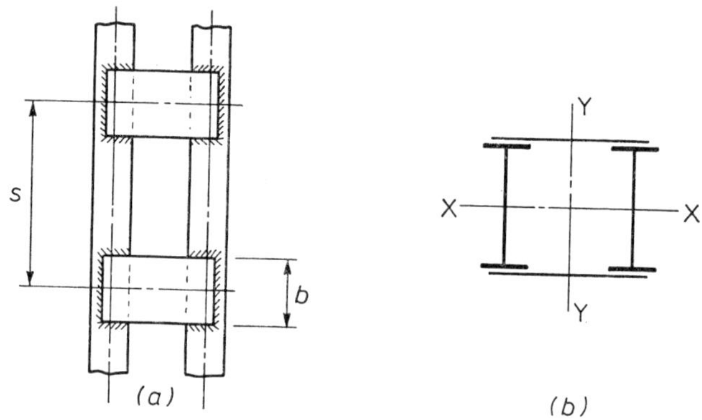

Figure 6.13

(a) The member should have two components preferably of the same cross section. Battens should be opposite each other at the ends of the member and intermediate battens should divide the member into not less than three bays.

(b) If the slenderness ratio about the YY axis, l_{YY}/r_{YY}, is not more than 0.8 times the slenderness ratio about the XX axis, l_{XX}/r_{XX}, the spacing of battens centre to centre of end fastenings shall be such that the slenderness ratio l/r of the lesser main component over that distance shall be not greater than 50 or greater than 0.7 times the slenderness ratio of the member as a whole about its XX axis. If the slenderness ratio about the YY axis is more than 0.8 times the slenderness ratio about the XX axis, the spacing of battens centre to centre of end fastenings shall be such that the slenderness ratio of the lesser main component over that distance shall be not greater than 40 or greater than 0.6 times the slenderness ratio of the member as a whole about its weaker axis.

(c) The battens shall be designed to carry the bending moments and shears arising from a transverse shear force F_q of $2\frac{1}{2}\%$ of the total axial force on the whole compression member at any point in the length of the member divided equally between parallel planes of battens,

(d) Battens shall be plates, channels or I sections riveted or welded to the main components to resist simultaneously:

a longitudinal shear force $F_1 = F_q d/na$

and a moment $M = F_q d/2n$

where d is the longitudinal distance centre to centre of battens, a is the minimum transverse distance between the centroids of the rivet groups or welding, F_q is the transverse shear force given in (c) and n is the number of parallel planes of battens.

(e) End battens... shall have an effective length, longitudinally, of not less than the perpendicular distance between the centroids of the main members and intermediate battens shall have an effective length of not less than three-quarters of this distance but in no case shall the effective length of any batten be less than twice the width of one member in the plane of the battens.

(f) The effective length of a batten shall be taken as the longitudinal distance between... end welds.

(g) Batten plates shall have a thickness of not less than 1/50 of the minimum distance... between welds... . Where channels... are used as battens... this requirement does not apply.

(h) The length of weld connecting each longitudinal edge of the batten plate to a member shall, in the aggregate, be not less than half the length of the batten plate, and at least one third of the weld shall be placed at each end of the longitudinal edge. In addition, the welding shall be returned along the ends of the plate for a length equal to at least four times the thickness of the plate.

Where tie or batten plates are connected between main members they shall be connected to each member by fillet welds on each side of the plate, equal in length to at least that specified in the preceding paragraph, or by complete-penetration butt welds.

The application of these rules will be shown in the example that follows.

6.5.4 Example—Battened Stanchion

A battened compression member is made up of 2 No. 381×152×52 kg/m UB spaced at 300 mm centre to centre. The effective length of the member is 6.5 m. Determine the safe load for the member and the size and spacing of the batten plates.

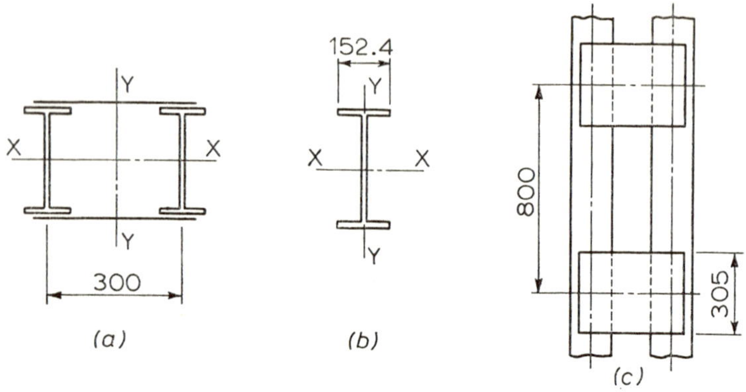

Figure 6.14

The section of the member is shown on Figure 6.14(a). The properties for the universal beam shown in Figure 6.14(b) are as follows.

$$A = 66.4 \text{ cm}^2$$
$$r_{YY} = 3.21 \text{ cm}$$
$$r_{XX} = 15.5 \text{ cm}$$
$$I_{YY} = 685 \text{ cm}^4$$

For the stanchion section:

$$I_{YY} = 2\times685\times10^4+2\times66.4\times10^2\times150^2$$
$$= 31.17\times10^7 \text{ mm}^4$$
$$r_{YY} = [(31.17\times10^7)/(2\times66.4\times10^2)]^{1/2} = 153 \text{ mm}$$
$$r_{XX} = 155 \text{ mm}$$
$$l/r_{YY} = 6500/153 = 42.5$$
$$p_c = 137.5 \text{ N/mm}^2 \quad \text{from Table 17 (BS 449)}$$
$$\text{Safe load} = (2\times66.4\times10^2\times137.5)/10^3 = 1822 \text{ kN}$$

l/r_{YY} is more than $0.8l/r_{XX}$, so the spacing of battens shall be such that l/r of the member between the battens shall not exceed:

(i) 40

(ii) $0.6\times42.5 = 25.5$

Spacing $l = 25.5\times32.1 = 818$ mm
Adopt a spacing of 800 mm.

End battens—The length longitudinally is to be not less than the perpendicular distance between the centroids of the main members length = 300 mm.

Intermediate batten—The length is to be not less than $\frac{3}{4}$ of the length of the end batten = 225 mm or length of any batten is to be not less than twice the width of one member in the plane of the battens:

$$\text{Length} = 2\times152.4 = 304.8 \text{ mm}$$

Make length of all batten plates 305 mm. The batten plate arrangement is shown in Figure 6.14(c).

A detail of the batten plate is shown in Figure 6.15(a) and the weld is shown in (b).

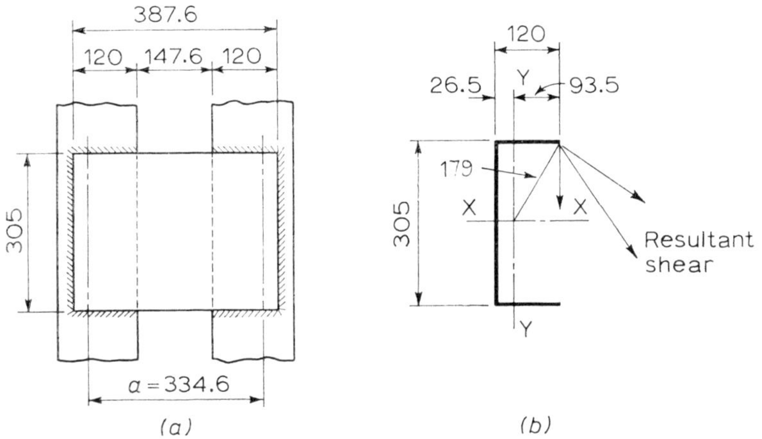

Figure 6.15

The properties of the weld group can be calculated as set out in section 2.3.2. These are:

$$L = 545 \text{ mm}$$

$$I_p = 8.72\times10^6 \text{ mm}^3$$

Forces on batten plate and weld:

Transverse shear $\quad F_q = 0.025\times1822 = 44.6 \text{ kN}$

Longitudinal shear $\quad F_l = F_q d/na$

$$= (44.6\times800)/(2\times334.6) = 58.6 \text{ kN}$$

Moment $\quad M = F_q d/2n$

$$= (44.6\times800)/(2\times2) = 8920 \text{ kN mm}$$

Forces on weld

Direct shear $= (58.6\times10^3)/545 = 107.8 \text{ N/mm}$

Shear due to moment $= (8920\times10^3\times179)/(8.72\times10^6)$

$$= 183 \text{ N/mm}$$

Resultant shear $= (107.8^2+183^2+2\times107.8\times183\times93.5/179)^{1/2}$

$= 256$ N/mm

Use 4 mm fillet weld, strength 322 N/mm

Batten plate thickness = 387.6/50 = 7.75 mm

Use 8 mm plate. This will be strong enough to resist the above forces.

Summarising the batten plate design:

Batten plates—305×390×8 plate at 800 mm centres

Weld—4 mm continuous fillet weld.

6.6 CASED STANCHIONS

6.6.1 Code Requirements

Solid concrete casing acts as fire protection for the steel stanchion and the casing assists in carrying the load and preventing the column from buckling if the design is in accordance with the recommendations in BS 449. The regulations governing design are set out in Clause 30(b) of the code. These requirements are as follows.

'Struts of single I section or of two channels back to back in contact or spaced apart not less than 20 mm or more than half their depth and battened or laced... may be designed as cased struts when the following conditions are met:

'1 The steel strut is unpainted and solidly encased in ordinary dense concrete with 10 mm aggregate (unless solidity can be obtained with a larger aggregate) and a works strength not less than 21 N/mm² at 28 days...

'2 The minimum width of solid casing is equal to $b+100$ mm where b is the overall width of the steel flange or flanges in millimetres.

'3 The surface and edges of the steel strut have a concrete cover of not less than 50 mm.

'4 The casing is effectively reinforced.... The wire shall be at least 5 mm in diameter and the reinforcement shall be in the form of stirrups or binding at not more than 150 mm pitch, so arranged to pass through the centre of the covering of the edges and outer faces of the flanges and supported by and attached to longitudinal spacing bars of not less than 4 in number.

'The radius of gyration of the cased strut about the axis in the plane of its web may be taken as $0.2(b+100)$ mm. The radius of gyration about its other axis shall be taken as that of the uncased section.

'In no case shall the axial load on a cased strut exceed twice that which would be permitted on the uncased section, nor shall the slenderness ratio of the uncased section measured over its full length exceed 250.

'In computing the allowable axial load on the cased strut the concrete shall be taken as assisting in carrying the load over its rectangular cross section, any cover in excess of 75 mm from the overall dimensions of the steel section of the cased strut being ignored.

'This section of concrete shall be taken as assisting in carrying the load on the basis of a stress equal to the allowable stress in the steel (as given in Table 17) divided by 0.19 times the numerical value of p_{bc} given in Table 2 for the grade of steel concerned.'

The requirements above regarding the casing are shown in Figure 6.16.

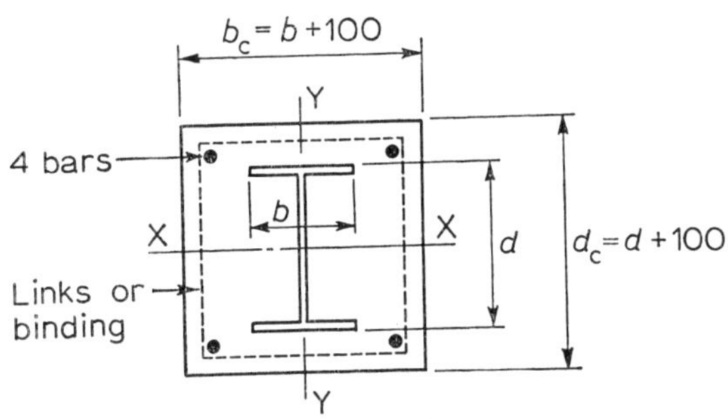

Figure 6.16

Radii of gyration:

$$r_{XX} = \text{radius of gyration of steel section}$$

$$r_{YY} = 0.2(b+100) \text{ mm}$$

$$\text{Safe load} = p_c A_S + p_c A_c / 0.19 p_{bc}$$

where p_c is the allowable compressive stress from Table 17 for the slenderness ratio of the cased strut, A_S is the area of steel section, A_c is the gross area of cased strut ($= b_c d_c$) and p_{bc} is 165 N/mm^2 for grade 43 steel.

Safe loads for cased struts for various effective lengths are given in the Handbook on Structural Steelwork.

6.6.2 Example

An internal stanchion in a multi-storey building has an actual length of 4.2 m centre to centre floor beams. The steel section is 203×203×52 kg/m UC. Calculate the safe axial load that this strut can carry if cased in accordance with Clause 30(b) of BS 449.

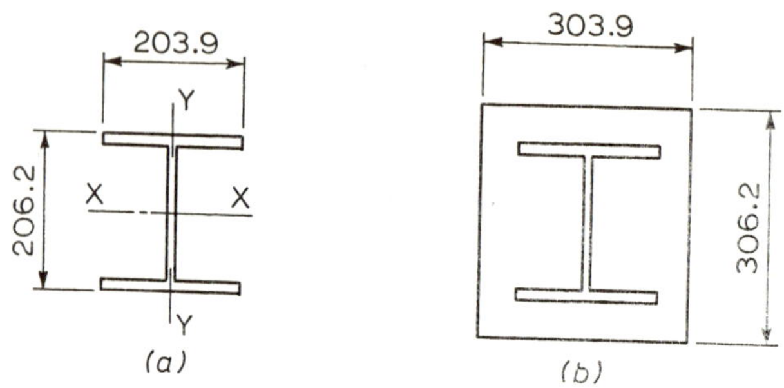

Figure 6.17

The steel core and cased section are shown in Figure 6.17. The properties of the steel section are:

$$A = 66.4 \text{ cm}^2$$
$$r_{XX} = 8.9 \text{ cm}$$
$$r_{YY} = 5.16 \text{ cm}$$

(a) Safe load on the steel section

$$l/r_{YY} = (0.7\times4200)/51.6 = 57$$
$$p_c = 128 \text{ N/mm}^2 \text{ from Table 17 (BS 449)}$$
$$\text{Safe load} = (66.4\times10^2\times128)/10^3 = 850 \text{ kN}$$

(b) Cased section

Slenderness ratio on the full length $= 4200/51.6 = 81.5$ less than 250.

For the cased section:

$$r_{YY} = 0.2\times303.9 = 60.7 \text{ mm}$$
$$l/r_{YY} = (0.7\times4200)/60.7 = 48.4$$
$$\left.\begin{array}{l} p_c = 134.5 \text{ N/mm}^2 \quad \text{from Table 17} \\ p_{bc} = 165 \text{ N/mm}^2 \quad \text{from Table 2} \end{array}\right\} \text{BS 449}$$
$$\text{Safe load} = (66.4\times10^2\times134.5)/10^3+(134.5\times306.2\times303.9)/(0.19\times165\times10^3)$$
$$= 891+400$$
$$= 1291 \text{ kN}$$

This value gives the safe load as it is not more than twice the safe load on the uncased section.

The safe load has been calculated for the minimum permitted section of cased strut. In practice the strut would be made 310 mm square. This section would be fully effective in carrying load, as only concrete cover in excess of 75 mm from the steel section is ignored.

6.7 CONSTRUCTION DETAILS

6.7.1 Splices

Recommendations for stanchion splices are given in Clause 32(b) of BS 449. These are:

> 'Where the ends of compression members are faced for complete bearing over the whole area, they shall be spliced to hold the connected members accurately in place and resist any tension where bending is present. Where such members are not faced for complete bearing the joints shall be designed to transmit all the forces to which they are subjected'

In multi-storey buildings splices are usually located just above floor level. Some typical splices are given in Figure 6.18. The ends of the stanchion lengths are usually machined for bearing and unless the

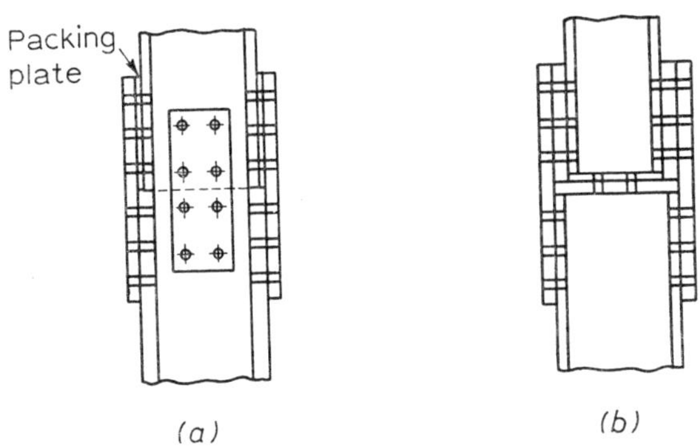

Figure 6.18 Typical splices

stanchion carries a moment sufficient to overcome the axial load to cause tension, the splice merely holds the stanchion lengths in place. Any particular stanchion splice must be checked and, if required, designed along the lines set out in Chapter 2. Combined bolted and welded splices using high-strength friction-grip bolts and full-strength welded splices are used where high moments have to be resisted as well as axial forces. Examples of this include the stanchions in rigid-frame buildings.

6.7.2 End and Connection Details

Some typical stanchion cap details are shown in Figure 6.19. Stanchion base details are given in Chapter 9. Refer to Chapters 3 and 4 for connections for beams and girders.

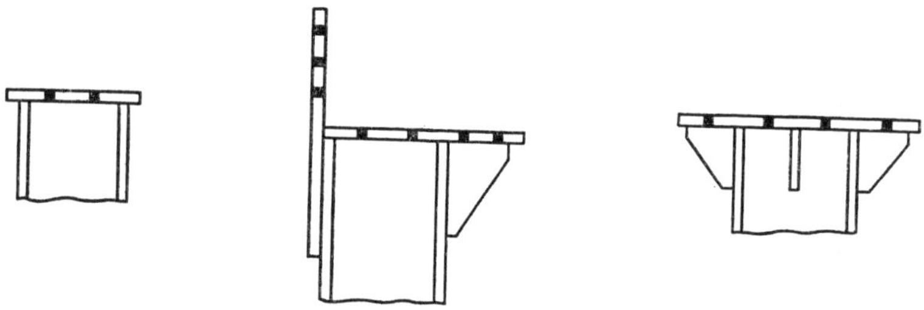

Figure 6.19 Stanchion cap details

6.7.3 Built-up Stanchions

These are H or box sections as shown in Figure 6.1. The shear in stanchions is low and is taken as $2\frac{1}{2}\%$ of axial force, so the web to flange welds are usually minimum-leg-length continuous fillet welds. Stiffeners in H section columns are provided and the rules given for intermediate stiffeners in plate girders can be used as a guide for design here. Load points may require local stiffening and this is sometimes a matter of experience. Comments given in section 4.2 regarding the fabrication of plate girders also apply to these members. Diaphragms are provided in box members.

6.8 PROBLEMS

1 The lower length of an internal stanchion in a multi-storey building carries an axial load including a self weight of 1730 kN. The stanchion has a substantial base and four floor beams frame into it at first-floor level. The height from base to centreline of the connections of the first-floor beams is 6.2 m. Design the stanchion length.

2 A stanchion carrying a floor load is shown in Figure 6.20. The stanchion can be considered as pinned at the top and the base and propped by a strut near the mid height as shown. Design the stanchion for the load at the base shown on the figure.

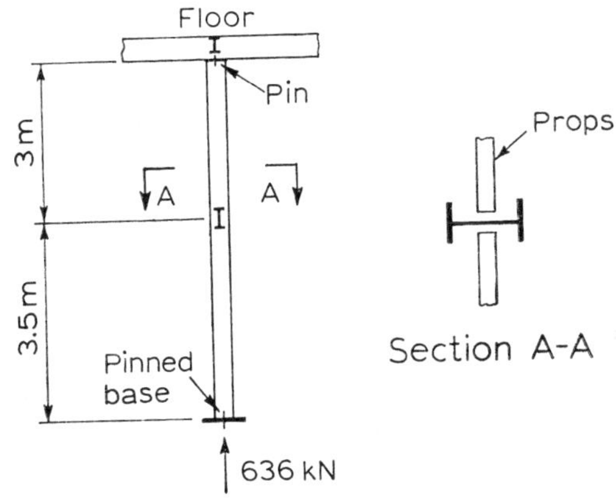

Figure 6.20 Stanchion carrying a floor load

3 A stanchion has an effective height of 5.2 m and has to carry an axial load of 137 kN. This includes an allowance for self weight. Make designs using the following sections:

(a) Universal column
(b) Circular hollow section
(c) Rectangular hollow section

4 A built-up stanchion of section shown in Figure 6.21 has an effective length for buckling about each axis of 7.4 m. Determine the safe axial load that the stanchion can carry.

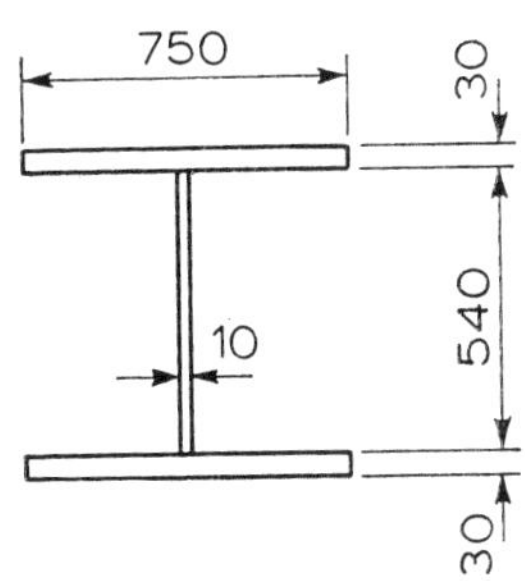

Figure 6.21

5 A battened stanchion consists of two No. 229×76×26.06 kg/m channels spaced at 140 mm back to back as shown in Figure 6.22. If the effective length of the stanchion is 6 m, determine the safe axial load. Determine also the size and spacing of batten plates required and design the batten-plate-to-channel weld.

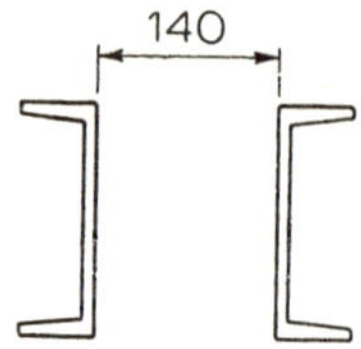

Figure 6.22

6 A universal beam, 305×165×54 kg/m, is cased in accordance with the provisions of Clause 30 of BS 449. The effective length of the stanchion is 6 m. Check that the section can carry a load of 480 kN.

7 Part of an operating floor in an industrial plant is shown in Figure 6.23.

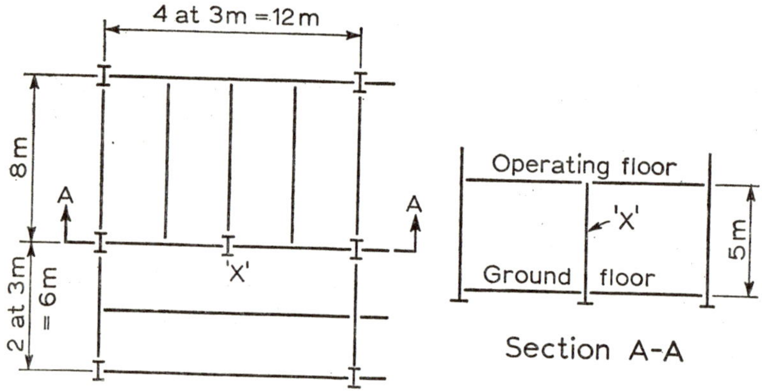

Figure 6.23 Part of an operating floor in an industrial plant

The floor loading is as follows:

Dead load—steel beams, slab, screed and services

$$= 815 \text{ kg/m}^2 \qquad = 8 \text{ kN/m}^2$$

Imposed load $= 12 \text{ kN/m}^2$

(a) Determine the load on stanchion 'X'
(b) Design the stanchion

7

Eccentrically loaded stanchions

Most practical stanchions are subjected to axial load and bending moment. The combination of stresses can have a severe effect on the strength of a stanchion. It is very important to calculate the moments or make appropriate allowance for eccentric loads.

7.1 DESIGN CONSIDERATIONS

7.1.1 Types of Loading

Various types of loading and construction cause bending moments as well as axial loads in stanchions. Some of these cases are shown in Figure 7.1.

Eccentric loads are due to floor beam reactions, crane loads, pipe loads, etc. Lateral loads from crane surge and wind cause moments in stanchions. In rigid-frame construction, moments are transmitted through the joints into the stanchions. Crane stanchions are dealt with in the next chapter. Rigid-frame construction is outside the scope of this book. In dealing with side stanchions in single-storey buildings, the action of the whole frame must be taken into account.

7.1.2 Stresses due to Axial Load and Moment

An axial load P and bending moment M acting at a section are equivalent to the axial load P at an eccentricity e, where

$$e = M/P$$

Refer to Figure 7.2(a) and (b).

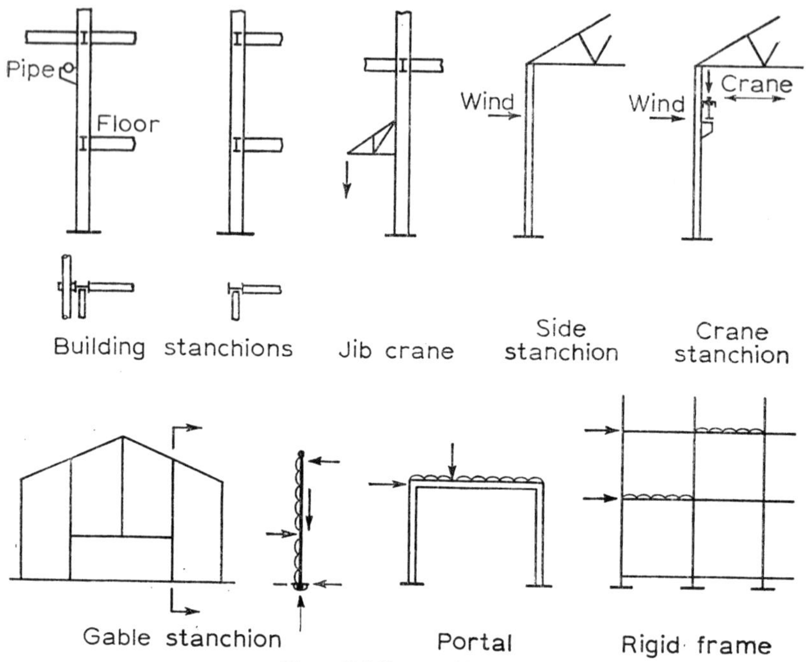

Figure 7.1 Types of loading

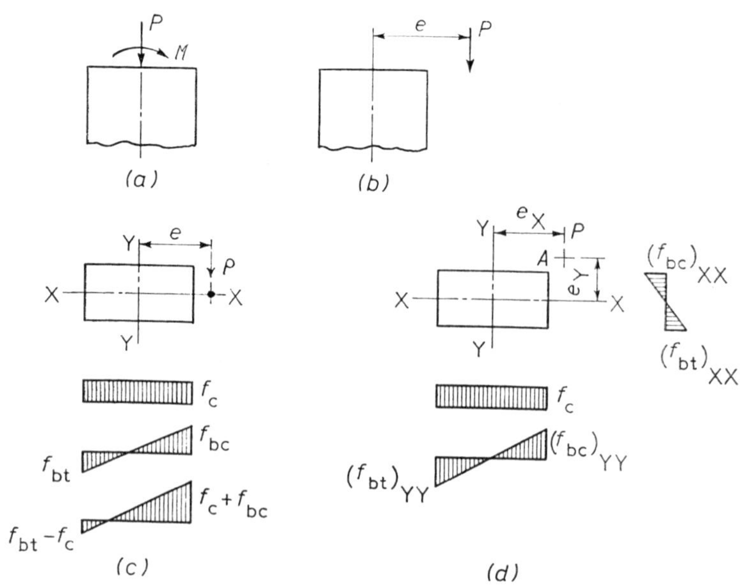

Figure 7.2

First consider bending about the YY axis only. The section properties are:

A = area of section

Z_{YY} = modulus of section about YY axis

Axial stress $f_c = P/A$

Bending stress $f_b = M/Z_{YY}$

Maximum compressive stress $= f_c + f_b$

If f_c is greater than f_b there will be compression over the whole section. If f_b is greater than f_c there will be a tensile stress. The stresses for this case are shown in Figure 7.2(c).

In the case of bending about two axes, where

Z_{XX} = modulus of section about the XX axis,

$M_{XX} = Pe_Y$ = moment about XX axis

$M_{YY} = Pe_X$ = moment about YY axis,

Axial stress $f_c = P/A$

Bending stress $(f_b)_{XX} = M_{XX}/Z_{XX}$

$(f_b)_{YY} = M_{YY}/Z_{YY}$

These stresses are shown in Figure 7.2(d). The maximum compressive stress occurs at point A and is equal to

$$f_c + (f_{bc})_{XX} + (f_{bc})_{YY}$$

Stresses at any other point on the cross section can be found from values on the stress diagrams.

7.1.3 Combined stresses

The method of determining whether a section subjected to bending and axial compression is overstressed is given in BS 449 in Clause 14(a). This clause states:

'Members subject to both axial compression and bending stresses shall be so proportioned that the quantity

$$\frac{f_c}{p_c} + \frac{f_{bc}}{p_{bc}}$$

does not exceed unity at any point, where:

f_c = the calculated average axial compressive stress

p_c = the allowable compressive stress in axially loaded struts (see Table 17)

f_{bc} = the resultant compressive stress due to bending about both rectangular axes

p_{bc} = the appropriate allowable compressive stress for members subject to bending (see Clause 19)

'In cased struts for which allowance is made for the load carried by the concrete in accordance with Sub-clause 30(b), the ratio in the above expression shall be replaced by the ratio of calculated axial load on the strut to the allowable axial load determined from Sub-clause 30(b).'

Application of the above principles and the clause from BS 449 will be shown in the design examples that follow.

7.2 DESIGN EXAMPLES

7.2.1 A Steel Stanchion Subjected to Axial Load and Moment

A steel stanchion has an effective length of 6.1 m and carries an axial load of 500 kN in addition to a bending moment of 36.6 kN m at the base of the stanchion. The compression flange of the stanchion in bending has an unrestrained length of 6.1 m. Design the stanchion using a universal column section. Check if a lighter section could be obtained if a universal beam were used. The stanchion is shown in Figure 7.3(a).

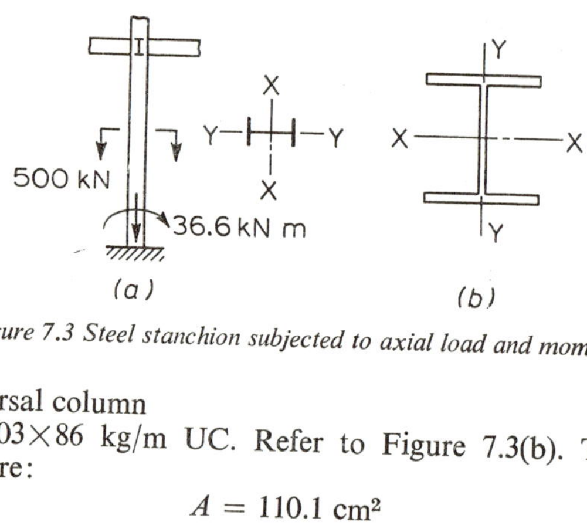

Figure 7.3 Steel stanchion subjected to axial load and moment

(a) Universal column

Try 203×203×86 kg/m UC. Refer to Figure 7.3(b). The section properties are:

$$A = 110.1 \text{ cm}^2$$

$$Z_{XX} = 851.5 \text{ cm}^3$$

$$r_{YY} = 5.32 \text{ cm}$$

$$D/T = 10.8$$

$$l/r_{YY} = 6100/53.2 = 114$$

Allowable axial stress $p_c = 65$ N/mm² (from Table 17)
Allowable bending stress $p_{bc} = 165$ N/mm² (from Table 3)

Actual stresses:

$$f_c = (500\times10^3)/(110.1\times10^2) = 45.5 \text{ N/mm}^2$$
$$f_{bc} = (36.6\times10^6)/(851.5\times10^3) = 43 \text{ N/mm}^2$$

Combined: $45.5/65+43/165 = 0.7+0.26 = 0.96$

The trial section is satisfactory.

(b) Universal beam

Try 533×210×82 kg/m UB.

$$A = 104.3 \text{ cm}^2$$
$$Z_{XX} = 1793 \text{ cm}^3$$
$$r_{YY} = 4.18 \text{ cm}$$
$$D/T = 40$$
$$l/r_{YY} = 6100/41.8 = 146$$
$$p_c = 42 \text{ N/mm}^2 \quad \text{(from Table 17)}$$
$$f_c = (500\times10^3)/(104.3\times10^2) = 47.9 \text{ N/mm}^2$$

In this case the universal beam required would be heavier than the universal column section.

7.2.2 A Cased Stanchion Subjected to Axial Load and Moment

Redesign the stanchion in the previous example as a cased stanchion.

Referring to Clause 14(a) of BS 449 which is given above in section 7.1.2, no provision is made for the concrete casing to assist in resisting the bending moment. The casing assists in carrying the axial load in accordance with Clause 30(b).

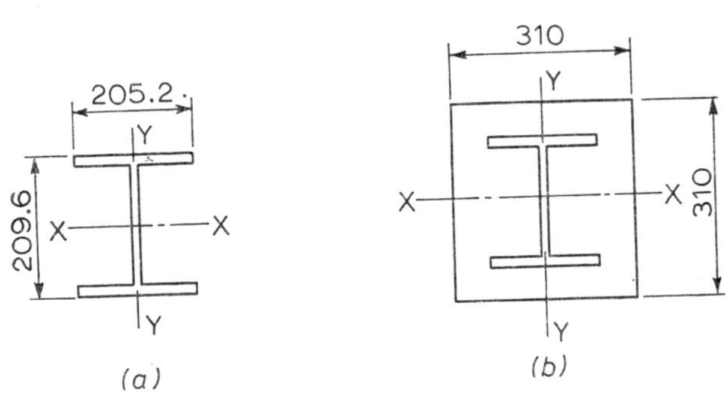

Figure 7.4

Try 203×203×60 kg/m UC. The steel section and cased section are shown in Figure 7.4. The properties of the steel section are:

$$A = 75.8 \text{ cm}^2$$
$$Z_{XX} = 581.1 \text{ cm}^3$$
$$r_{YY} = 5.19 \text{ cm}$$
$$r_{XX} = 8.96 \text{ cm}$$
$$D/T = 14.8$$

For the uncased section:

$$l/r_{YY} = 6100/51.9 = 117.5$$
$$p_c = 61.5 \text{ N/mm}^2 \quad \text{(from Table 17)}$$
$$p_{bc} = 163 \text{ N/mm}^2 \quad \text{(from Table 3)}$$
$$\text{Safe axial load} = (75.8\times10^2\times61.5)/10^3 = 466 \text{ kN}$$
$$f_{bc} = (36.6\times10^6)/(581.1\times10^3) = 63 \text{ N/mm}^2$$

For the cased section (refer to Figure 7.4(b)):

$$r_{YY} = 0.2\times305.2 = 61 \text{ mm}$$
$$l/r_{YY} = 6100/61 = 100$$
$$p_c = 79 \text{ N/mm}^2 \quad \text{(from Table 17)}$$
$$p_{bc} = 165 \text{ N/mm}^2 \quad \text{(from Table 2)}$$

This is the allowable bending stress for grade 43 steel in the formula for the safe axial load.

$$\text{Safe load} = (75.8\times10^2\times80.5)/10^3+(79\times310^2)/(0.19\times165)$$
$$= 610+246 = 856 \text{ kN}$$

This is the safe axial load for the strut since it does not exceed twice the safe load on the uncased section.

Check the combined bending and axial loading in accordance with BS 449:

$$500/856+63/163 = 0.584+0.387 = 0.971$$

The steel section chosen is satisfactory.

7.3 ECCENTRICITY OF LOADING FROM FLOOR BEAMS

7.3.1 Code Regulations

The eccentricities of beam and truss reactions and the methods for evaluating the bending moments due to these reactions are given in Clause 34 of BS 449. This clause states:

'(a) For the purpose of determining the stress in a stanchion or column section, the beam reactions or similar loads shall be assumed to be applied 100 mm from the face of the section or at the centre of the bearing, whichever dimension gives the greater eccentricity and with the exception of the following two cases:

'1 In the case of cap connections, the load shall be assumed to be applied at the face of the stanchion section or edge of the packing if used, towards the span of the beam.
'2 In the case of a roof truss bearing on a cap, no eccentricity need be taken for simple bearings without connections capable of developing an appreciable moment.

'(b) In effectively jointed and continuous stanchions the bending moments due to eccentricities of loading at any one floor or horizontal frame level may be taken as being:

'1 Ineffective at the floor or frame level above and below that floor.
'2 Divided equally between the stanchion lengths above and below that floor or frame level, provided the moment of inertia of either stanchion section, divided by its actual length, does not

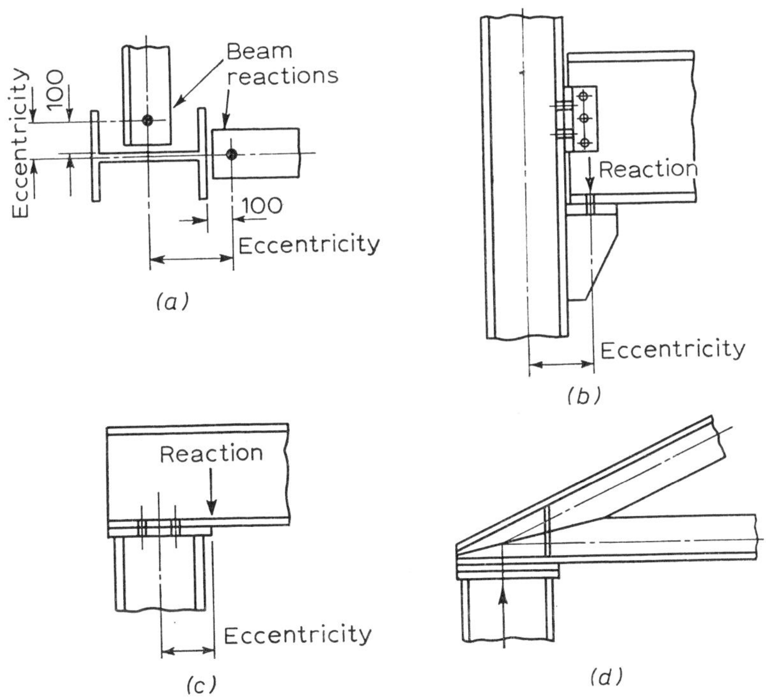

Figure 7.5 Eccentricities for various connections

exceed 1.5 times the corresponding value for the other length. In cases exceeding this ratio the bending moment shall be divided in proportion to the moments of inertia of the stanchion sections, divided by their respective actual lengths.'

Eccentricities for various connections in accordance with Clause 34 of BS 449 are shown in Figure 7.5. In (a) in the figure, for normal web connections for beam to stanchion, the reaction is taken to act at a minimum of 100 mm from the face of the column. In this case stresses caused by bending about the weak axis of the stanchion can be very important in design. In (b) all the reaction is taken by the lower bracket and the actual eccentricity is used in calculating the bending moment in the stanchion. Cap connections are shown in (c) and (d) in the figure. Note that there is no eccentricity in the roof truss connection.

7.3.2 Design of a Corner Stanchion of a Steel-Frame Building

The part plan of the floor and roof steel for an office building is shown in Figure 7.6(a). An elevation of the corner stanchion is shown in (b) on the figure. The roof and floor loading is as follows:

Roof—dead load (total) $= 510\ \text{kg/m}^2 = 5\ \text{kN/m}^2$
imposed load $= 1.5\ \text{kN/m}^2$
Floors—dead load (total) $= 715\ \text{kg/m}^2 = 7\ \text{kN/m}^2$
imposed load $= 3\ \text{kN/m}^2$

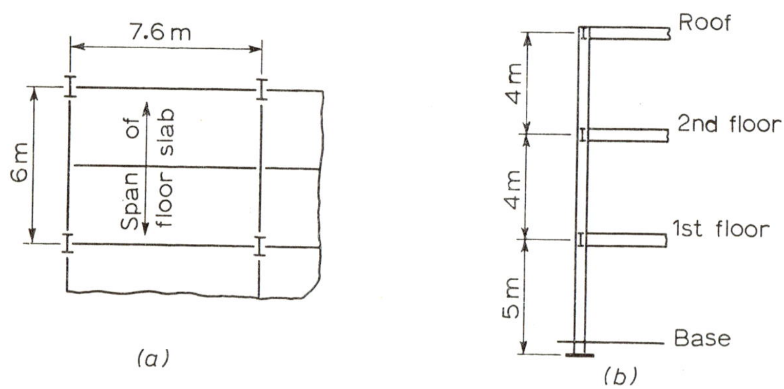

Figure 7.6

The self weight of the stanchion may be taken as 153 kg/m, that is, 1.5 kN/m. In addition, all external beams carry the following loading due to brick walls and concrete casing:

Roof beams—parapet and casing $= 204\ \text{kg/m} = 2\ \text{kN/m}$
Floor beams—walls and casing $= 612\ \text{kg/m} = 6\ \text{kN/m}$

The reinforced concrete slabs for the roof and floors are one-way slabs spanning in the direction shown on the figure.

Design the corner stanchion of the building. The stanchion is to be designed on the basis of a steel section only.

In accordance with Table 2 of CP3, Chapter V, 1967, the imposed loading may be reduced as follows:

One floor carried by member—no reduction

Two floors carried by member—10% reduction

Three floors carried by member—20% reduction

The roof is counted as a floor. Note that the reduction is only taken into account in the axial load on the stanchion. The full imposed load at that section is taken in calculating the moments due to eccentric beam reactions.

(a) Loading and reactions of floor beams

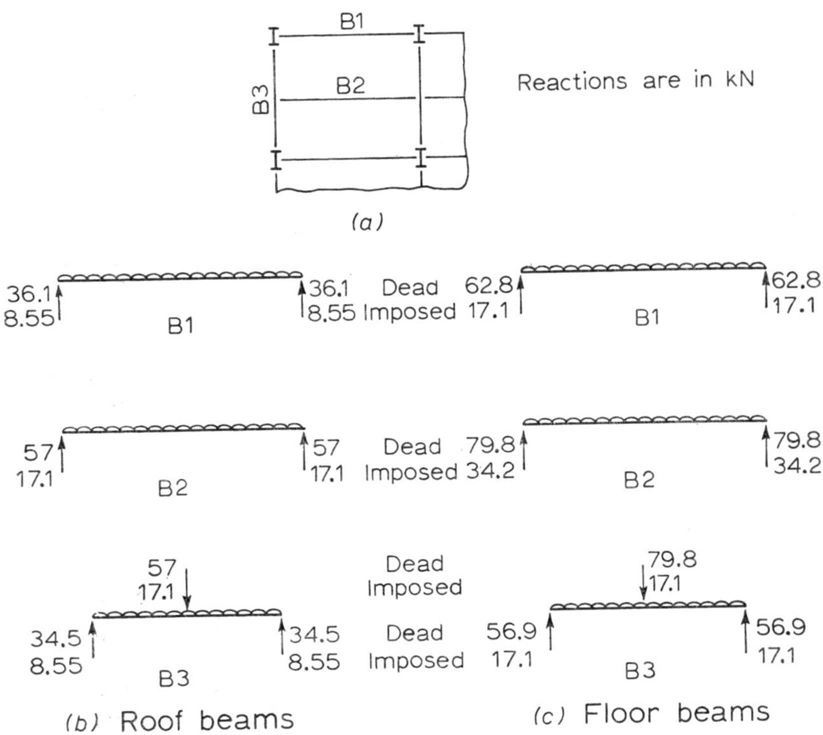

Figure 7.7

Mark numbers for the floor beams as shown in Figure 7.7(a). Beam end reactions are calculated below:

Roof

B1	Dead load	$= (5\times3.8\times1.5)+(2\times3.8)$	= 36.1 kN
	Imposed load	$= (1.5\times3.8\times1.5)$	= 8.55 kN
B2	Dead load	$= (5\times3.8\times3)$	= 57 kN
	Imposed load	$= (1.5\times3.8\times3)$	= 17.1 kN
B3	Dead load	$= 28.5+(2\times3)$	= 34.5 kN
	Imposed load	$= 8.55$	= 8.55 kN

The roof beam reactions are shown on Figure 7.7(b).

Floors

B1	Dead load	$= (7\times3.8\times1.5)+(6\times3.8)$	= 62.8 kN
	Imposed load	$= (3\times3.8\times1.5)$	= 17.1 kN
B2	Dead load	$= (7\times3.8\times3)$	= 79.8 kN
	Imposed load	$= (3\times3.8\times3)$	= 34.2 kN
B3	Dead load	$= 39.9+(6\times3)$	= 57.9 kN
	Imposed load	$= 17.1$	= 17.1 kN

The floor beam reactions are shown in Figure 7.7(c).

(b) Loading and moments at floor levels

The loading at the roof, second floor, first floor and the base is totalled from values taken from Figure 7.7. The values for imposed loads are given separately so that the reductions permited can be calculated and the net axial load found. The reductions allowed at various positions on the stanchion are shown in Figure 7.8.

The moments due to the eccentricities of the roof and floor beam reactions are based on the following assumed sizes for the stanchion lengths:

Roof to second floor—152×152 UC

Second floor to first floor—203×203 UC

First floor to base—203×203 UC

Further it will be assumed initially that the moments at the floor levels can be divided between the upper and lower column lengths in proportion to the stiffnesses which are based on inertia ratios shown on the stanchion lengths in Figure 7.8. The actual values of the moments of inertia are not required. The division of moments is made as follows:

1 Joint at second-floor level

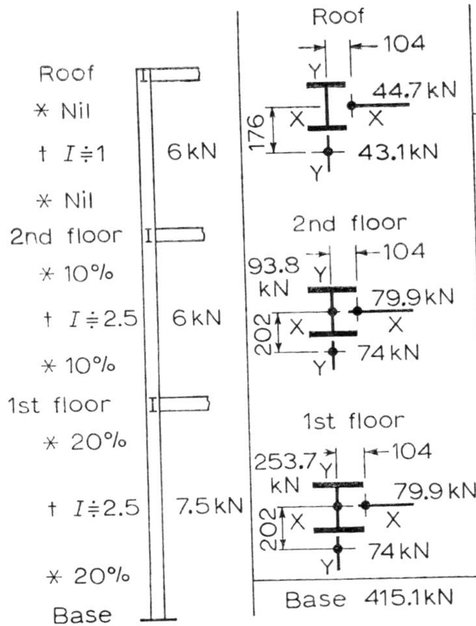

Position	Total axial load (kN)	Imposed load (kN)	Reduction in imposed load %	Reduction in imposed load Load	Design axial (kN)	Moment M_{XX} (kN m)	Moment M_{YY} (kN m)
Roof	87.8	17.1	Nil	–	87.8	7.59	4.65
Above 2nd floor	93.8	17.1	Nil	–	93.8	4.24	2.38
Below 2nd floor	247.7	51.3	10%	5.1	242.6	10.7	5.94
Above 1st floor	253.7	51.3	10%	5.1	248.6	7.47	4.16
Below 1st floor	407.6	85.5	20%	17.1	390.5	7.47	4.16
Base	415.1	85.5	20%	17.1	398	–	–

† $I \doteqdot 1$ assumed value of the rates for the moment of inertia

* 10% permitted value for reduction in imposed load

Figure 7.8 Eccentricities of beam reactions and loads and moments for design

Upper stanchion length

$$\text{Stiffness } I/l = \tfrac{1}{4} \qquad = 0.25$$

Lower stanchion length

$$\text{Stiffness } I/l = 2.5/4 = 0.625$$

If M is the moment due to the eccentric floor beam reaction then the moment in the upper stanchion length is given by

$$M_U = [0.25/(0.25+0.625)]M \;\; = 0.286M$$

Moment in the lower stanchion length

$$M_L = [0.625/(0.25+0.625)]M = 0.714M$$

2 Joint at first-floor level

It will be assumed that the same stanchion section will be used for the two lower lengths. Hence, the moments of inertia are the same and the stiffnesses are inversely proportional to the stanchion lengths. Upper stanchion length

$$\text{Stiffness} = \tfrac{1}{4} = 0.25$$

Lower stanchion length

$$\text{Stiffness} = \tfrac{1}{5} = 0.2$$

The stiffness of the upper length does not exceed 1.5 times the stiffness of the lower length. Thus the moments may be divided equally between the upper and lower lengths.

These assumed values for the division of the moments will be checked after the stanchion sections have been designed. The eccentricities of the beam reactions and the loads and moments for design are shown in Figure 7.8. The maximum value of the imposed load has been used in calculating the moments.

(c) Stanchion design

(i) Roof to second floor

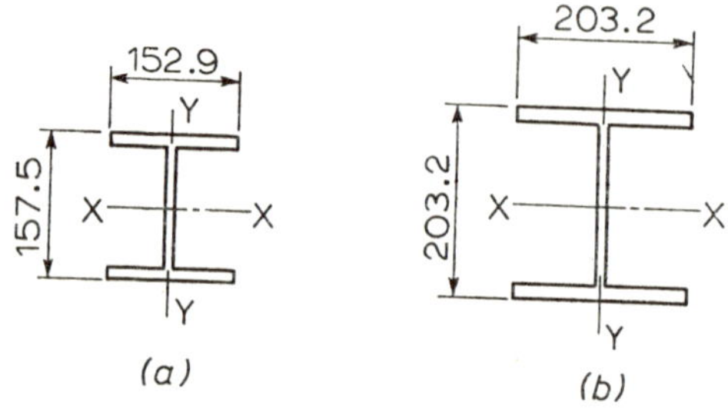

Figure 7.9

Try 152×152×30 kg/m UC. Refer to Figure 7.9(a). Properties from steel tables are:

$$A = 38.2 \text{ cm}^2 \qquad I_{YY} = 558 \text{ cm}^4$$
$$Z_{XX} = 221.2 \text{ cm}^3 \qquad r_{YY} = 3.82 \text{ cm}$$
$$Z_{YY} = 73.06 \text{ cm}^3 \qquad D/T = 16.8$$
$$I_{XX} = 1742 \text{ cm}^4$$
$$l/r_{YY} = (0.7\times4000)/38.2 = 73.4$$
$$\left.\begin{aligned} p_c &= 111.5 \text{ N/mm}^2 \quad \text{from Table 17} \\ p_{bc} &= 165 \text{ N/mm}^2 \quad \text{from Table 3} \end{aligned}\right\} \text{BS 449}$$

Stresses at the top of the stanchion:

$$f_c = (87.8\times10^3)/(38.2\times10^2) = 23 \text{ N/mm}^2$$
$$(f_{bc})_{XX} = (7.59\times10^6)/(221.2\times10^3) = 34.4 \text{ N/mm}^2$$
$$(f_{bc})_{YY} = (4.65\times10^6)/(73.06\times10^3) = 63.8 \text{ N/mm}^2$$
$$\text{Combined bending stress} = 98.2 \text{ N/mm}^2$$
$$\text{Combined:} = 23/111.5+98.2/165$$
$$= 0.206+0.595 = 0.801$$

The next lightest universal column section is overstressed, so that the section checked is the most economical one.

Stanchion 152×152×30 kg/m UC

(ii) Second floor to first floor

Try 203×203×46 kg/m UC. Refer to Figure 7.9(b). Properties from steel tables are:

$$A = 58.8 \text{ cm}^2 \qquad I_{YY} = 1539 \text{ cm}^4$$
$$Z_{XX} = 449.2 \text{ cm}^3 \qquad r_{YY} = 5.11 \text{ cm}$$
$$Z_{YY} = 151.5 \text{ cm}^3 \qquad D/T = 18.5$$
$$I_{XX} = 4564 \text{ cm}^4$$
$$l/r_{YY} = (0.7\times4000)/51.1 = 55$$
$$\left.\begin{aligned} p_c &= 130 \text{ N/mm}^2 \quad \text{from Table 17} \\ p_{bc} &= 165 \text{ N/mm}^2 \quad \text{from Table 3} \end{aligned}\right\} \text{BS 449}$$

Stresses just below second floor level are:

$$f_c = (242.6\times10^3)/(58.8\times10^2) = 41.2 \text{ N/mm}^2$$
$$(f_{bc})_{XX} = (10.7\times10^6)/(449.2\times10^3) = 24 \text{ N/mm}^2$$
$$(f_{bc})_{YY} = (5.94\times10^6)/(151.5\times10^3) = 39.3 \text{ N/mm}^2$$
$$\text{Combined bending stress} = 63.3 \text{ N/mm}^2$$

$$\text{Combined: } = (41.2/130)+(63.3/165)$$
$$= 0.318+03.384 = 0.702$$

The next lighter universal section, 152×152×37 kg/m UC, would be overstressed.

Stanchion 203×203×46 kg/m UC

(iii) First floor to base

Try 203×203×46 kg/m UC. The properties for this section are given above.

$$l/r_{YY} = (0.7\times5000)/51.1 = 68.5$$

p_c = 117.5 N/mm² from Table 17 } BS 449

p_{bc} = 165 N/mm² from Table 3 } BS 449

Stresses at the underside of first-floor beams:

$$f_c = (390.5\times10^3)/(58.8\times10^2) = 66.5 \text{ N/mm}^2$$
$$(f_{bc})_{XX} = (7.47\times10^6)/(449.2\times10^3) = 16.7 \text{ N/mm}^2$$
$$(f_{bc})_{YY} = (4.16\times10^6)/(151.5\times10^3) = 27.5 \text{ N/mm}^2$$

Combined bending stress $= 44.2$ N/mm²

$$\text{Combined: } = (66.5\times117.5)+(44.2/165)$$
$$= 0.565+0.268 = 0.833$$

Stanchion 203×203×46 kg/m UC

Use one length of stanchion from the base to a splice above second-floor level. The stanchion length is then about 9.3 m. This is spliced to the smaller section of the top-storey length.

(d) Check the stiffness ratios

Stiffnesses

Roof to second floor	$I_{XX}/l = 1742/400 =$	4.38
	$I_{YY}/l = 558/400 =$	1.39
Second to first floor	$I_{XX}/l = 4564/400 =$	11.41
	$I_{YY}/l = 1539/400 =$	3.85
First floor to base	$I_{XX}/l = 4564/500 =$	9.13
	$I_{YY}/l = 1539/500 =$	3.08

Stiffness Ratios

1 Second-floor joint

$$\text{XX axis } \frac{(I/l)\text{ lower}}{(I/l)\text{ upper}} = \frac{11.41}{4.38} = 2.6 : 1$$

$$\text{YY axis } \frac{(I/l)\text{ lower}}{(I/l)\text{ upper}} = \frac{3.85}{1.39} = 2.77 : 1$$

The ratio of 2.5 assumed for the inertia of the lower length relative to that of the upper length is satisfactory.

2 First-floor joint
The stiffness ratios for both axes are inside the limit of 1.5 stated in BS 449 for this joint. Thus the assumed distribution, that is dividing the moment equally between the upper and lower stanchion lengths, is satisfactory.

7.4 THE SIDE STANCHION FOR A SINGLE-STOREY INDUSTRIAL BUILDING

7.4.1 Arrangement and Loading

The cross section and side elevation of a single-storey industrial building are shown in Figure 7.10(a) and (b). The stanchions are assumed to be fixed at the base and pinned at the top and act as par-

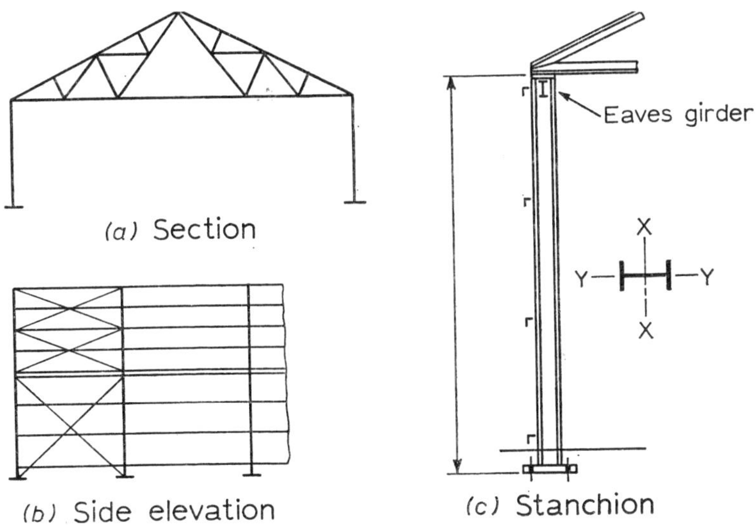

Figure 7.10 Cross section and side elevation of a single-storey industrial building

tially propped cantilevers in resisting lateral loads. The top of the stanchion is held in the longitudinal direction by the eaves girder and bracing as shown on the side elevation.

The loading on the stanchion is due to:

1 Dead and imposed load from the roof and dead load from the walls and steel frame
2 Wind loading on roof and walls

The loading on the roof consists of:

1 Dead load due to sheeting, insulation board, purlins, and weight of roof truss and bracing. This may be taken as 30 to 60 kg/m^2, that is, approximately 0.3 to 0.6 kN/m^2 of roof area.
2 Imposed load due to snow, erection and maintenance loads. This is given in CP3, Chapter V, Part 1, 1967. The load is 0.75 kN/m^2 on plan area.

The loading from the walls is due to sheeting, insulation board, sheeting rails and the weight of stanchion and bracing. This load may be taken as 30 to 60 kg/m^2, that is, approximately 0.3 to 0.6 kN/m^2 of wall area.

In calculating the wind load, the dimensions of the whole building must be taken into account amongst other factors. The method of calculating the wind load is taken from CP3, Chapter V, Part 2, 1970. This method is shown in the example that follows.

The breakdown and diagrams for the calculation of the loading are shown in Figure 7.11. The following comments are made on these figures.

1 Dead and imposed loads give an axial reaction R at the base. See (a) on the figure.
2 Wind on the roof and walls is shown in (b). On the windward slope, there may be a pressure or suction depending on the angle of slope. The reactions from wind on the roof only are shown in (c). The vertical uplift results in reactions R_1 and R_2. The net horizontal reaction is assumed to be divided equally between the two stanchions; this is $\frac{1}{2}(H_2-H_1)$.
3 The wind on the walls causes the frame to deflect as shown in (d). The top of each stanchion moves by the same amount δ. The wind loads p_1 and p_2 on each wall, taken as uniformly distributed, will have different values and this results in a force P in the bottom chord of the truss as shown in (e). The value of P may be found by equating deflections at the top of each stanchion. Thus for the case shown where p_1 is greater than p_2 there is a compression P in the bottom chord:

$$\frac{p_1L^4}{8EI}-\frac{PL^3}{3EI}=\frac{p_2L^4}{8EI}+\frac{PL^3}{3EI}$$

This gives $$P=\frac{3L}{16}(p_1-p_2)$$

where I is the moment of inertia of the stanchion about the XX axis (same for each stanchion) and E is Young's modulus.

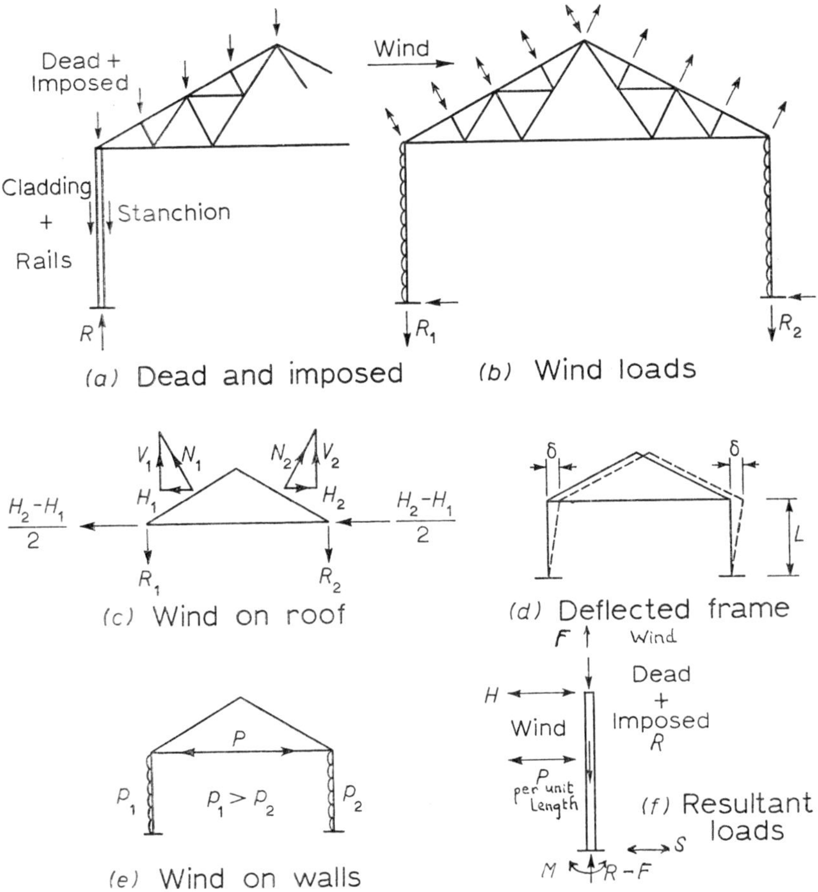

Figure 7.11

4 The resultant loading on the stanchion is shown in (f) where the horizontal point load at the top is:

$$H = P \pm \left(\frac{H_2 - H_1}{2}\right)$$

The stanchion is designed for moment and axial load at the base. The bending moment is due entirely to wind load so the allowable stress for bending may be increased by 25% in accordance with Clause 13 of BS 449.

The deflection at the stanchion cap due to horizontal loads is limited by Clause 31(b) of BS 449 to 1/325 of the stanchion height. The deflection is δ shown in Figure 7.11(d).

7.4.2 Effective Length of Stanchion for Axial Load

The effective lengths for the stanchion shown in Figure 7.10(c) are taken from Figure 15 of Appendix D of BS 449. These are:

$$\text{XX axis} \quad l = 1.5L$$

$$\text{YY axis} \quad l = 0.75L$$

It is further stated in Clause 31(b) that these lengths apply when:

'1 The bases of the stanchions are properly held in position and restrained in direction; and

'2 The caps of the stanchions are held in position by provision of diagonal or portal bracing in a vertical plane in at least one longitudinal bay in each line, or by other adequate means.'

The requirements are shown in Figure 7.10(a) and (b).

Universal beams are often used for these lightly loaded stanchions. This section is strong with regard to bending about the XX axis, but is very weak with regard to buckling about the YY axis. It is often economical to introduce a strut at mid height to halve the slenderness ratio about the YY axis. A laced strut composed of two angles, one of which also serves as a sheeting rail, may be used as shown in Figure 7.12(a). In light buildings the bracing may also be provided direct from the sheeting rails as shown in (b) in the figure.

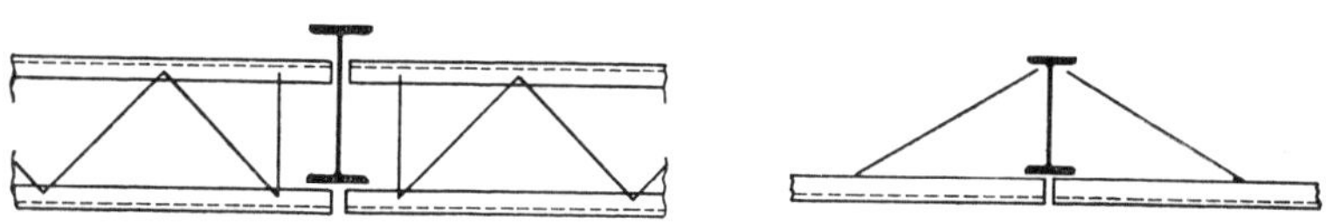

(a) (b)

Figure 7.12

7.4.3 Effective Length of the Compression Flange in Bending

The stanchion shown in Figure 7.10(c) is subjected to bending as well as axial load. The stanchion is a partially restrained vertical cantilever with the free end held in position longitudinally. The effective length of the compression flange is given in Clause 26(c) of BS 449 which states:

> 'For cantilever beams of projecting length L, the effective length to be used... shall be...:
>
> 2 Built-in at the support, restrained against torsion at the end by contiguous construction ... $l = 0.75\ L$'

Bracing the stanchion at mid height, as set out above, reduces the effective length of the compression flange in bending.

7.4.4 Design of a Side Stanchion

A section through a single-storey building is shown in Figure 7.13. The frames are at 5 m centres and the length of the building is 30 m. The stanchions are pinned at the top and fixed at the base. The loading is as follows:

Roof: dead load—measured on plan
sheeting, insulation board, truss = 51 kg/m² = 0.5 kN/m²
imposed load—measured on plan = 0.75 kN/m²
Walls: sheeting, insulation board, sheeting rails = 0.5 kN/m²
Stanchion: (estimate) = 410 kg = 4 kN

Wind load—This is taken from CP3, Chapter V, Part 2, 1970. The wind loading is set out below where all necessary data regarding the building are given. The student should consult the code for further information.

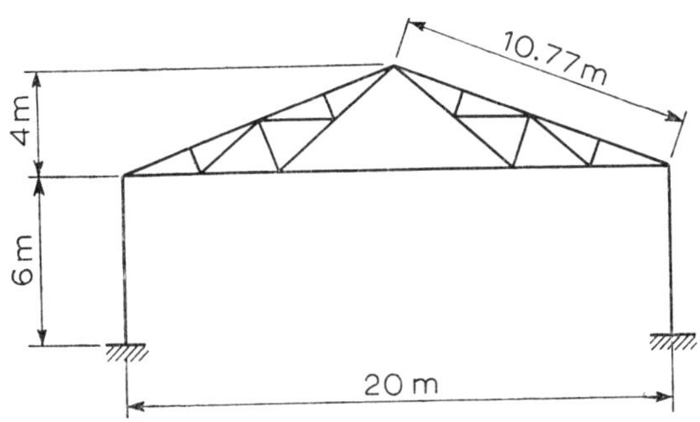

Figure 7.13

(a) Determine the loads and moments at the base of the stanchions.
(b) Design the side stanchion. Note: sheeting rails do not provide adequate lateral support to the stanchion.

(a) Dead and imposed load

Roof—dead load	$= 10 \times 5 \times 0.5$	$= 25.0$ kN
imposed load	$= 10 \times 5 \times 0.75$	$= 37.5$ kN
Walls	$= 6 \times 5 \times 0.5$	$= 15.0$ kN
Stanchion		$= 4.0$ kN
Total load at stanchion base		$= 81.5$ kN

(b) Wind load

Location—North-East England

Basic wind speed		= 45 m/s
Topography factor	S_1	= 1

Ground roughness 3. The location is on the outskirts of a city with obstructions up to 10 m in height.

Building size—Class B. The greatest horizontal or vertical dimension does not exceed 50 m.

Height to the top of the roof	H	= 10 m
Height to the top of the walls	H	= 6 m
From Table 3		
Roof	S_2	= 0.74
Walls	S_2	= 0.67
Statistical factor	S_3	= 1

Design wind speed:

Roof $V_S = 0.74 \times 45 = 33.3$ m/s

Walls $V_S = 0.67 \times 45 = 30.3$ m/s

Dynamic pressure from Table 4 (CP3, Chap. V, Part 2):

Roof $q = 682$ N/m^2

Walls $q = 558$ N/m^2

1 *Roof*

The external pressure coefficients C_{pe} are shown in Figure 7.14. These are taken from Table 8.

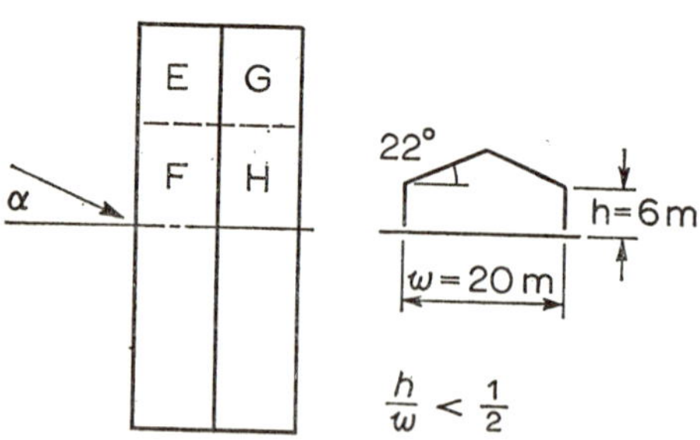

Roof angle	Wind angle 0°	
	EF	GH
22°	−0.32	−0.4

Figure 7.14

The interal pressure coefficients are taken from Appendix E. The case will be considered where there is only a negligible probability of a dominant opening occurring during a severe storm. Here C_{pi} shall be taken as the more onerous of $+0.2$ and -0.3.

The wind loading normal to the roof slope $= [(682 \times 10.77 \times 5)/10^3]$ $(C_{pe} - C_{pi})$
The pressure coefficients and wind loads are shown in Figure 7.16.

2 *Walls*
The external pressure coefficients are taken from Table 7. These are shown in Figure 7.15.

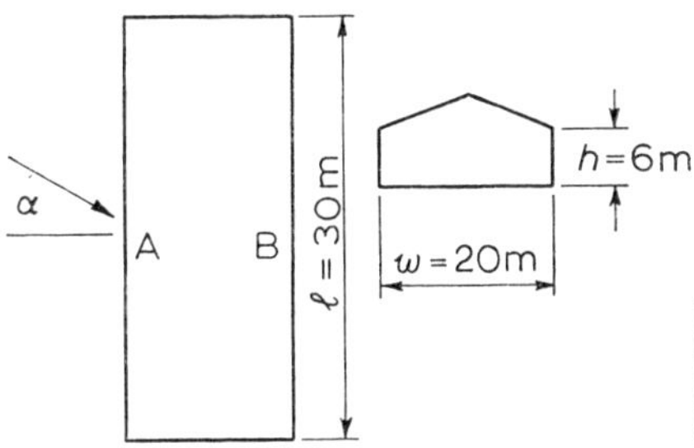

$$\frac{h}{w} = \frac{6}{20} = 0.3 < \frac{1}{2}$$

$$\frac{\ell}{w} = \frac{30}{20} = 1.5$$

$$1 < \frac{\ell}{w} \leq 1.5$$

Wind angle	C_{pe} for surface	
	A	B
0°	+0.7	−0.2

Figure 7.15

The internal pressure coefficients are +0.2 and −0.3 as stated above.
The wind loading normal to the walls $= [(558 \times 5 \times 6)/10^3]\,(C_{pe} - C_{pi})$
The pressure coefficients and wind loads are shown in Figure 7.16.

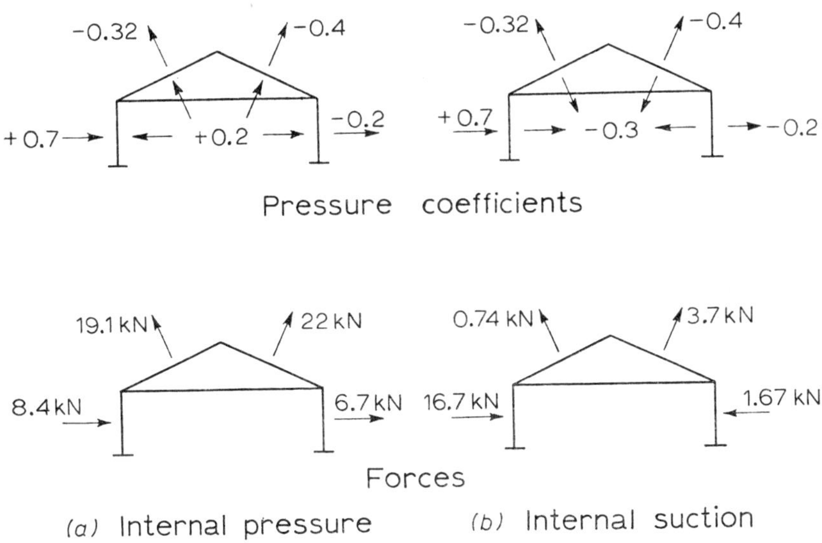

Figure 7.16

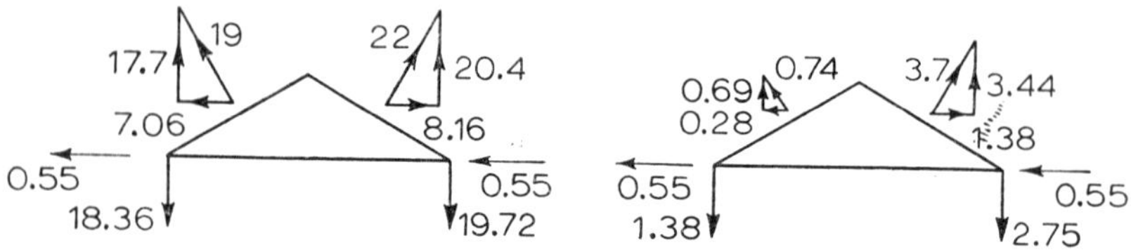

(a) Internal pressure (b) Internal suction

Figure 7.17

The reactions due to the loads on the roof truss are shown in Figure 7.17.

The horizontal reaction is divided equally between each support. The reactions are applied to the top of each stanchion. The wind loading on the walls requires the analysis set out above where each stanchion top is assumed to deflect by an equal amount and a force

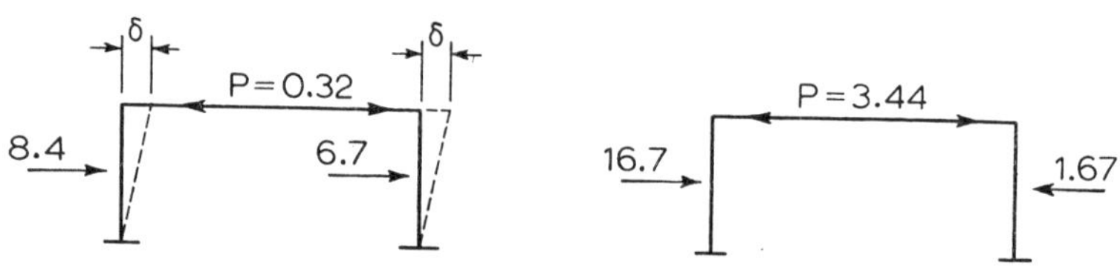

(a) Internal pressure (b) Internal suction

Figure 7.18

is transmitted through the bottom chord of the roof truss. The calculations are given for the internal pressure case. The loading is shown in Figure 7.18(a) and (b). Equating deflections at the top of each stanchion in (a):

$$\frac{PL^3}{3EI} + \frac{6.7L^3}{8EI} = \frac{-PL^3}{3EI} + \frac{8.4L^3}{8EI}$$

$$P = 0.32 \text{ kN}$$

A similar calculation can be made for the internal suction case.

(c) Summary of loads and moments

The loading and moments on the stanchions are summarised in Figure 7.19. The maximum design conditions can be chosen from this figure. These are for the case of internal suction on the windward stanchion. Hence for the case of dead + imposed + wind loads:

$$W = 80.1 \text{ kN}$$
$$M = 32.8 \text{ kN m}$$

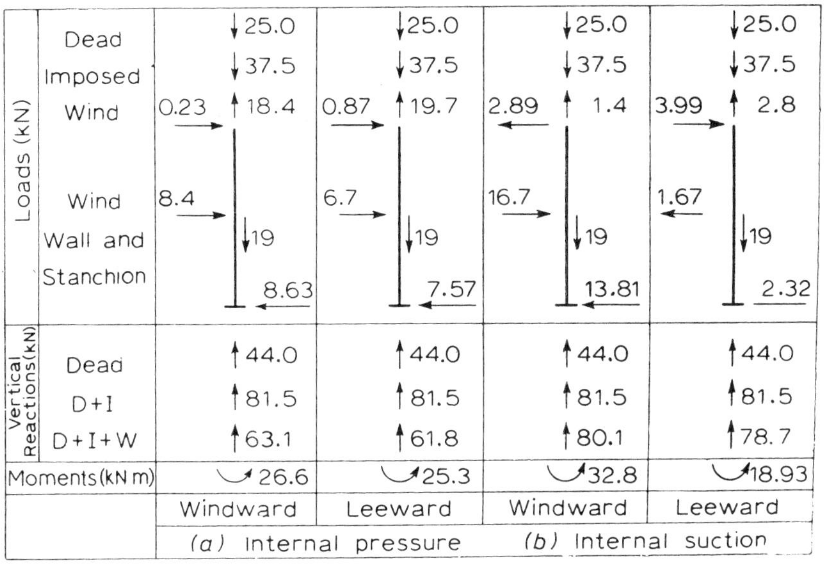

Figure 7.19

(d) Design of the stanchion

Try 406×140×39 kg/m UB. The properties are:

$$A = 49.3 \text{ cm}^2 \qquad Z_{XX} = 624.7 \text{ cm}^3$$
$$r_{XX} = 15.9 \text{ cm} \qquad I_{XX} = 12\,408 \text{ cm}^4$$
$$r_{YY} = 2.75 \text{ cm} \qquad D/T = 46.2$$
$$l/r_{XX} = (1.5\times 6000)/159 = 56.6$$
$$l/r_{YY} = (0.75\times 6000)/27.5 = 164$$

The effective length for obtaining the allowable bending stress is the same as that for axial compression in this case. Hence

$$p_c = 34 \text{ N/mm}^2\text{—Table 17}$$
$$p_{bc} = 88+25\% \text{ for wind} = 109 \text{ N/mm}^2\text{—Table 3}$$

The stresses at the stanchion base are

$$f_c = (80.1\times 10^3)/(49.3\times 10^2) = 16.3 \text{ N/mm}^2$$
$$f_{bc} = (32.8\times 10^6)/(624.7\times 10^3) = 52.5 \text{ N/mm}^2$$
$$\text{Combined: } (16.3/34)+(52.5/109) = 0.48+0.482 = 0.962$$

This is less than 1 and so is satisfactory.

The deflection at the stanchion cap is given by

$$\delta = (16\,700\times 6000^3)/(8\times 2\times 10^5\times 12\,408\times 10^4)$$
$$-(2890\times 6000^3)/(3\times 2\times 10^5\times 12\,408\times 10^4)$$
$$= 18.1-8.4 = 9.7 \text{ mm}$$

$$\delta/\text{Height} = 9.7/6000 = 1/620$$

The deflection may be checked for the case of internal pressure:

$$\delta = (230\times 6000^3)/(3\times 2\times 10^5\times 12\,408\times 10^4)$$
$$+(8400\times 6000^3)/(8\times 2\times 10^5\times 12\,408\times 10^4)$$
$$= 0.67+9.1$$
$$= 9.77 \text{ mm}$$

The stanchion section chosen is satisfactory for stress and deflection.

7.5 GABLE STANCHIONS

7.5.1 Arrangement and Loading

A typical gable framing arrangement is shown in Figure 7.20(a). The sheeting rails and bracing in the end gable are not shown on this figure. The gable columns such as AB and CD carry vertical loads from

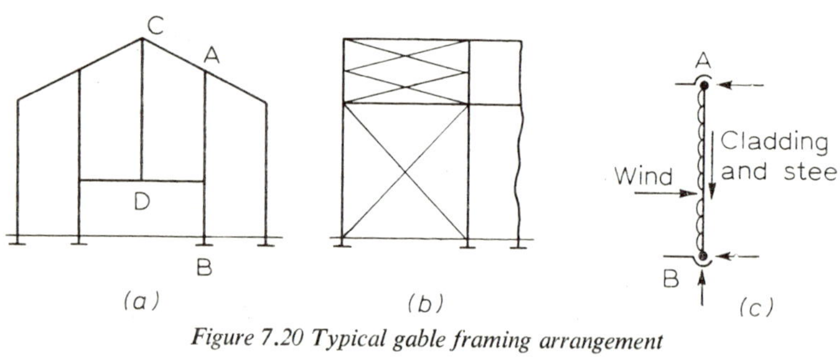

Figure 7.20 Typical gable framing arrangement

the wall cladding and weight of steel and horizontal loads from wind on the gable. The wind load may act in either direction and is taken as uniformly distributed. The loading depends on the arrangement of the framing.

The stanchion is generally assumed to have pin supports. The top support is formed by the roof and sidewall bracing as shown in the figure in (b). The bottom of the stanchion is carried on a light base. The loading and supports for the stanchion are shown in (c) on the figure.

The bending moment due to wind is the most important loading and thus universal beams are often used for gable stanchions. It is important to provide lateral support for the compression flange in order to increase the allowable stress. These points in design will be illustrated in the following example.

7.5.2 Design of a Gable Stanchion

Design the gable stanchion AB shown in Figure 7.21. The loading on this stanchion is as follows.

Dead load—sheeting, insulation board, rails, self weight of steel and door frame = 41 kg/m^2 = 0.4 kN/m^2

Wind load—can act in either direction = 0.5 kN/m^2

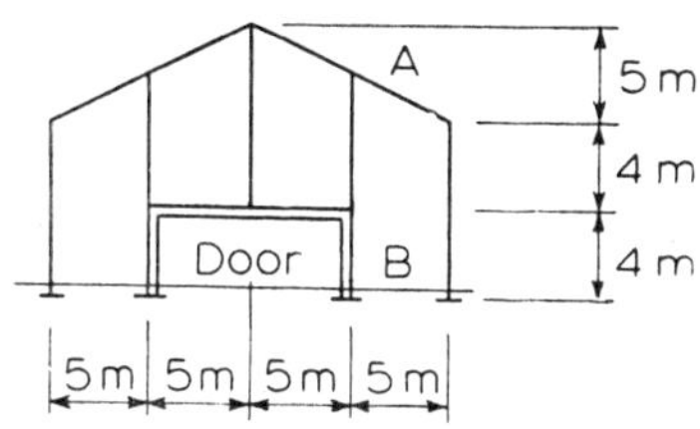

Figure 7.21

The stanchion may be taken as simply supported at the top and the bottom. Wind loading on doors is set out below. There are two doors running on wheels at the bottom and supported at the top by the door frame.

(a) *Loading on the stanchion*

Referring to Figure 7.22, the loading on the stanchion AB may be estimated as follows.

Dead load to point C on the stanchion

$= (0.4\times6.5\times5)+(8.87\times2.5\times0.4)$ = 21.87 kN

Dead load between C and B = $0.4\times4\times2.5$ = 4 kN

Stanchion base reaction = 25.87 kN

Wind load:

Uniform load on AC = 5×0.5 = 2.5 kN/m

Uniform load on BC = 2.5×0.5 = 1.25 kN/m

Point load at C = $(0.5\times8.87\times2.5)+(2\times5\times0.5)$ = 16.1 kN

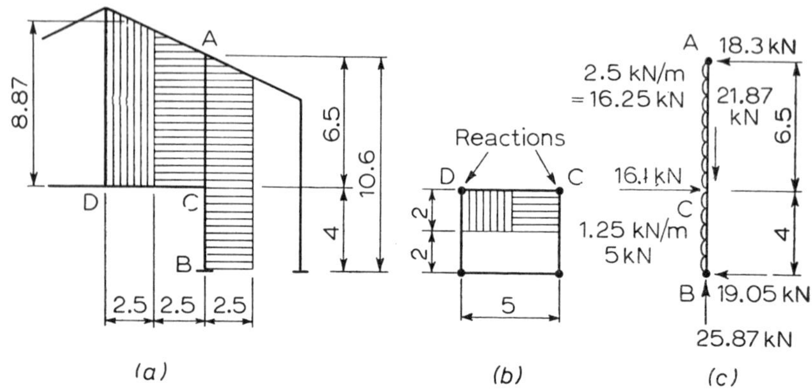

Figure 7.22

In the above, each door has been assumed to be 5 m wide although the actual door would be less than this. The wind on the door is resisted by four reactions at the corners. The two top reactions C and D as shown in (b) in the figure will result in a concentrated load at C on stanchion AB. The final loading on AB is shown in (c) in the figure. The moments and reactions in the stanchion are now calculated:

$$R_A = (16.1\times4)/10.5+(16.25\times7.25)/10.5+(5\times2)/10.5 = 18.3 \text{ kN}$$
$$R_B = 16.1+16.25+5-18.3 = 19.05 \text{ kN}$$
$$M_C = (18.3\times6.5)-(16.25\times3.25) = 66.3 \text{ kN m}$$

(b) *Design of the stanchion*

Assume that the compression flange of the stanchion is supported laterally at C so that the effective length for axial load and bending is 6.5 m.

Try 356×171×51 kg/m UB. The properties for this section from steel tables are:

$A = 64.5 \text{ cm}^2$

$Z_{XX} = 794 \text{ cm}^3$

$r_{YY} = 3.71 \text{ cm}$

$D/T = 30.9$

$l/r_{YY} = 6500/37.1 = 175$

$p_c = 30 \text{ N/mm}^2$

$p_{bc} = 82+25\%$ allowance for wind stress

$= 102 \text{ N/mm}^2$

Stresses at point C on the stanchion are:

$f_c = (21.87\times10^3)/(64.5\times10^2) = 3.4\ \text{N/mm}^2$

$f_{bc} = (66.3\times10^6)/(794\times10^3) = 83.5\ \text{N/mm}^2$

Combined: $3.4/30+83.5/102 = 0.113+0.82 = 0.933$

The section chosen is satisfactory.

7.6 PROBLEMS

1 A steel stanchion of effective length 3.6 m carries an axial load of 400 kN and a moment about the major axis of 150 kN m at the base. Check that 254×254×107 kg/m UC is satisfactory.

2 If the stanchion in problem 1 is cased in accordance with the provisions given in Clause 30(b) of BS 449, determine the lightest section universal column of serial size 254×254 that can be used.

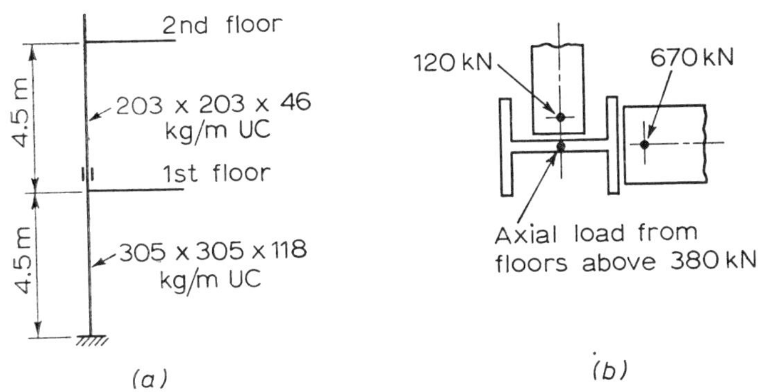

Figure 7.23 Corner stanchion of a multi-storey building

3 A corner stanchion of a multi-storey building is shown in Figure 7.23(a). The loading at the underside of the first floor is shown in (b) in the figure. Check that the lower stanchion section shown is satisfactory.

4 The side stanchion of a building is subjected to the loading shown in Figure 7.24. The horizontal loads on the stanchion are due to wind. The vertical reaction at the column base is the net value due to dead and imposed load and the uplift due to the wind. The stanchion is held in position longitudinally by an eaves girder at the top.

(a) If no lateral support is provided at mid height, check that 356×171×45 kg/m UB will be satisfactory for stress at the base and deflection at the stanchion cap.

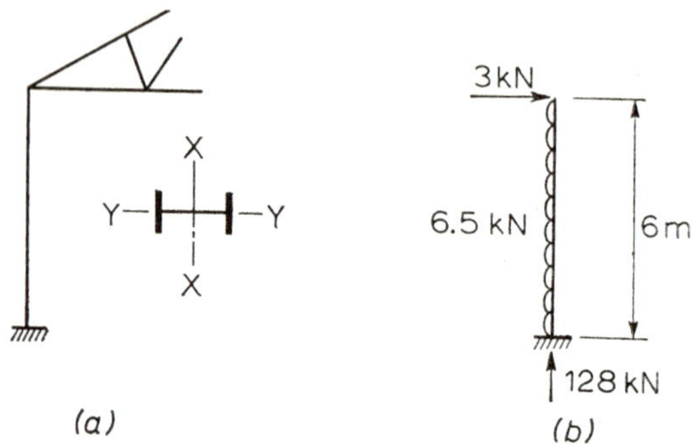

Figure 7.24

(b) If the stanchion is supported laterally at mid height, check that the section could be reduced to 406×140×39 kg/m UB.

5 An extension to an existing building is shown in Figure 7.25. The frames are at 6 m centres. The loading is as follows:

Roof—total dead load = 61 kg/m² = 0.6 kN/m²
imposed load = 0.75 kN/m²

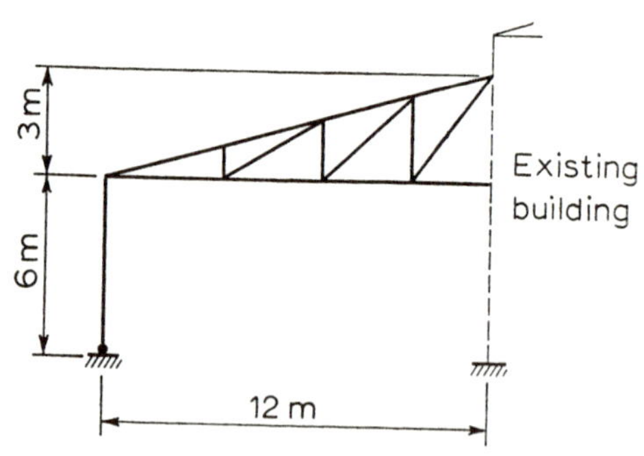

Figure 7.25

Walls—cladding, rails, stanchion etc. = 61 kg/m² = 0.6 kN/m²
Wind on the wall of the extension = 0.58 kN/m²

Design the stanchion for the extension assuming that the joints at the top and bottom are pinned. The lateral wind load will be resisted by the existing building.

6 Design the gable stanchions AB and CD on the end framing shown in Figure 7.26. The loading is as follows.

Total dead load—sheeting and steel = 51 kg/m^2 = 0.5 kN/m^2

Wind load—acts in either direction = 0.5 kN/m^2

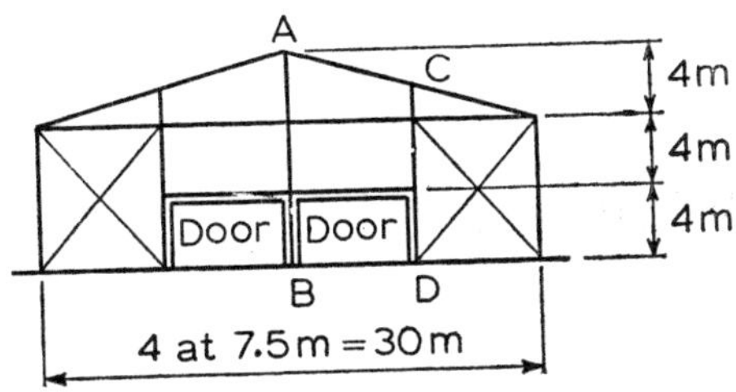

Figure 7.26

8

Crane stanchions

Stanchions carrying cranes in single-storey industrial buildings are considered. The truss and stanchion frame must be taken into account to determine the loads and moments in each stanchion.

8.1 TYPES AND CODE REQUIREMENTS

8.1.1 Types

Three common types of crane stanchions used in single-storey industrial buildings are shown in Figure 8.1. These are:

1 A stanchion of uniform section carrying the crane girder on a bracket. These are used for light cranes.
2 The laced stanchion composed of two universal beams.
3 The compound stanchion either built up from plate or fabricated from two universal beams.

Types 2 and 3 are used for heavy cranes and the crane girder is directly supported either by one member or by the flange of the stanchion. Only the design of a uniform crane stanchion will be given here. Reference should be made to other books for the design of laced crane stanchions.

8.1.2 Effective Lengths for Axial Load

The effective lengths for the stanchion shown in Figure 8.1(a) are taken from Figure 17 of Appendix D of BS 449. These are:

$$\text{XX axis} \quad l = 1.5L$$

$$\text{YY axis} \quad l = 0.75L_1$$

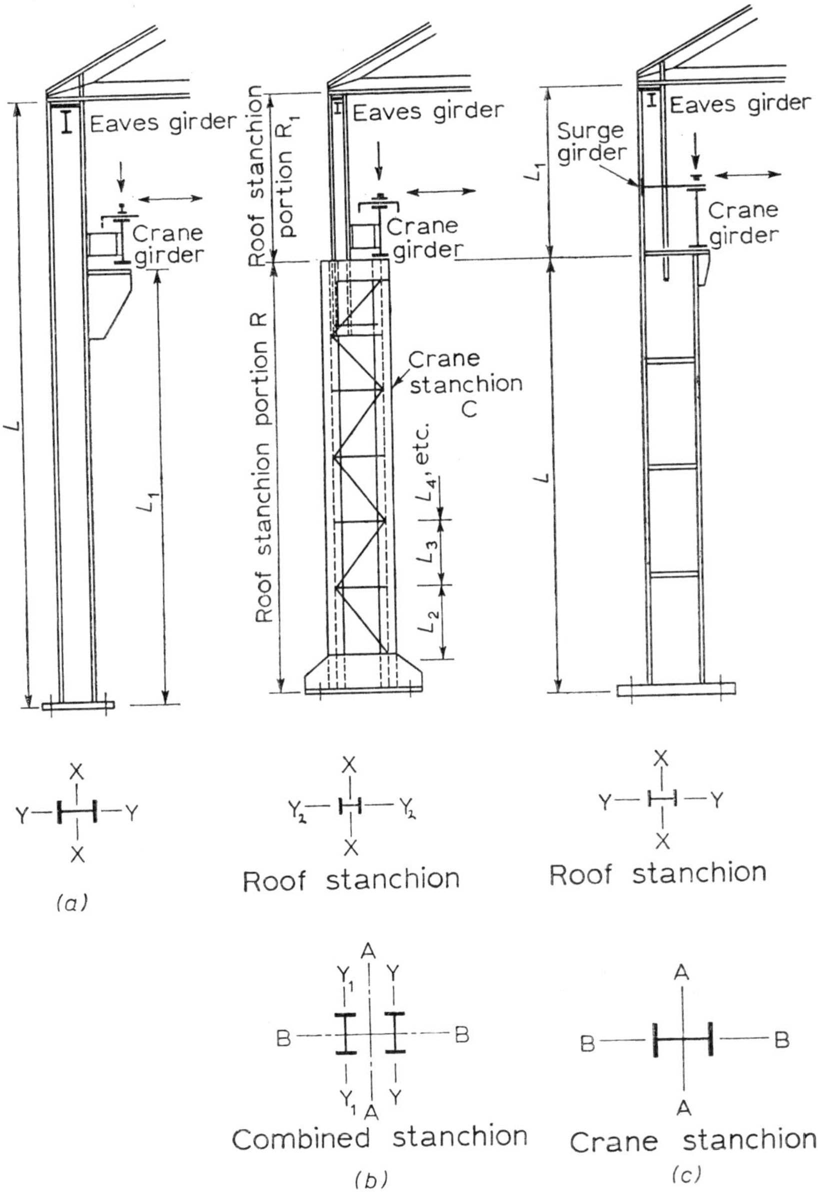
Eaves girder
Crane girder
L
L_1
(a)
Roof stanchion portion R_1
Eaves girder
Crane girder
Roof stanchion portion R
Crane stanchion C
L_4, etc.
L_3
L_2
Roof stanchion
Combined stanchion
(b)
Eaves girder
Surge girder
Crane girder
L_1
L
Roof stanchion
Crane stanchion
(c)

Figure 8.1

For the stanchions shown in Figure 8.1(b) and (c) the effective lengths are taken from Figure 19 of Appendix D of BS 449. These are:

(a) Laced stanchion

Roof stanchion portion R_1

XX axis $l = 1.5\ L_1$

Y_2Y_2 axis $l = L_1$

Roof stanchion portion R

BB axis $l = 0.85\ L$

Y_1Y_1 axis $l = L_2, L_3$, etc., whichever is the greatest

Crane stanchion C

BB axis $l = 0.85\ L$

YY axis $l = L_2, L_3$, etc., whichever is the greatest

Combined roof stanchion portion R and crane stanchion C

AA axis $l = 1.5\ L$

BB axis $l = 0.85\ L$

(b) Built-up stanchion

Roof stanchion

XX axis $l = 1.5\ L_1$

YY axis $l = L_1$

Combined roof and crane stanchion

AA axis $l = 1.5\ L$

BB axis $l = 0.85\ L$

8.1.3 Effective Length of Compression Flange in Bending

With regard to bending, these stanchions are vertical cantilevers with restraints against torsion at two levels as follows:

1 Restraint at the stanchion cap supplied by the eaves strut
2 Restraint at crane girder level

The effective lengths are controlled by Clause 26(c) of BS 449. This clause was discussed in section 7.4.3. The effective lengths for the crane stanchions then are:

1 The stanchion shown in Figure 8.1(a)

$$l = 0.75\ L_1$$

2 The stanchion shown in Figure 8.1(c)

Roof leg	$l = 0.75\ L_1$
Combined crane and roof stanchion	$l = 0.75\ L$

8.2 FRAMES WITH UNIFORM STANCHIONS

8.2.1 Loading

A building frame carrying a crane is shown in Figure 8.2(a). The hook load is placed as far as possible to the left to give the maximum load on that stanchion. The building, crane and wind loads on the stanchion are shown on the figure in (b), (c) and (d) respectively.

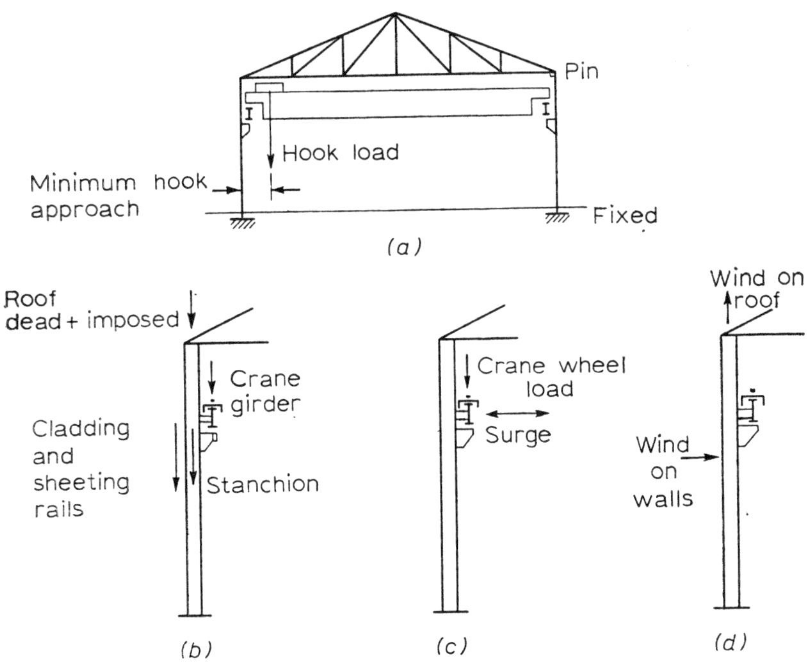

Figure 8.2

8.2.2 Frame Action and Analysis

In order to determine the values of the loading and moments on a stanchion, the frame must be considered as a whole. It will be necessary to adopt a convention for the sign of the bending moments. This will be:

Moments causing tension on the outside of the stanchions of the frame are positive.

Consider the frame shown in Figure 8.2(a) where the stanchions are of uniform section pinned at the top and fixed at the base. The loading is discussed as follows.

1 *Dead and imposed loads*

The dead and imposed loads from the roof, walls and self weight of the steel frame are taken as acting axially on the stanchion. The dead load from the crane girder causes moments as well as axial load in the stanchion. Refer to the crane wheel loads below.

2 *Wind loads*

Wind loading for this type of frame was dealt with in Chapter 7. The analysis is summarised again here. A lateral load causes the frame to sway sideways so that the deflection at the top of each stanchion is the same. This is used to determine the loading on each stanchion. The frame action, loading and moments are shown in Figure 8.3.

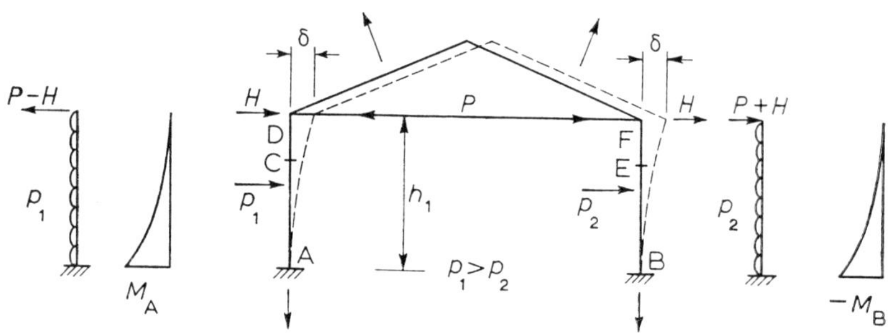

Figure 8.3

The forces H at the tops of the stanchions are the horizontal reactions from the wind on the roof. Equating deflections at the top of each stanchion gives:

$$\frac{p_1h_1^4}{8EI}-\frac{Ph_1^3}{3EI}=\frac{p_2h_1^4}{8EI}+\frac{Ph_1^3}{3EI}$$

This gives

$$P=\tfrac{3}{16}h_1(p_1-p_2)$$

The moments at the bases of the stanchion are:

$$M_A=\tfrac{1}{2}p_1h_1^2-(P-H)h_1$$

$$M_B=-[\tfrac{1}{2}p_2h_1^2+(P+H)h_1]$$

3 *Vertical crane wheel loads*

The vertical crane wheel loads set up moments as well as vertical loads in each stanchion. The weight of the crane girder is added to the

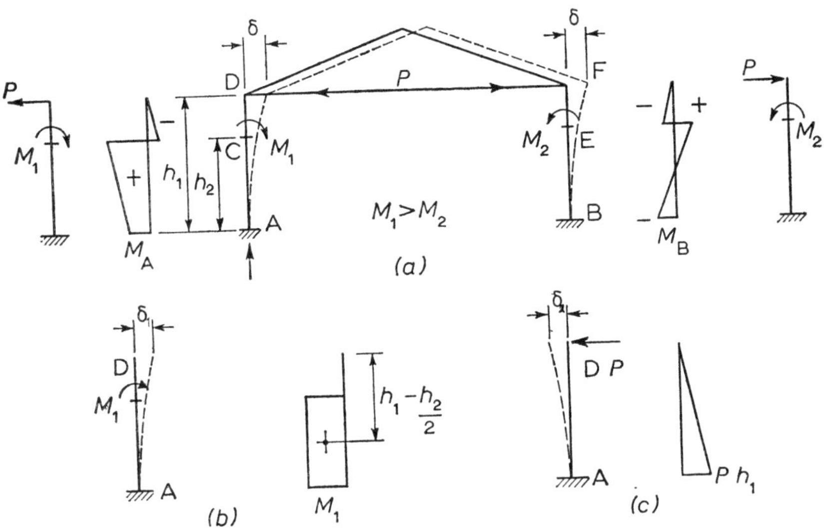

Figure 8.4

reaction from the crane wheels to give the total eccentric load. The moments are unequal so the frame sways as shown in Figure 8.4(a) and a force P is set up in the bottom chord of the truss.

In order to calculate the deflection at the top of the stanchion, consider the moment M_1 and force P for stanchion AD separately, as shown in (b) and (c) in the figure. The deflections may be calculated by the moment area method. This is stated:

> 'The deflection at D relative to the tangent at A is numerically equal to the moment of the M/EI diagram between A and D taken about D'

(Reference should be made to text books on the theory of structures for the moment area method.) Then

$$\delta_1 = \frac{M_1 h_2}{EI}\left(h_1 - \frac{h_2}{2}\right)$$

$$\delta_2 = \frac{Ph_1^3}{3EI}$$

where I is moment of inertia of the stanchions and E is Young's modulus.

Then the deflection of the frame

$$\delta = \delta_1 - \delta_2$$

and equating deflections at the top of each stanchion gives

$$\frac{M_1 h_2}{EI}\left(h_1-\frac{h_2}{2}\right)-\frac{Ph_1^3}{3EI} = -\frac{M_2 h_2}{EI}\left(h_1-\frac{h_2}{2}\right)+\frac{Ph_1^3}{3EI}$$

whence

$$P = \frac{3h_2}{2h_1^3}\left(h_1-\frac{h_2}{2}\right)(M_1+M_2)$$

The moments in the stanchions can now be found.

The moments at the base are:

$$M_A = M_1 - Ph_1$$
$$M_B = M_2 - Ph_1$$

4 *Crane surge*

If the surge load S is the same each side, then each stanchion acts as a free cantilever. The loads and moments for this are shown in

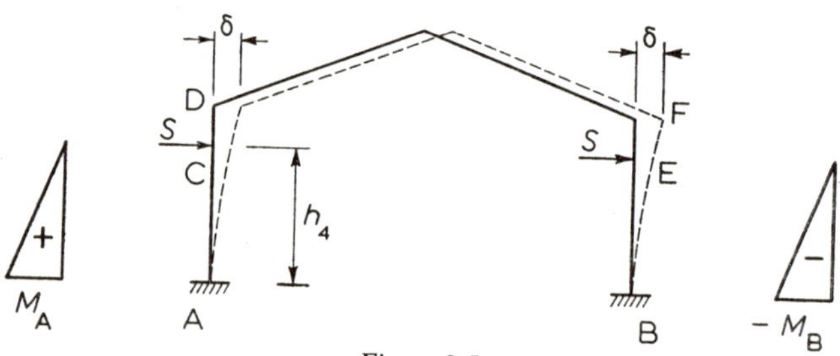

Figure 8.5

Figure 8.5. The moment at the base of the stanchion is

$$M_A = Sh_4 = -M_B$$

If all the surge load is applied to one stanchion, then load is transmitted through the truss to the other stanchion. In this case an analysis similar to that above for wind load would be made.

8.2.3 Design of a Uniform Crane Stanchion

The single-storey building frame shown in Figure 8.6(a) carries a 50 kN electric overhead crane. The frames are at 5 m centres and the length of the building is 50 m. The static crane wheel loads are shown in (b). The end carriage has two wheels at 3 m centres. The crane girders are simply supported, spanning 5 m between stanchions and the weight of one girder is approximately 815 kg, that is 8 kN. The arrangement of the stanchion and crane girder showing the end clearance and eccentricity are shown in (c).

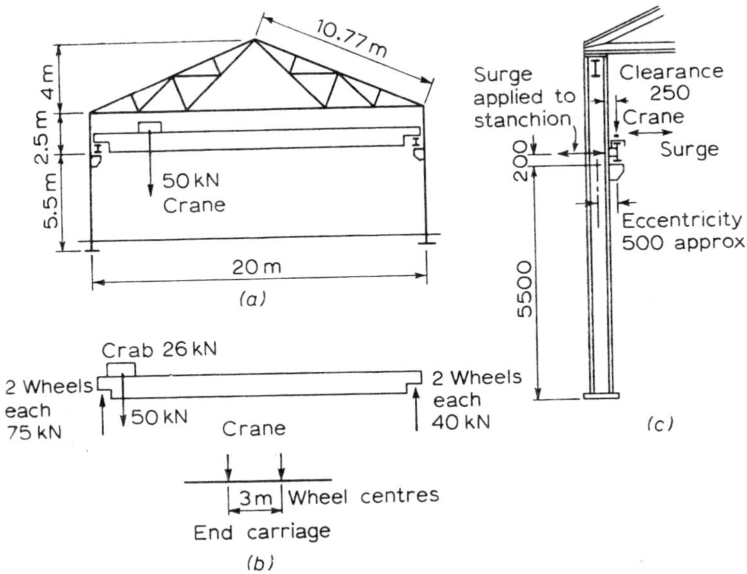

Figure 8.6

The loads on the building are:

Roof—dead load—cladding, purlins, truss	$= 61\ kg/m^2$	$= 0.6\ kN/m^2$
imposed load		$= 0.75\ kN/m^2$
Walls—cladding and rails	$= 61\ kg/m^2$	$= 0.6\ kN/m^2$
Stanchion (estimate)	$= 1020\ kg$	$= 10\ kN$

Wind loads are taken from CP3, Chapter V, Part 2, 1970. The wind loading is set out below.

(a) Determine the loading on the stanchion.
(b) Design the stanchion.

(a) Loading and moments

1 *Dead and imposed loads*

The load on one stanchion is:

Roof—dead	$= 10 \times 5 \times 0.6$	$= 30\ kN$
imposed	$= 10 \times 5 \times 0.75$	$= 37.5\ kN$
Total load at stanchion cap		$= 67.5\ kN$
Wall	$= 8 \times 5 \times 0.6$	$= 24\ kN$
Stanchion		$= 10\ kN$
Total		$= 34\ kN$

These loads act axially on the stanchion.

2 *Wind loads*

Basic wind speed $= 45$ m/s

Topography factor $S_1 = 1$

Ground roughness $= 3$

Building size Class B

Roof—height = 12 m $S_2 = 0.78$

Walls—height = 8 m $S_2 = 0.7$

Factor $S_3 = 1$

Design wind speed

Roof $V_S = 45 \times 0.78 = 35.1$ m/s

Walls $V_S = 45 \times 0.7 = 31.6$ m/s

Dynamic pressure

Roof $q = 760$ N/m²

Walls $q = 613$ N/m²

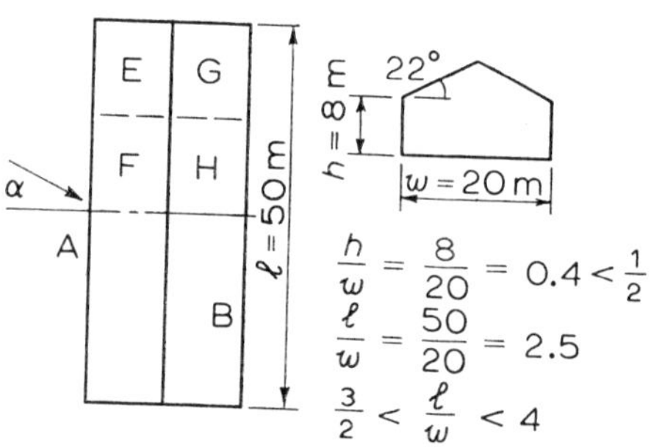

Roof		
Roof angle	Wind angle 0°	
	EF	GH
22°	−0.32	−0.4
Walls		
Wind angle	C_{pe} for surface	
	A	B
0°	+0.7	−0.25

Figure 8.7

The external pressure coefficients C_{pe} for the roof and walls are shown in Figure 8.7. The internal pressure coefficients C_{pi} are taken as the more onerous of +0.2 and −0.3. The wind load on the roof slopes or walls is

$$\frac{q \times \text{Area}}{10^3}(C_{pe} - C_{pi}) \quad \text{kN}$$

The pressure coefficients and wind loads for the frame for the cases of internal pressure and suction are shown in Figure 8.8(a). The loads and reactions from the wind on the roof are shown in (b) and the loads on the walls and in the bottom chord are shown in (c). The

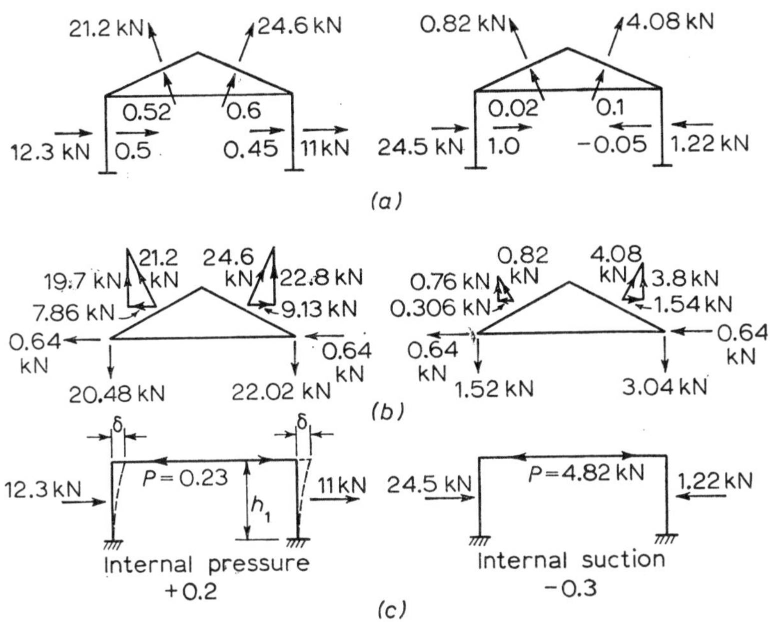

Figure 8.8

calculation for the load in the bottom chord for the case of internal pressure is:

$$-\frac{Ph_1^3}{3EI}+\frac{12.3h_1^3}{8EI}=\frac{Ph_1^3}{3EI}+\frac{11h_1^3}{8EI}$$

$$P = 0.23 \text{ kN}$$

The moments at the stanchion bases are:

Internal pressure $M_A = (12.3\times4)+(0.64-0.23)8 = 52.5$ kN m

$M_B = -[(11\times4)+(0.64+0.23)8] = -50.9$ kN m

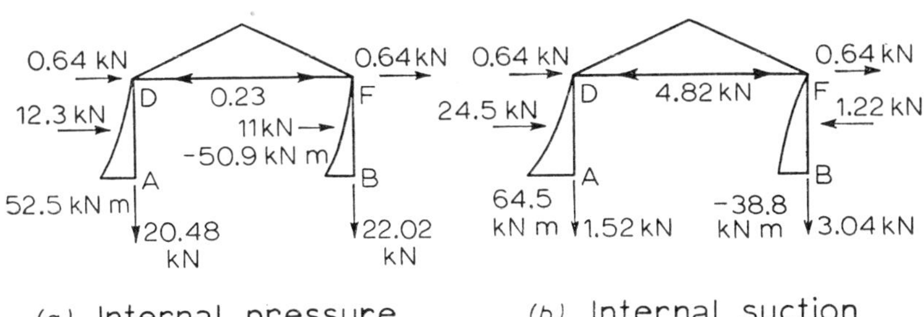

Figure 8.9

Internal section $M_A = (24.5\times4)-(4.82-0.64)8 \quad = \quad 64.5$ kN m

$M_B = (1.22\times4)-(4.82+0.64)8 \quad = -38.8$ kN m

The loads and moments in the stanchions are shown in Figure 8.9.

3 *Vertical Crane Wheel Loads*

The crane wheel loads including impact are:

Maximum wheel loads $= 75+25\% = 93.75$ kN

Light side wheel loads $= 40+25\% = 50$ kN

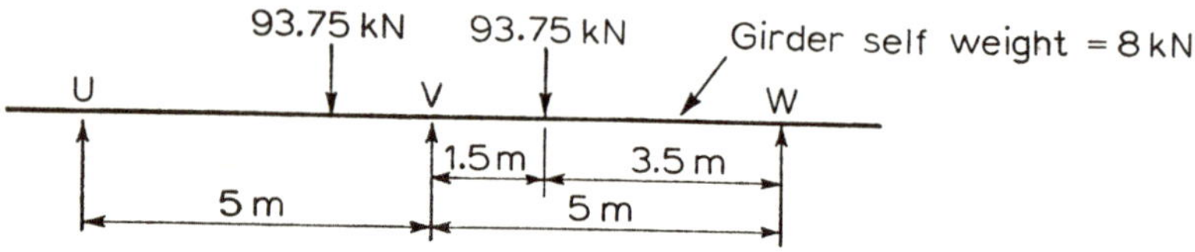

Figure 8.10

To determine the maximum stanchion reaction, place the wheel loads equidistant about the stanchion V as shown in Figure 8.10. Then the stanchion reaction for the maximum wheel loads is:

Maximum $R_V = (2\times93.75\times3.5)/5+8 = 131+8 = 139$ kN

The same value would be obtained if one wheel load is placed over the stanchion V. The light side stanchion reaction is found by proportion.

Light side $R_V = (50/93.75)\ 131+8 = 69.8+8 = 77.8$ kN

The moments on the stanchions caused by these loads are:

Maximum moment $M_1 = 139 \times 0.5 = 69.5$ kN m

Light side moment $M_2 = 77.8\times0.5 = 38.9$ kN m

The loads on the frame are shown in Figure 8.11. Referring to the crane wheel load case in the section above:

$$P = \frac{(3\times5.5)}{2\times8^3}\left(\frac{8-5.5}{2}\right)(69.5+38.9) = 9.15 \text{ kN}$$

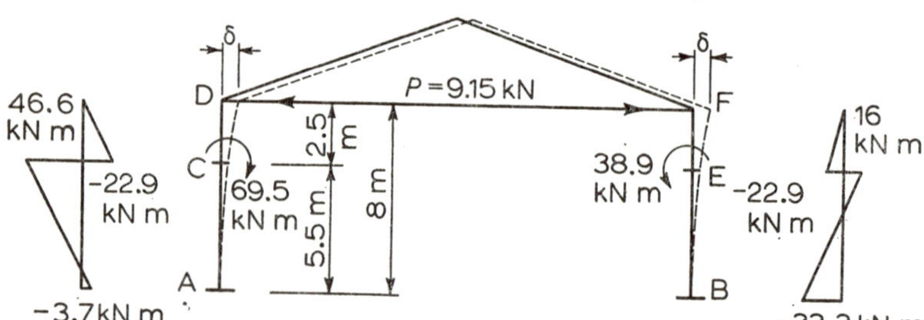

Figure 8.11

The moments in the stanchions are:

Stanchion AD	$M_{CD} = -9.15 \times 2.5$	$= -22.9$ kN m
	$M_{CA} = 69.5 - 22.9$	$= 46.6$ kN m
	$M_A = 69.5 - (9.15 \times 8)$	$= -3.7$ kN m
Stanchion BF	$M_{EF} =$	$= -22.9$ kN m
	$M_{EB} = 38.9 - 22.9$	$= 16.0$ kN m
	$M_B = 39.9 - 73.2$	$= -33.3$ kN m

The bending moment diagrams for the stanchions are shown in Figure 8.11.

4 *Crane surge*

The horizontal surge load is 10% of hook load plus the weight of the crab

$$= 0.1(50+26) = 7.6 \text{ kN}$$

This is divided between the four wheels, so the load per wheel is 1.9 kN. The column reaction from the horizontal wheel loads may be found in a similar way to the vertical wheel loads above:

$$S = (2 \times 1.9 \times 3.5)/5 = 2.66 \text{ kN}$$

The moments at the base of the stanchions are:

$$M_A = -M_B = 2.66 \times 5.7 = 15.2 \text{ kN m}$$

The loads and moments are shown in Figure 8.12.

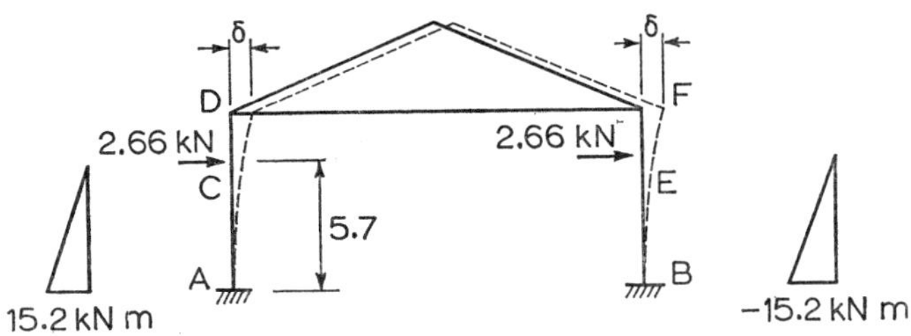

Figure 8.12

(b) Maximum design conditions

Several cases may require investigation to find the maximum design conditions. The loading and base reactions for three cases are set out in Figure 8.13. Wind loads cause more than 25% of the base moments in all cases.

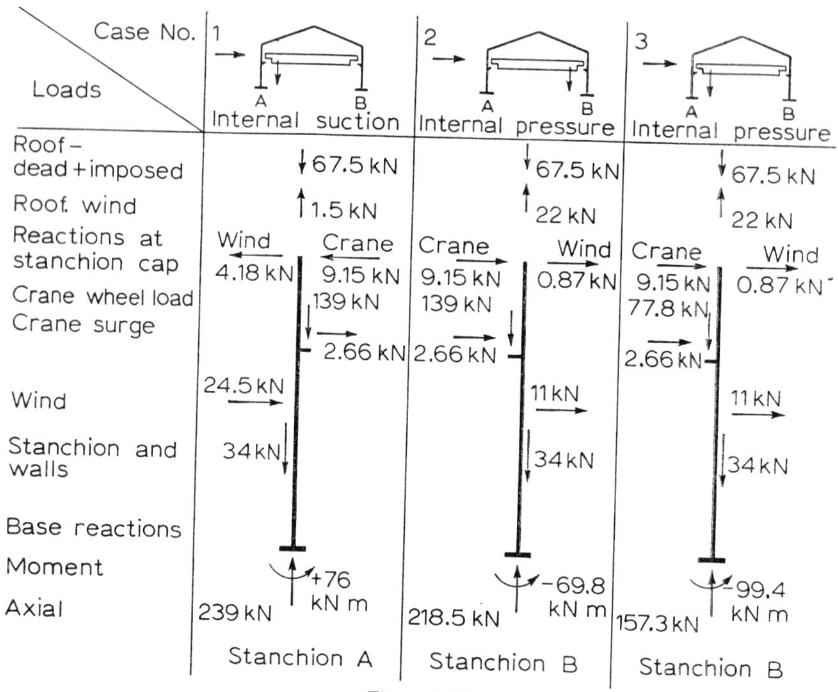

Figure 8.13

(c) Design of the stanchion

Try 457×152×67 kg/m UB. The properties are:

$$A = 85.3 \text{ cm}^2$$
$$Z_{XX} = 1248 \text{ cm}^3$$
$$r_{XX} = 18.3 \text{ cm}$$
$$r_{YY} = 3.12 \text{ cm}$$
$$D/T = 30.5$$

Permissible compressive stress:

$$l/r_{XX} = (1.5\times8000)/183 = 65.5$$
$$l/r_{YY} = (0.75\times5500)/31.2 = 132.5$$
$$p_c = 50.5 \text{ N/mm}^2 \text{ from Table 17 (BS 449)}$$

Permissible bending stress in compression:

$$l/r_{YY} = 132.5;\ D/T = 30.5. \text{ From Table 3:}$$
$$p_{bc} = 123+25\% \text{ wind allowance}$$
$$= 154 \text{ N/mm}^2$$

The actual stresses for Case 1 loading are:

$$f_c = (239\times10^3)/(85.3\times10^2) = 28\ \text{N/mm}^2$$
$$f_{bc} = (76\times10^6)/(1248\times10^3) = 61\ \text{N/mm}^2$$

Combined:

$$(28/50.5)+(61/154) = 0.555+0.397$$
$$= 0.952$$

The section selected is satisfactory. This is the most severe loading case.

8.3 PROBLEMS

1 Design the stanchion for the building frame shown in Figure 8.14(a). The frames are of 8 m centres and the length of the building is 48 m.

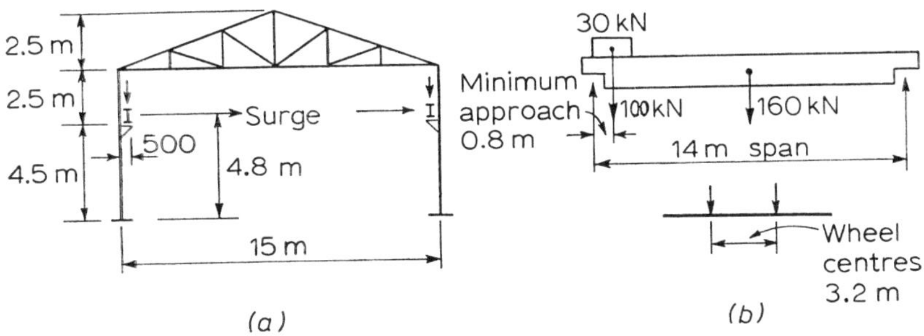

Figure 8.14

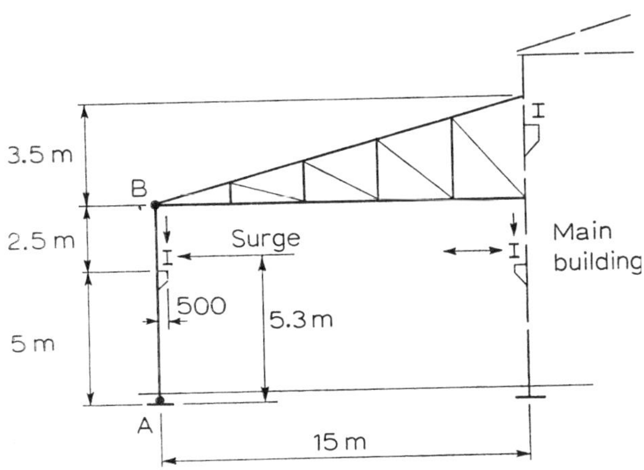

Figure 8.15

The data for the 100 kN crane are shown in (b). Assume suitable values for the dead loads. The imposed and wind loads are to be in accordance with the relevant British Standards.

2 The side bay of a factory is shown in Figure 8.15. The frames are at 6 m centres. The stanchion AB is pinned at the eaves and at the base. The crane in the side bay is 100 kN capacity and the data are as shown in Figure 8.14(b). Assume suitable values for the dead loads. The imposed and wind loads are in accordance with CP3, Chapter V, Parts 1 and 2 respectively. The lateral loading on the side bay is taken on the main building. Design stanchion AB.

9

Stanchion bases

Stanchion bases transmit the loads and moments from the steel stanchion to the concrete foundation. The main purpose is to distribute the loads safely to the weaker material. The concrete base transmits the loads to the earth.

Steel grillages are used to spread very heavy loads on to a concrete base or through a concrete casing on to the earth.

The design of bases is approximate and is conservative.

9.1 DESIGN CONSIDERATIONS

9.1.1 Types and Loading

The types of bases used are illustrated in Figure 9.1. These are:

1 Slab or gusseted base shown in (a), (b) and (c) on the figure
2 Pocket base shown in (d)
3 Grillage shown in (e)

The gusseted bases dealt with are of welded construction.

Bases are required to transmit axial load only or axial load combined with bending moments and horizontal loads. Bending moments may act about both rectangular axes. Depending on the values of axial load and moment, there may be compression over the whole base or compression over part of the base and tension in the holding-down bolts. Horizontal loads are resisted by shear in the holding-down bolts and friction between the base and the concrete. These loads are small.

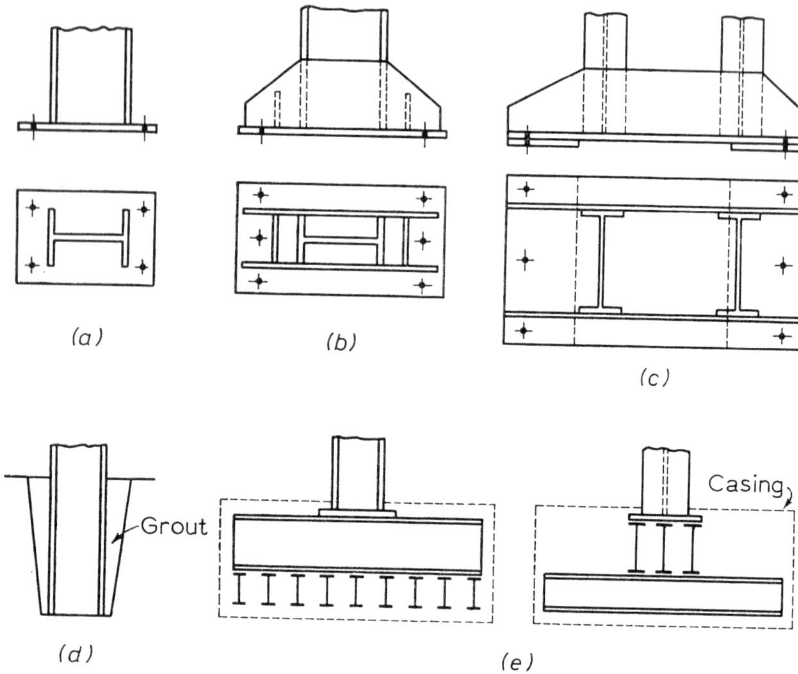

Figure 9.1 Types of stanchion base

9.1.2 Permissible Stresses

(a) Stanchion base plates

The permissible bending stress in slab bases is given in Table 2 and Clause 38 of BS 449. This is 185 N/mm² for all steels. For gusseted bases, a bending stress of 165 N/mm² for grade 43 steel is used in the gussets and stiffeners, but a stress of 185 N/mm² is used in the base plate.

(b) Grillage foundations

The stresses are governed by Clause 40 of BS 449. All permissible working stresses may be increased by $33\frac{1}{3}\%$ provided that the grillage is set out and encased in accordance with this clause. This will be given in detail later.

(c) Holding-down bolts

The permissible stresses for holding-down bolts are given in Table 20 of BS 449. Tensile stresses are calculated on the net area at the root of the thread. Net areas and safe tension values for bolts of strength grade designation 4.6 are given in the Appendix in *Table* 2.

(d) Concrete
The permissible stress for the concrete in bearing is taken from CP 114, Part 2, 1969, 'The Structural Use of Concrete.' All calculations in this chapter will be based on a nominal 1 : 2 : 4 mix, where

$$p_c = \text{permissible stress in direct compression}$$
$$= 5.3 \text{ N/mm}^2$$

This stress will be used where axial load and bending occur together. A stress of 7 N/mm² is allowed in CP 114 for concrete in compression in bending in reinforced concrete beams. This stress will be used in the design of pocket bases.

Stronger mixes are specified in CP 114 and designed mixes can be used. The student should refer to the code of practice and to text books on reinforced-concrete design.

9.2 SLAB BASES

9.2.1 Code Requirements and Theory

This type of base is used extensively in modern practice. On very heavily loaded stanchions, slabs 100 mm and more in thickness are used. The slab base is free from pockets where corrosion may start and maintenance is simpler than with gusseted bases.

Rules for the design of slab bases are given in Clause 38(b) of BS 449. This clause states:

> 'Stanchions with slab bases need not be provided with gussets, but fastenings shall be provided to retain the parts securely in place and to resist all moments and forces, other than direct compression.... When the slab alone will distribute the load uniformly, the minimum thickness of a rectangular slab shall be:
>
> $$t = \left[\frac{3\omega}{p_{bct}}\left(A^2 - \frac{B^2}{4}\right)\right]^{1/2}$$
>
> where t is the slab thickness in mm, A is the greater projection of the plate beyond the stanchion in mm, B is the lesser projection of the plate beyond the stanchion in mm, ω is the pressure or loading on the underside of the base in N/mm² and p_{bct} is the permissible bending stress in the steel (= 185 N/mm²).
> When the slab will not distribute the load uniformly or when the slab is not rectangular, special calculations shall be made to show that the stresses are within the special limits... .'

The above equation takes into account plate bending in two directions at right angles. The plate is stronger than if it were composed of a series of independent strips set side by side. Poisson's ratio is introduced to allow for this effect. This is stated as:

A stress produces a strain in its own direction and an opposite strain in every direction at right angles to its own. The ratio

$$\frac{\text{Lateral strain}}{\text{Longitudinal strain}} = \text{Poisson's ratio} \quad \text{(a constant)}$$

$$= \tfrac{1}{4} \text{ for steel} \quad \text{(0.3 is generally used now)}$$

Thus a moment in the base slab is reduced by the coexistent moment at right angles. To show this, the equation from the code is now derived.

Consider an element at W and the two cantilever strips 1 mm wide shown in Figure 9.2. The pressure under the base is ω N/mm². The moments at W are

$$M_{XX} = \tfrac{1}{2}\omega A^2$$

$$M_{YY} = \tfrac{1}{2}\omega B^2$$

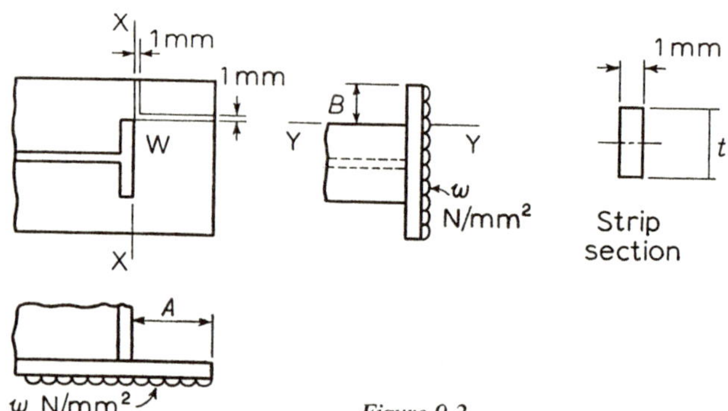

Figure 9.2

The projection A is greater than B, so the net moment

$$M_{XX} = \tfrac{1}{2}\omega A^2 - \tfrac{1}{8}\omega B^2$$

$$= \frac{\omega}{2}\left(A^2 - \frac{B^2}{4}\right)$$

$$= p_{bct}\frac{t^2}{6}$$

where the modulus of section for the cantilever strip is $Z = t^2/6$. Thus

$$t = \left[\frac{3\omega}{p_{bct}}\left(A^2 - \frac{B^2}{4}\right)\right]^{1/2}$$

In axially loaded bases, the stanchion end and slab are usually machined for bearing. In this case the weld between stanchion and slab only holds the slab in position; the axial load is transmitted by bearing. With light struts, the surfaces may not be machined for bearing and in such a case the weld must be designed to transmit the load.

In a case where the stanchion is subjected to axial load and bending moment the weld may require designing if tension develops on one flange of the stanchion. This will be considered later.

9.2.2 Slab Base for an Axially Loaded Stanchion

A stanchion consisting of a 305×305×198 kg/m UC carries an axial load of 2400 kN at the base. The safe bearing pressure on the concrete is 5.3 N/mm². Adopting a square slab, determine the size and thickness required.

The area required for the base is calculated first.

$$\text{Area} = (2400\times10^3)/5.3 \qquad = 45.2\times10^4\ \text{mm}^2$$

Make base 680 mm square

$$\text{Pressure } \omega = (2400\times10^3)/680^2 = 5.2\ \text{N/mm}^2$$

The arrangement of the stanchion on the base slab is shown in Figure 9.3. From this:

A—greater projection of base = 182.9 mm

B—lesser projection of base = 170 mm

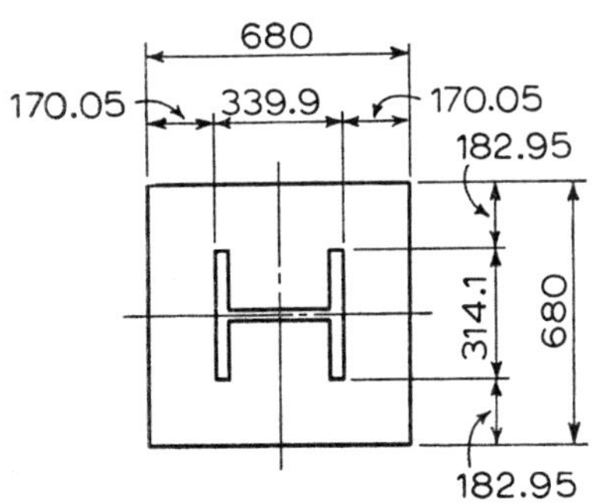

Figure 9.3

The thickness of the base plate is given by:

$$t = \left[\frac{3\times5.2}{185}\left(182.9^2 - \frac{170^2}{4}\right)\right]^{1/2}$$

$$= 47.1\ \text{mm}$$

Make the base slab 50 mm thick. Use 8 mm fillet weld all round to hold the base slab in place. The surfaces are to be machined for direct bearing. The holding-down bolts required are nominal, but four No. 24 mm diameter bolts would be used.

Base slab: 680 mm×680 mm×50 mm thick

9.3 GUSSETED BASES

With this type of base, part of the load is transmitted from the stanchion through the gussets to the base slab. The gussets and stiffeners support the base slab against bending and hence a thinner plate can be used than with the slab base.

These bases are more difficult to maintain than slab bases and corrosion is more likely to occur.

Design considerations are set out in Clause 38(a) of BS 449. This states:

> 'For stanchions with gusseted bases, the gusset plates, angle cleats, stiffeners, fastenings, etc., in combination with the bearing area of the shaft shall, where all the parts are fabricated flush for bearing be sufficient to take loads, bending moments and reactions to the base plate without exceeding the specified stresses.
> Where the ends of the stanchion shaft and the gusset plates are not faced for complete bearing, the fastenings connecting them to the base plate shall be sufficient to transmit all the forces to which the base is subjected.'

The method generally adopted for the design of gusseted bases is set out in the example given in section 9.4.4.

9.4 ECCENTRICALLY LOADED BASES

Consider bases subject to axial load and bending about one axis. The case of bending about two axes will be considered later. Two separate cases can be considered. These are:

(a) Pressure over the whole of the base
(b) Pressure over part of the base and tension in the holding-down bolts

Which case will occur in a given instance depends on the relative values of the moment and axial load.

9.4.1 Compression over the Whole of the Base

A stanchion base and loading is shown in Figure 9.4 with the pressure distribution under the base.

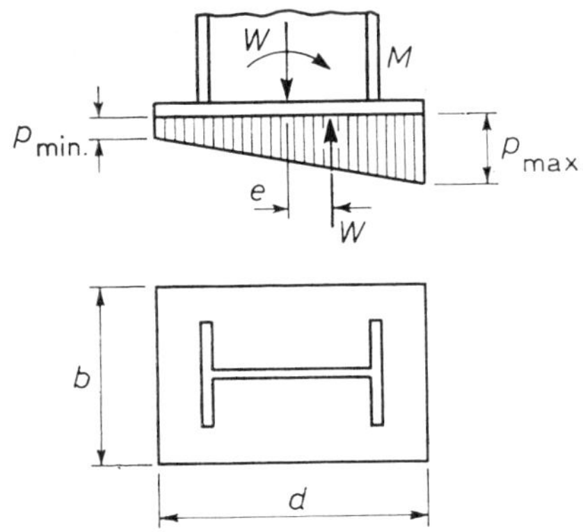

Figure 9.4

The following symbols are defined:

W = total load on base

M = moment at base

b = breadth of base

d = length of base

e = eccentricity

$= M/W$

p_{max}, p_{min} = maximum and minimum pressures under the base, respectively

The middle-third rule of structural mechanics applies and if the eccentricity e is less than $d/6$, that is the resultant load lies within the middle third of the length of the base, pressure occurs over the whole of the base.

A = area of base

$= bd$

Z = modulus of base

$= bd^2/6$

Maximum pressure on the concrete:

$$p_{max} = (W/A)+(M/Z)$$

Minimum pressure on the concrete:

$$p_{min} = (W/A)-(M/Z)$$

A special case occurs where the pressure varies from a maximum at one edge to zero at the other. In either case, the maximum pressure must not exceed the allowable bearing pressure on the concrete.

The size chosen for the base depends on practical considerations. Both the length and the breadth can be varied. The length must not be less than $6e$ for the condition of pressure over the whole base to apply. If the length of the base is fixed, the breadth may be determined so that the maximum allowable pressure is not exceeded.

9.4.2 Compression Over Part of the Base and Tension in the Holding-Down Bolts

The design method adopted here is that used for a reinforced concrete section subjected to axial force and bending moment, using elastic theory. A stanchion base is shown in Figure 9.5. The pressure distribution on the concrete and the tension in the holding-down bolts are shown in the elevation on the figure. The following symbols require definition:

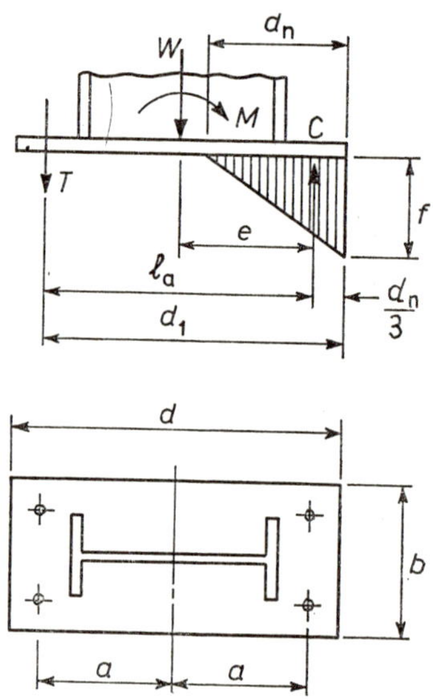

Figure 9.5

p_c = permissible stress in the concrete in bearing
p_{st} = permissible stress in the steel holding-down bolts in tension
m = modular ratio = Young's modulus for steel / Young's modulus for concrete
= 15

W = axial load on the base
M = moment on the base
d = length of base
b = breadth of the base
e = eccentricity = M/W
d_n = depth to the neutral axis
d_1 = distance from the centreline of the bolts in tension to the edge of base plate in compression
l_a = lever arm, the distance between the compressive force in the concrete and the tensile force in the bolts
A_{st} = net area of the holding-down bolts in tension
$2a$ = distance between the centres of the bolts in the base plate
C = compressive force in the concrete
T = tensile force in the bolts
f = maximum compressive stress in the concrete

An approximate design method is given first and this is followed by the exact elastic method which is used for checking a section.

(a) Design method
Assume that the maximum permissible stresses occur simultaneously in the concrete and steel. From the elastic theory of reinforced concrete design, the depth to the neutral axis is given by

$$d_n = \left(\frac{mp_c}{mp_c+p_{st}}\right) d_1$$

The lever arm $l_a = d_1 - d_n/3$

Take moments about the centreline of the bolts in tension:

$$M_1 = M + Wa$$

The compressive force in the concrete

$$C = M_1/l_a$$

The maximum compressive stress in the concrete

$$f = 2C/bd_n$$

This stress must not exceed the allowable compressive stress for the concrete. Alternatively, if the length of the base is assumed first, the breadth can be determined from the above equation, so that the allowable concrete stress is not exceeded. The tensile force in the holding-down bolts is

$$T = C - W$$

From this, the area at the root of the thread can be determined:

$$A_{st} = T/p_{st}$$

Suitable bolts may then be selected from *Table* 2 in the Appendix.

Thus an assumed size of base can be checked and the method gives accurate results if the final stresses in the concrete and steel bolts are close to the permissible values taken to determine the depth of the neutral axis.

(b) Elastic theory

Bases often have to be checked for different load conditions and the stresses in the steel and concrete will not be the maximum allowable values. For example, in designing the base for the side stanchion in a single-storey building, the maximum pressure on the concrete occurs when dead, imposed and wind load are taken together. However, the maximum tension in the bolts is found when dead and wind loads are acting. Here, the general elastic theory can be used to find the stresses in the given base. This is now outlined.

Consider the base loaded with an eccentric load as shown in Figure 9.6.

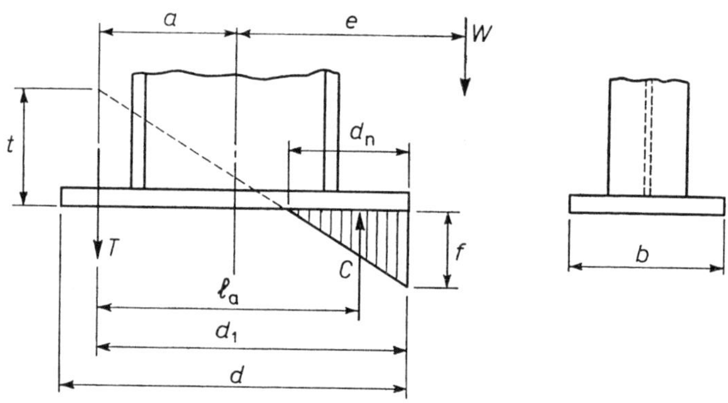

Figure 9.6

The following terms are defined in addition to those above:

t = tensile stress in the holding-down bolts

$$= 15f\left(\frac{d_1-d_n}{d_n}\right)$$

C = compressive force in the concrete

$$= \tfrac{1}{2}fbd_n$$

T = tensile force in the holding-down bolts

$$= t\,A_{st}$$

The load on the stanchion is

$$W = C-T$$

The moment on the base is

$$M = We$$
$$= C[(d/2)-(d_n/3)]+Ta$$
$$= (C-T)e$$

From the two equations for the moment M, a cubic equation is obtained in d_n which can be solved. Then by back substitution f and t can be found. These values should not exceed the permissible stresses.

Instead of making the above calculations, it is often sufficiently accurate to assume the value for the lever arm, l_a, from balanced design as set out in the design method above, and to calculate the stresses using this value. These methods will be demonstrated in an example that follows.

9.4.3 Slab Base—Compression Over the Whole of the Base

A stanchion base is subjected to a moment of 36.6 kN m and an axial load of 500 kN. The stanchion section is a 203×203×86 kg/m UC. The permissible bearing pressure on the concrete is 5.3 N/mm². Design a slab base and the weld from the stanchion shaft to the base plate assuming that the parts are machined for bearing.

(a) Size of base
Eccentricity of load

$$e = 36.6\times10^6/(500\times10^3) = 73.2 \text{ mm}$$

If the base is made $6e$ in length, there will be pressure over the whole of the base:

$$6e = 439.2 \text{ mm}$$

The breadth of the base b to limit the stress on the concrete to 5.3 N/mm² is given by

$$b = (2\times500\times10^3)/(439.2\times5.3) = 431 \text{ mm}$$

Thus a square base 440 mm×440 mm could be used. However, a rectangular base 500 mm long by 360 mm wide will be checked. The arrangement of the base is shown in Figure 9.7.

Area A	$= 500\times360$	$= 1.8\times10^4$ mm²
Modulus Z	$= (360\times500^2)/6$	$= 1.5\times10^7$ mm³
Maximum pressure	$= (500\times10^3)/(1.8\times10^4)+(36.6\times10^6)$ $/(1.5\times10^7)$	
	$= 2.78+2.44$	$= 5.22$ N/mm²
Minimum pressure	$= 2.78-2.44$	$= 0.34$ N/mm²

The pressure distribution on the concrete under the base is shown on the elevation of the base in Figure 9.7.

The maximum pressure is less than the allowable pressure and the size chosen for the base is satisfactory.

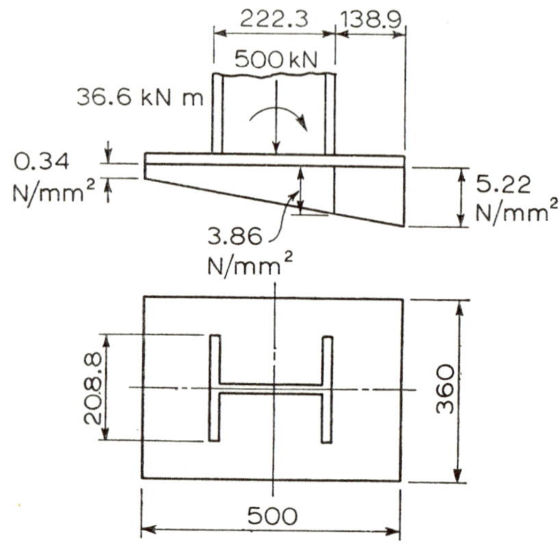

Figure 9.7

(b) Thickness of the base plate

Consider a 1 mm wide strip as shown on Figure 9.8. This acts as a cantilever from the face of the stanchion and the loading on the

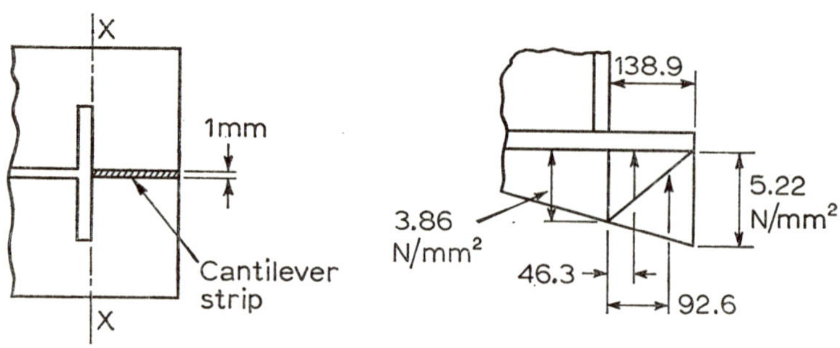

Figure 9.8

strip is shown on the figure. This treatment gives a conservative result for the thickness of base plate. Plate action due to bending in two directions at right angles is not taken into account.

Base pressure at section XX

$$= 0.34+[(500-138.9)/500]\,(5.22-0.34) = 3.86\ \text{N/mm}^2$$

The trapezoidal pressure diagram loading the cantilever strip is divided into two triangles as shown on the figure. The moment may be calculated as follows:

$$M = (3.86\times 139\times 46.3)/2+(5.22\times 139\times 92.6)/2$$
$$= 46.1\times 10^3\ \text{N mm}$$

The section of the base plate is 1 mm wide $\times$ t mm thick and

$$\text{Modulus } Z = t^2/6$$

whence the thickness can be calculated:

$$t = (6\times 46.1\times 10^3/185)^{1/2} = 38.7\ \text{mm}$$

A slab 40 mm thick would be used.

(c) Weld–stanchion to base plate
Stanchion 203×203×86 kg/m UC

$$A = 110.1\ \text{cm}^2$$
$$Z_{XX} = 851.5\ \text{cm}^3$$

Stresses at the stanchion-to-base plate joint:

Axial stress $f_c = (500\times 10^3)/(110.1\times 10^2) = 45.5\ \text{N/mm}^2$

Bending stress $f_{bc} = (36.6\times 10^6)/(851.5\times 10^3) = 43\ \text{N/mm}^2$

There is compressive stress over the whole of the stanchion section. The slab and stanchion are machined for bearing, so the weld is required to hold the base slab in position. Use, 8 mm fillet weld. A continuous weld around the stanchion profile will be specified to give a full seal and maximum protection against corrosion.

9.4.4 Gusseted Base—Compression Over the Whole of the Base

Redesign the base in the previous example using a gusseted base. The parts are not faced for bearing so the plates and welds are to be designed to transmit the whole of the stanchion load and moment to the base plate.

The arrangement of the base is shown in Figure 9.9.

(a) Gusset plates
The pressure under the base is shown in Figure 9.10(a) and the load on one gusset is shown in (b). The values for the pressures are taken from the previous example.

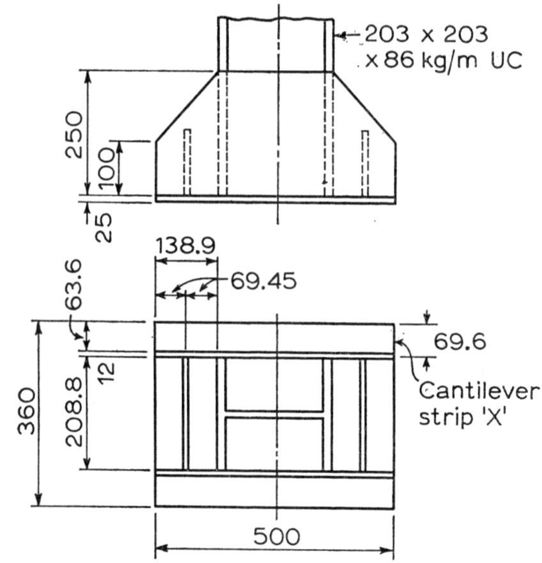

Figure 9.9

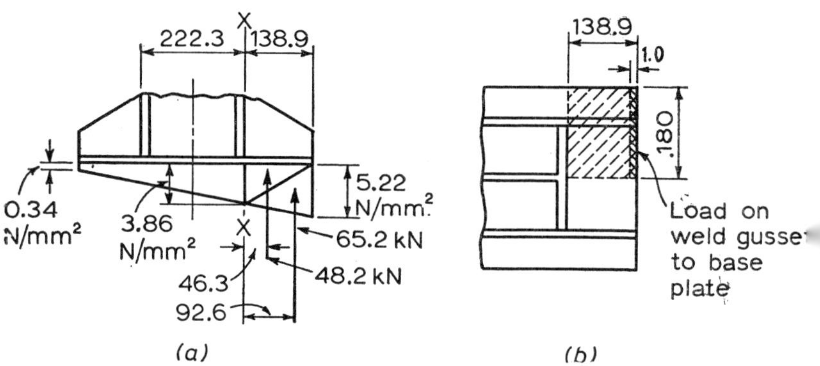

Figure 9.10

At section XX the shear on one gusset is

$$S = (3.86 \times 138.9 \times 180)/(2 \times 10^3) + (5.22 \times 138.9 \times 150)/(2 \times 10^3)$$
$$= 48.2 + 65.2$$
$$= 113.4 \text{ kN}$$

and the moment at the same section is

$$M = (48.2\times46.3)/10^3+(65.2\times92.6)/10^3$$
$$= 2.23+6.04$$
$$= 8.27 \text{ kN m}$$

Make the gusset plates 12 mm thick and determine the depth d to give a bending stress of 165 N/mm².

Modulus $Z = 12d^2/6 = 2d^2$ mm³

Depth $d = [(8.27\times10^6)/(165\times2)]^{1/2} = 159$ mm

Shear stress $= (113.4\times10^3)/(159\times12) = 59.5$ N/mm²

The weld from the gusset plate to the stanchion will have to be designed before the depth of the gusset plate can be determined finally.

(b) Gusset plate-to-stanchion weld

The welds between the stanchion, gussets and base plate have to transmit all of the load to the base plate. These welds are shown in Figure 9.11.

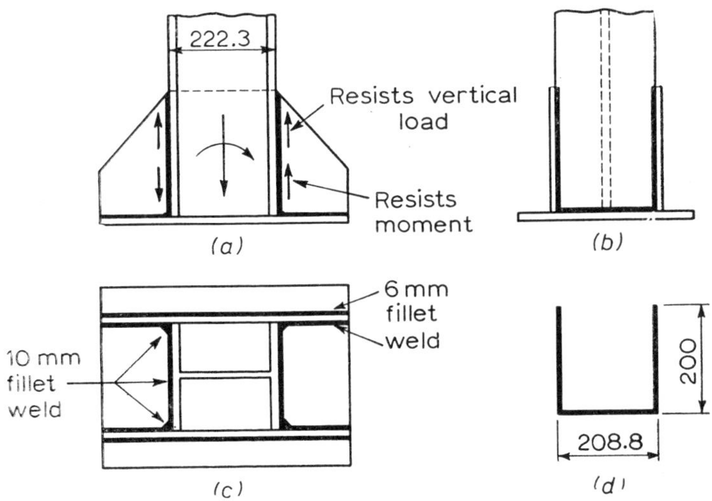

Figure 9.11

The stanchion-to-gusset and base plate weld is shown in (d).

Load per weld $= (500/2)+(36.6/0.22)$

$= 250+165 = 415$ kN

Use 10 mm fillet weld, strength 800 N/mm.

Length of weld $= 415/0.8 = 519$ mm

Use 200 mm depth of gusset plate. This gives a total length of 608.8 mm.

The welds between one gusset plate and the base plate must support the maximum pressure under the base. Refer to Figure 9.10(b) and consider a strip 1 mm wide at the edge of the base.

$$\text{Load per weld} = (180 \times 5.22)/2 = 469.8 \text{ N}$$

6 mm continuous fillet weld, strength 480 N/mm is satisfactory. The weld is shown in Figure 9.11.

(c) Base plate thickness

A secondary stiffener has been introduced as shown on Figure 9.9 mid way between the stanchion face and the edge of the base plate. Consider cantilever strip 'X', 1 mm wide between the gusset and the edge of the plate. This is the longest cantilever strip. For this cantilever:

$$M = (5.22 \times 69.5)/2 = 1.36 \times 10^4 \text{ N mm}$$

$$= 185t^2/6$$

$$t = (6 \times 1.36 \times 10^4/185)^{1/2} = 21 \text{ mm}$$

Use 25 mm thick plate.

9.4.5 Slab Base—Tension in Holding-Down Bolts

Design a slab for the side stanchion of a single-storey building for the following conditions (See Figure 7.19):

1 Dead+imposed+wind loads:

Vertical load	W =	80.1 kN
Horizontal load	H =	13.81 kN
Moment	M =	32.8 kN m

2 Dead+wind loads:

Vertical load	W =	42.6 kN
Horizontal load	H =	13.81 kN
Moment	M =	32.8 kN m

In each case the total moment on the base is due to wind load. The stanchion is a 406×140×39 kg/m UB.

(a) Size of the base

The arrangement of the base is shown in Figure 9.12. Consider case 1 loading first as this will determine the size of base. A length of 560 mm

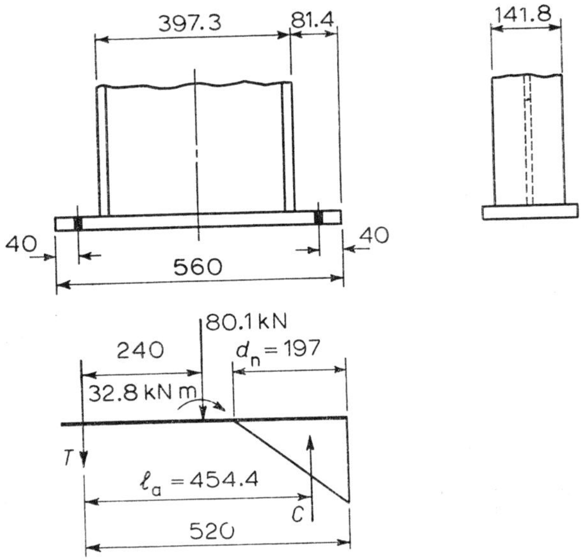

Figure 9.12

has been selected and the breadth will be calculated. The permissible stresses are:

Concrete = 5.3+25% wind allowance = 6.63 N/mm^2

Bolts = 130+25% wind allowance = 162.5 N/mm^2

The depth to the neutral axis is

$$d_n = [(15\times6.63)/(15\times6.63+162.5)]520 \quad = 197 \text{ mm}$$

The lever arm

$$l_a = 520-(197/3) \quad = 454.4 \text{ mm}$$

Take moments about the centreline of the holding-down bolts:

$$M_1 = 32.8+(80.1\times0.24) \quad = 52 \text{ kN m}$$

The compressive force in the concrete is

$$C = (52\times10^6)/(454.4\times10^3) \quad = 114.5 \text{ kN}$$

The base width to limit the concrete stress to 6.63 N/mm^2 is

$$b = (2\times114.5\times10^3)/(6.63\times197) = 175 \text{ mm}$$

Adopt a base width of 200 mm, the maximum concrete stress is

$$f = (175/200)\times6.63 \quad = 5.8 \text{ N/mm}^2$$

The tension in the holding-down bolts is

$$T = 114.5 - 80.1 \qquad = 34.4 \text{ kN}$$

The area at the root of the thread for two bolts is

$$(34.4 \times 10^3)/(162.5 \times 2) \qquad = 106 \text{ mm}^2$$

Try two No. 22 mm bolts—net area 303 mm². This size of bolt has been selected for practical reasons and the steel stress will be very much less than the permissible stress:

$$\text{Approximate steel stress} = (106/606) \times 162.5 = 28.4 \text{ N/mm}^2$$

The stresses will be checked for case 2 loading to determine the maximum tension in the bolts. The elastic method will be used first and the

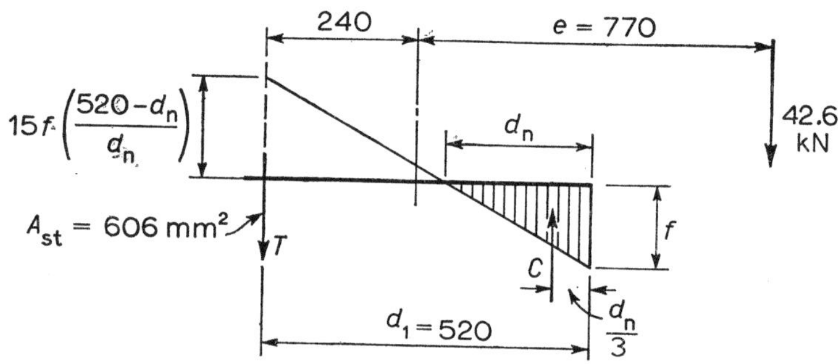

Figure 9.13

approximate stresses using the data derived in case 1 will be calculated for comparison. The loading and stress distribution are shown in Figure 9.13.

The eccentricity $e = (32.8 \times 10^6)/(42.6 \times 10^3) = 770$ mm

$$W = (\tfrac{1}{2} \times f \times d_n \times 200) - 15f \times 606\ [(520 - d_n)/d_n]$$
$$= 100fd_n + 9090f - 47.2 \times 10^5 f/d_n$$
$$M = 770W$$
$$= 7.7 \times 10^4 fd_n + 6.98 \times 10^6 f - 3.64 \times 10^9 f/d_n \qquad (1)$$
$$= 100fd_n\ [280 - (d_n/3)] + 15f \times 606\ [(520 - d_n)/d_n]\ 240$$
$$= 2.8 \times 10^4 fd_n - 33.3fd_n^2 + (1.13 \times 10^9 f/d_n) - 2.18 \times 10^6 f \qquad (2)$$

Equations (1) and (2) are equal. Collecting terms and reducing gives:

$$d_n^3 + 1.47 \times 10^3 d_n^2 + 2.75 \times 10^5 d_n - 1.43 \times 10^8 = 0$$

Solving this cubic equation by trial and error gives:

$$d_n = 221 \text{ mm},$$

The stress f may be found by back substitution in the expression for W:

$$42.6\times 10^3 = 100\times 221f + 9090f - (47.2\times 10^5 f)/221$$

Concrete stress $f = 4.32\ \text{N/mm}^2$

Bolt stress $t = 15\times 4.32\times(299/221) = 87.5\ \text{N/mm}^2$

Check the combined stress on the bolt shank. The horizontal load is assumed to be taken by shear on the shanks of the four holding-down bolts.

Area of bolt shank $= 380\ \text{mm}^2$

$$f_t = (87.5\times 303)/380 = 69.7\ \text{N/mm}^2$$

$$f_q = (13.81\times 10^3)/(4\times 380) = 9.1\ \text{N/mm}^2$$

$$\text{Combined tensile stress} = \frac{69.7}{2} + \left[\left(\frac{69.7}{2}\right)^2 + 9.1^2\right]^{1/2}$$

$$= 71.1\ \text{N/mm}^2$$

The bolt is understressed but the size would not be reduced.

If the data derived in case 1 are used, the following stresses are derived:

$$M_1 = 32.6 + 42.6\times 0.24 = 42.8\ \text{kN m}$$

$$C = (42.8\times 10^3)/454.4 = 94.2\ \text{kN}$$

$$f = (2\times 94.2\times 10^3)/(200\times 197) = 4.77\text{N/mm}^2$$

$$T = 94.2 - 42.6 = 51.6\ \text{kN}$$

$$t = (51.6\times 10^3)/606 = 85.5\ \text{N/mm}^2$$

These stresses are of the same order as the values found above by more accurate means.

(b) Thickness of base plate

The pressures under the edge of the base plate are shown in Figure 9.14. Consider a cantilever strip 1 mm$\times$81.4 mm.

$$M = (3.4\times 81.4^2)/(2\times 3) + (5.8\times 81.4^2\times 2)/(2\times 3)$$

$$= 16.61\times 10^3\ \text{N mm}$$

Permissible bending stress $= 185 + 25\%$ wind allowance

$= 231\ \text{N/mm}^2$

Base plate thickness:

$$t = (6\times 16.61\times 10^3/231)^{1/2} = 20.8\ \text{mm}$$

Use 22 mm thick plate.

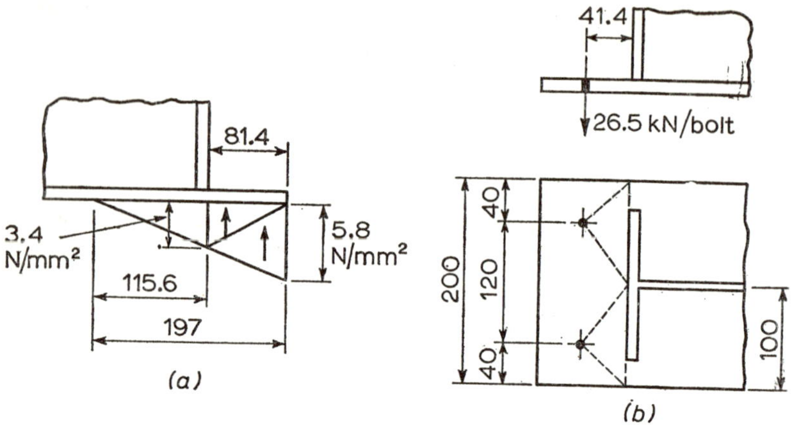

Figure 9.14

The moment in the base slab due to bolt tension is checked. Refer to Figure 9.14(b).

Bolt Load $= (303\times 87.5)/10^3 = 26.5$ kN

The moment at the stanchion face due to the load in one bolt:

$$M = 26.5\times 41.4 = 1100 \text{ kN mm}$$

The section at the stanchion face resisting the bolt load in bending is generally considered to subtend an angle of 120° with the bolt. However, for the narrow base width here a section 100 mm wide is assumed to resist the moment. This is half the base width. Spreading the load at 120° would give a width of section of 143 mm.

The bending stress on a section 100 mm wide and 25 mm deep is

$$f_b = (1100\times 10^3\times 6)/(100\times 22^2) = 136.2 \text{ N/mm}^2\text{—safe}$$

(c) Stanchion-to-slab weld

The weld will be designed to transmit the whole of the load and moment between the stanchion and the base slab. The weld is shown in

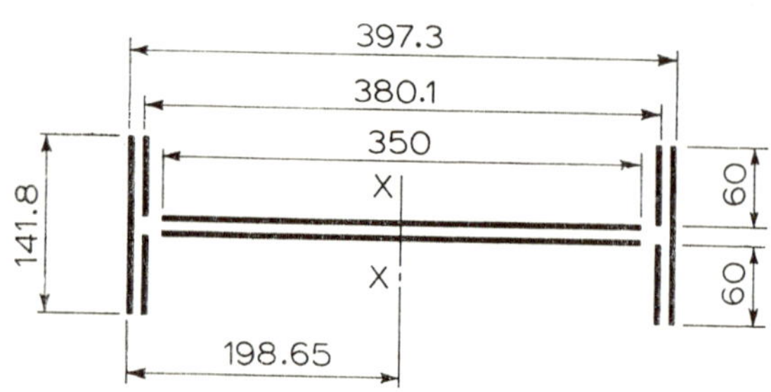

Figure 9.15

Figure 9.15. Weld on the edge of the flanges and at the web flange junction has been neglected.

$$L = 2(141.8+120+350) = 1223.6$$

$$I_{XX} = 2\times 141.8\times 198.65^2+2\times 120\times 190.05^2+(2\times 350^3)/12$$
$$= 27.1\times 10^6$$

The maximum load on the weld is

$$(80.1\times 10^3)/1223.6+(32.8\times 10^6\times 198.65)/(27.1\times 10^6)$$
$$= 305.6 \text{ N/mm}$$

Use 6 mm fillet weld with a strength of 483 N/mm. Weld all round the stanchion profile to seal the joint.

The above method of designing the joint is conservative. An analysis can be made in a section subjected to axial load and moment where bearing on the compression flange and part of the web is taken into account, with tension supplied by shear in fillet welds on the remaining part of the section. This analysis presents some difficulty and is not considered to be necessary. The effective section in such a case is shown in Figure 9.16(a). In sections subjected to large moments a full-strength weld between the flange and slab may be required as shown in Figure 9.16(b).

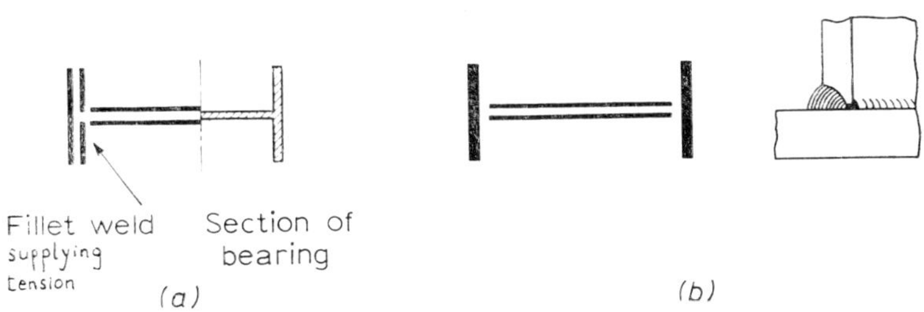

Figure 9.16

9.5 BASES SUBJECTED TO BENDING ABOUT TWO AXES

9.5.1 Compression Over the Whole Base

The base shown in Figure 9.17 is subjected to bending about two axes as well as to axial thrust. In the case where only compressive stresses occur, the analysis is straightforward and the stress at any point is the algebraic sum of the stresses due to:

(a) axial load

(b) moments about the two principal axes XX and YY

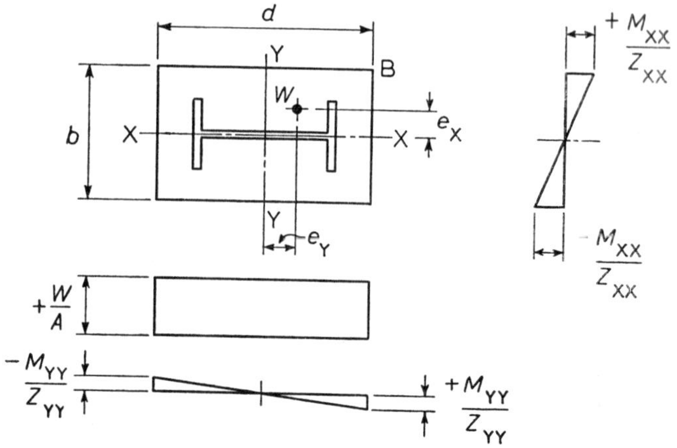

Figure 9.17

The maximum stress can be expressed as follows:

W = axial load

$M_{XX} = We_X$ = moment about XX axis

$M_{YY} = We_Y$ = moment about YY axis

$A = bd$ = area of base

$Z_{XX} = db^2/6$ = modulus of section about XX axis

$Z_{YY} = bd^2/6$ = modulus of section about YY axis

The maximum stress occurs at B and is

$$f_{max} = (W/A)+(M_{XX}/Z_{XX})+(M_{YY}/Z_{YY})$$

This should not exceed the allowable stress for concrete. Several trials may be necessary to establish a suitable size for the base.

9.5.2 Compression over Part of the Base and Tension in the Holding-Down Bolts

The case where tensile and compressive stresses occur, and where the holding-down bolts provide the tension, presents difficulty.

The base shown in Figure 9.18(a) is subjected to an axial load W and moments M_{XX} and M_{YY} about the XX and YY axes respectively. The resultant moment vector M_R is shown in (b). The neutral axis NA of the section is at right angles to the plane of action of M_R. The stress distribution is shown in (c).

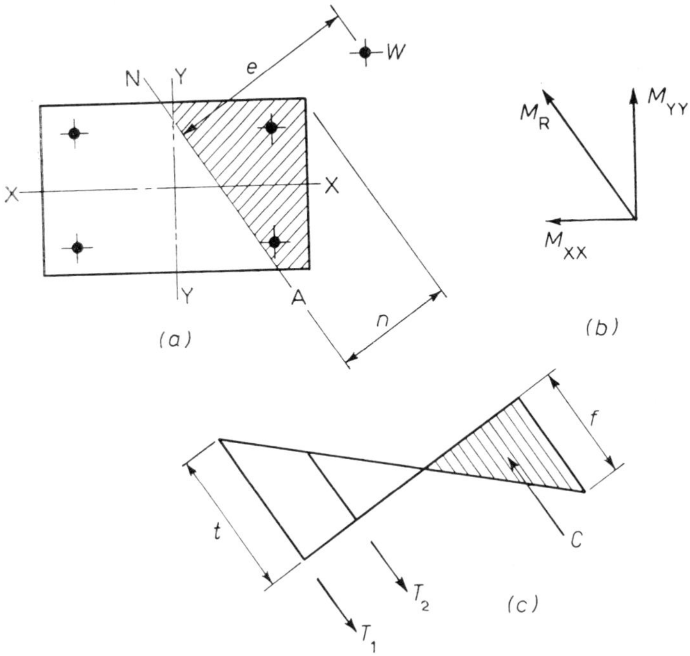

Figure 9.18

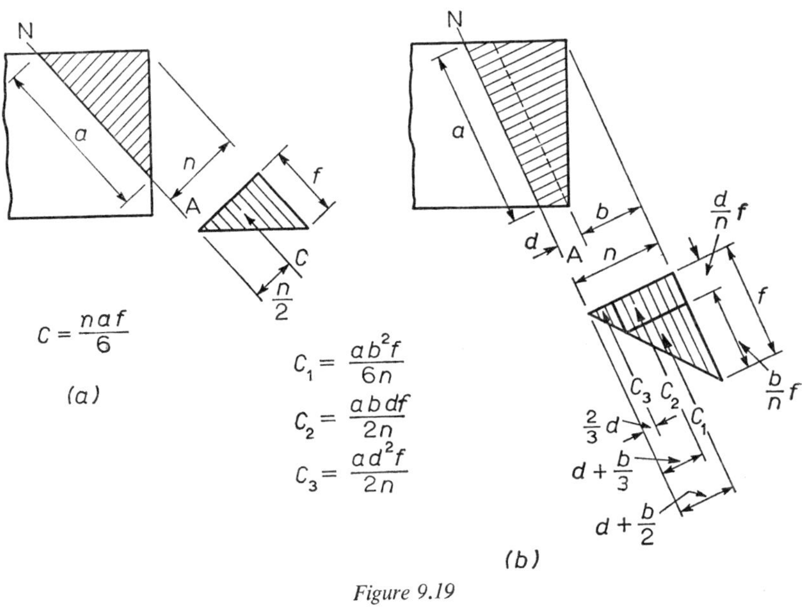

Figure 9.19

An equation can be formed connecting the external and internal actions. The forces in the concrete and steel are expressed in terms of the concrete stress f and the depth of the neutral axis n. This can be solved for n and the concrete and steel stresses can be found.

The neutral axis NA can lie in either of the two positions shown in Figure 9.19(a) and (b). A trial calculation is required to establish this. The expressions for the value and position of the compressive force C in the concrete for the triangular stress block are shown in Figure 9.19(a). The breakdown and expressions for the complex stress block are given in (b). The method is demonstrated in a simple example.

9.5.3 Example. Compression Over the Whole Base

The base shown in Figure 9.20 is subjected to the following loading:

Axial load $W = 360$ kN

Moment $M_{XX} = 6$ kN m

Moment $M_{YY} = 9$ kN m

Find the bearing stress on the concrete.

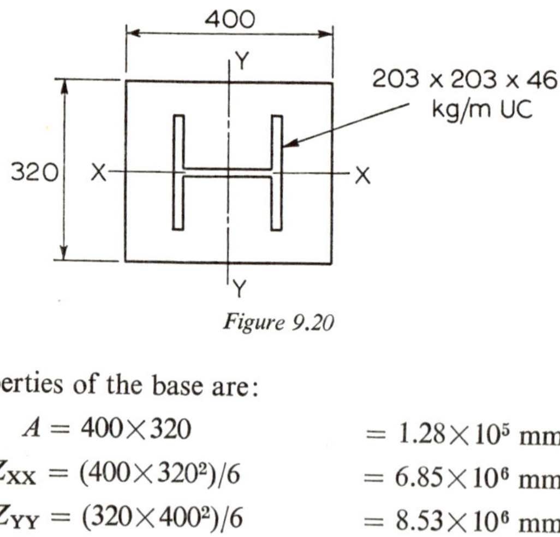

Figure 9.20

The properties of the base are:

$$A = 400 \times 320 \qquad = 1.28 \times 10^5 \text{ mm}^2$$

$$Z_{XX} = (400 \times 320^2)/6 \qquad = 6.85 \times 10^6 \text{ mm}^3$$

$$Z_{YY} = (320 \times 400^2)/6 \qquad = 8.53 \times 10^6 \text{ mm}^3$$

The stresses are:

Axial $f_c = (360 \times 10^3)/(1.28 \times 10^5) = 2.81$ N/mm²

XX axis $f_b = (6 \times 10^6)/(6.85 \times 10^6) = 0.88$ N/mm²

YY axis $f_b = (9 \times 10^6)/(8.53 \times 10^6) = 1.06$ N/mm²

Maximum bearing stress $= 4.75$ N/mm²

9.5.4 Example. Tension in the Holding-Down Bolts

A stanchion consisting of 203.2×203.2×6.3 RHS is subjected to an axial load of 28 kN and equal moments of 18.7 kN m about the rectangular axes XX and YY at the base. Check that the base size and holding-down bolts shown in Figure 9.21 are satisfactory.

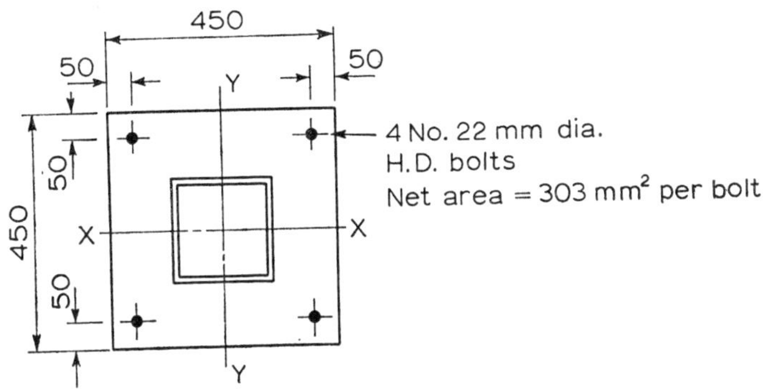

Figure 9.21

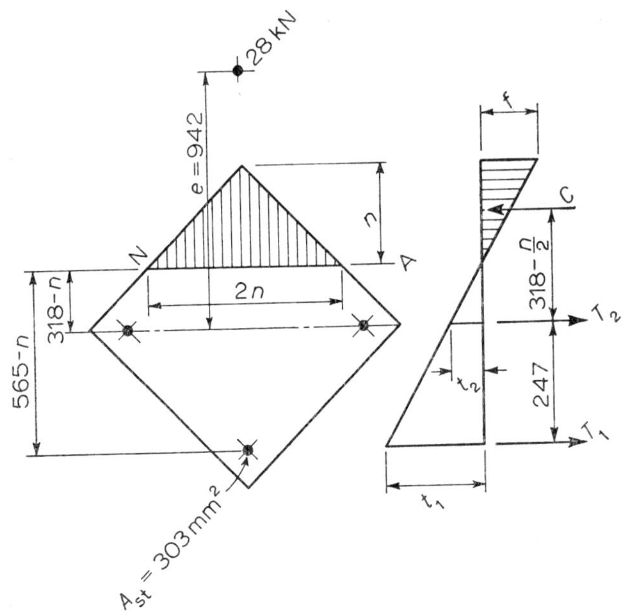

Figure 9.22

The stress diagram for the base is shown in Figure 9.22. The steel stresses expressed in terms of the concrete stresses are:

$$t_1 = 15f[(565-n)/n]$$
$$t_2 = 15f[(318-n)/n]$$

The resultant moment and eccentricity are:

$$M = 18.7\sqrt{2} \quad = \quad 26.4 \text{ kN m}$$
$$e = (26.4\times 10^3)/28 = 942 \text{ mm}$$

The axial load

$$W = \tfrac{1}{3}n^2f - 15f[(318-n)/n]606 - 15f[(565-n)/n]303$$
$$= \tfrac{1}{3}n^2f - 54.6\times 10^5(f/n) + 13\,635f$$

Take moments about the centre of the base:

$$We = 314\,n^2f - 51.5\times 10^8(f/n) + 12.85\times 10^6 f \qquad (1)$$

Moments of the internal forces about the centre of the base:

$$M = \tfrac{1}{3}n^2f[(318-n)/2] + 15f[(565-n)/n]303\times 247$$
$$= 106n^2f - (n^3f/6) + 6.32\times 10^8(f/n) - 1.12\times 10^6 f \qquad (2)$$

Equate expressions (1) and (2), collect terms and reduce:

$$n^4 + 1248n^3 + 83.82\times 10^6 n - 347.9\times 10^8 = 0$$

Solve by trial and error. This gives:

$$n = 222$$

Substitution in the expression for W above gives:

$$f = 5.25 \text{ N/mm}^2$$

and

$$t_1 = 122 \text{ N/mm}^2$$

These stresses are satisfactory. An approximate method would have to be adopted to find the thickness of the base plate.

9.6 GRILLAGE FOUNDATIONS

9.6.1 Code Requirements

The design of grillage foundations is governed by Clause 40 of BS 449. This states:

> 'Where grillage beams are enveloped in a solid block of dense concrete as specified in subclause 21.2, the permissible working stresses specified in this British Standard for uncased beams may be increased by $33\tfrac{1}{3}\%$ provided that:

'(a) The beams are spaced apart so that the distance between the edges of adjacent flanges is not less than 75 mm.
'(b) The thickness of the concrete cover on the top of the upper flanges, at the ends, and at the outer edges of the sides of the outermost beam is not less than 100 mm.
'(c) The concrete is properly compacted solid around all beams.

These increased stresses shall not apply to hollow compound girders.'
Subclause 21.2 states:

> 'The beams are to be unpainted and solidly encased in ordinary dense concrete, with 10 mm aggregate (unless solidity can be obtained with a larger aggregate), and of a works strength not less than 21 N/mm² at 28 days when tested in accordance with BS 1881, "Methods of testing Concrete".'

9.6.2 Design Theory

The grillage foundation is used under very heavily loaded columns. It is made up of two or more tiers of beams with the successive tiers running at right angles to each other. The whole is encased in concrete. Only grillages subjected to axial load will be considered here.

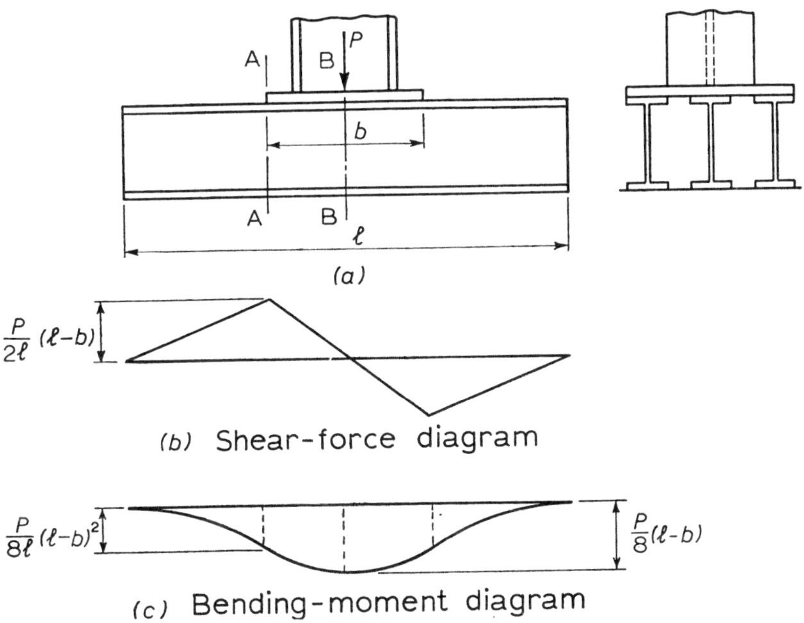

Figure 9.23

Consider the beams in one tier as shown in Figure 9.23. These are treated as a double cantilever and designed for moment, shear, web buckling, crushing and combined stresses. The slab can be designed by the method set out earlier in the chapter. It is assumed that the slab distributes the stanchion load uniformly to the top tier and these beams spread the load to the second tier.

$P =$ total load at column base

$G =$ weight of the grillage

$p =$ safe bearing pressure on the ground

$A =$ area of grillage

$= (P+G)/p$

Only the stanchion load P causes shear and bending in the grillage.

Uniform load on the top of beam $= P/b$

Uniform load on the underside of beam $= P/l$

Shear at A–A $= (P/l)(l/2-b/2) = (P/2l)(l-b)$

Shear at B–B $= 0$

Moment at A–A $= (P/2l)(l-b)[(l-b)/4]$

$= (P/8l)(l-b)^2$

Moment at B–B $= (P/l\times l/2\times l/4)-(P/6\times b/2\times b/4)$

$= (P/8)(l-b)$

The shear force and bending moment diagrams are shown in the figure.

9.6.3 Design of a Grillage Foundation

Design a two-tier square grillage foundation to carry the following loads:

Load at stanchion base	= 2800 kN
Approximate weight of grillage	= 300 kN
Total load	= 3100 kN

The safe bearing pressure on the ground is 220 kN/m^2

The stanchion section is 305×305×240 kg/m UC.

(a) Stanchion base plate

Assume a size for the base plate of 750 mm×750 mm. The pressure under the base plate is

$W = (2800\times 10^3)/750^2 = 4.98$ N/mm^2

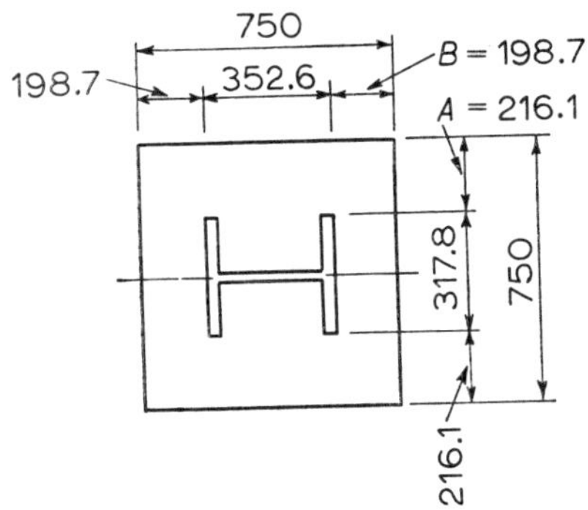

Figure 9.24

The base plate arrangement is shown in Figure 9.24. The thickness is given by

$$t = \left[\frac{3\times4.98}{185}\left(216.1^2-\frac{198.7^2}{4}\right)\right]^{1/2}$$

$$= 55.3 \text{ mm}$$

Use a 60 mm thick base slab.

(b) Grillage moments and shears

$$\text{Area} = 3100/220 = 14.2 \text{ m}^2$$

Use a square base 3.8 m by 3.8 m. The pressure on the lower tier is:

$$2800/3.8^2 = 195 \text{ kN/m}^2$$

Because the grillage is square the shears and moments are the same for each tier. Consider the top tier as shown in Figure 9.25(a).

Section A–A

Shear $= [2800/(2\times3.8)](3.8-0.75) = 1128$ kN

Moment $= 1128[(3.8-0.75)/4] = 860$ kN m

Section B–B

Shear $= 0$

Moment $= (2800/8)(3.8-0.75) = 1070$ kN m

The bending moment and shear force diagrams are shown in (b).

(c) Top tier

The grillage is encased in concrete in accordance with Clause 40 of BS 449. The permissible bending stress

$$p_{bc} = 165+33\tfrac{1}{3}\% = 220 \text{ N/mm}^2$$

Using three beams, the modulus required per beam is:

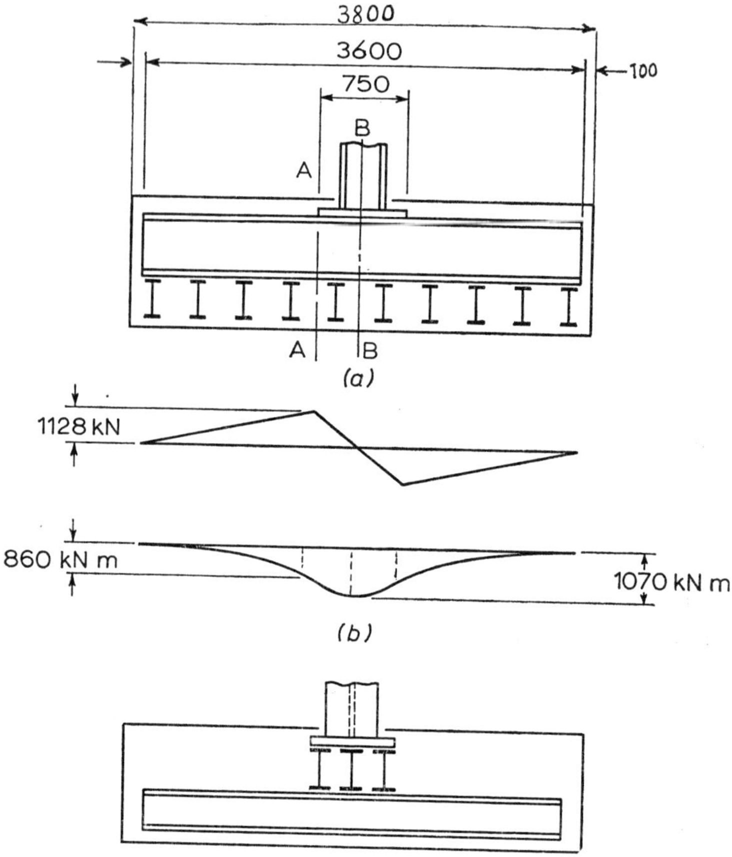

Figure 9.25

$$Z = (1070\times 10^6)/(220\times 3) = 1.62\times 10^6 \text{ mm}^3$$

Try three No. 457×191×89 kg/m UB where

$Z = 1767 \text{ cm}^3$

Shear stress $= (1128\times 10^3)/(463.6\times 10.6\times 3) = 76.6 \text{ N/mm}^2$

The universal beam is checked for web buckling and crushing. Refer to Figure 9.26(a) for web buckling.

$$l/r = (404.4\times \sqrt{3})/10.6 \qquad = 66$$

$$p_c = 120+33\tfrac{1}{3}\% \qquad = 160 \quad \text{N/mm}^2$$

$$f_c = (2800\times 10^3)/(3\times 1213.6\times 10.6) = 72.5 \text{ N/mm}^2$$

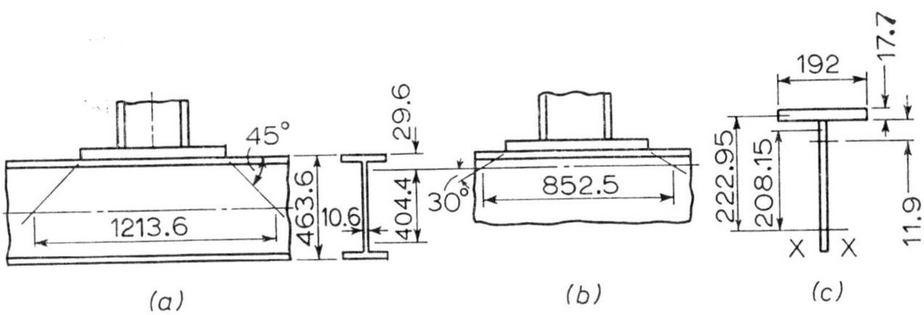

Figure 9.26

For web crushing refer to Figure 9.26(b).

Bearing stress $f_b = (2800\times10^3)/(3\times852.5\times10.6) = 102 \quad N/mm^2$

The allowable bearing stress

$$p_b = 190+33\tfrac{1}{3}\% \qquad = 253.3\ N/mm^2$$

The combined stress is checked at section AA at the edge of the base plate as shown in Figure 9.25(a). The bending stress in compression here is:

$$f_{bc} = (860\times10^6\times202.2)/(3\times1767\times10^3\times231.8) \qquad = 142 \quad N/mm^2$$

The dimensions for calculating the actual shear stress are shown in Figure 9.26(c).

$$f_q = \frac{(1128\times10^3)[(192\times17.7\times222.95)+(11.9\times10.6\times208.15)]}{3\times40\,956\times10^4\times10.6}$$

$$= 68\ N/mm^2$$

The combined stress

$$f_e = (142^2+102^2-162\times102+3\times68^2)^{1/2}$$
$$= 167\ N/mm^2$$

The allowable equivalent stress

$$p_e = 230+33\tfrac{1}{3}\% \qquad = 306.2\ N/mm^2$$

The beam section chosen is satisfactory.

(d) Lower tier

Using ten beams, the modulus required per beam is

$$Z = (1070\times10^6)/(10\times220) = 4.86\times10^5\ mm^3$$

Use $305\times165\times40$ kg/m UB where

$$Z = 559.6\ cm^3$$

These beams may be checked for web shear, buckling, crushing and combined stresses as above.

The sizes chosen for the two tiers will allow adequate spacing between flanges for placing and compaction of concrete as required in Clause 40 of BS 449.

9.7 POCKET BASES

9.7.1 Design Theory

In this type of base, the stanchion is grouted into a pocket in the concrete foundation as shown in Figure 9.27(a).

The axial load in the stanchion is resisted by direct bearing and bond between the steel and concrete. The moment is resisted by compression forces in the concrete acting on the flanges of the stanchion.

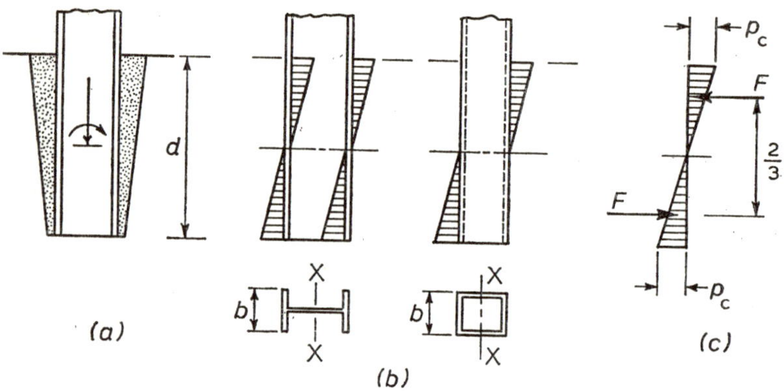

Figure 9.27

The forces act on both faces of the flanges of universal beams and columns and on one face of a rectangular hollow section. The action in resisting moment is shown in Figure 9.27(b) and (c).

The following terms are defined:

p_c = permissible stress in compression in bending in the concrete

p_s = permissible bending stress in the steel

Z_{XX} = modulus of section of the steel section

d = depth of embedment of the steel section

b = breadth of the stanchion flange

The depth to embed the stanchion to develop its strength in bending can be determined as follows:

The compressive force in the concrete

$$F = (p_c/2)\times(d/2)\times b = p_c db/4$$

For a universal beam or column, the moment of resistance of the concrete is

$$M = 2\times F\times \tfrac{2}{3}d = p_c d^2 b/3$$

$$= \text{moment of resistance of the steel section}$$

$$= p_s Z_{XX}$$

Thus the depth

$$d = (3p_s Z_{XX}/p_c b)^{1/2}$$

For a structural hollow section

$$M = p_c d^2 b/6$$

$$d = (6p_s Z_{XX}/p_c b)^{1/2}$$

9.7.2 Examples

1 Determine the depth that 305×127×48 kg/m UB must be embedded to develop its strength. The section dimensions are shown in Figure 9.28. The properties from steel tables are:

$$Z_{XX} = 611.1\ \text{cm}^3$$

$$A = 60.8\ \text{cm}^2$$

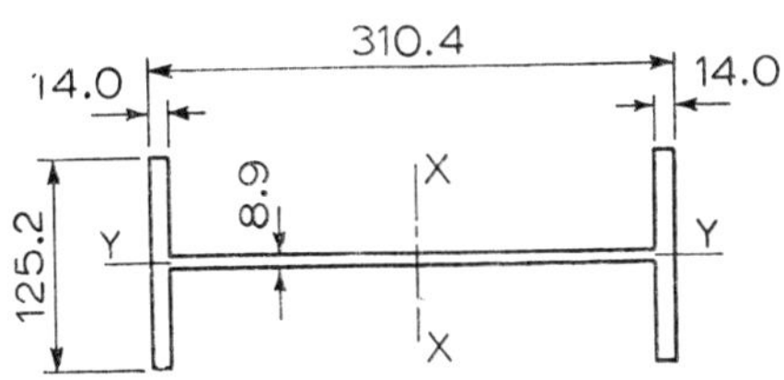

Figure 9.28

What is the safe axial load that the stanchion can trasmit to the concrete? The permissible stresses for 1 : 2 : 4 concrete taken from CP 114 are:

$$p_c = 7\ \text{N/mm}^2$$

Average bond stress $= 0.83\ \text{N/mm}^2$
The permissible bending stress in the steel

$$p_s = 165\ \text{N/mm}^2$$

The depth to embed the stanchion

$$d = [(3\times165\times611.1\times10^3)/(7\times125.2)]^{1/2} = 585 \text{ mm}$$

The safe axial load is calculated assuming that half the surface area of the flanges, that is, the area in compression, is effective in bond, together with the area of the web. For bearing on the concrete the permissible axial stress in direct compression in the concrete is 5.3 N/mm².

The net surface area in bond is

$$(125.2+116.3+2\times292.6)585$$
$$= 4.84\times10^5 \text{ mm}^2$$

$$\text{Safe load} = (4.84\times10^5\times0.83)/10^3+(60.8\times10^2\times5.3)/10^3$$
$$= 401+32$$
$$= 433 \text{ kN}$$

2 Determine the depth that 101.6×101.6×6.3 RHS must be set into concrete to develop its strength in bending.

Concrete: 1 : 2 : 4 mix

$$p_c = 7 \text{ N/mm}^3$$

Steel section: $Z = 71 \text{ cm}^3$

$$p_s = 165 \text{ N/mm}^2$$

Using the formula developed above:

$$d = (6\times165\times71\times10^3)/(7\times101.6)^{1/2}$$
$$= 314 \text{ mm}$$

9.8 PROBLEMS

1 Design a square slab base for a stanchion consisting of 203×203 ×86 kg/m UC that carries an axial load of 1050 kN. The permissible bearing stress on the concrete is 5.3 N/mm².

2 A stanchion has to resist an axial load of 700 kN and a moment of 50 kN m. The effective length of the stanchion is 3.6 m. Check that 254×254×73 kg/m UC is suitable for the stanchion and that the gusseted base shown in Figure 9.29 is adequate. Design the welds for the base.

3 A stanchion consisting of a 305×127×37 kg/m UB transmits an axial load of 50 kN and a moment of 32 kN m to the base. Check that a base plate 500×250×25 thick is suitable. Determine the size of

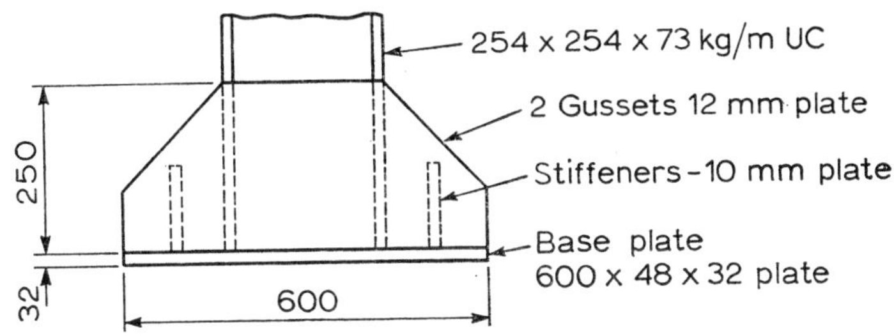

Figure 9.29

holding-down bolts and the size of weld between the stanchion and base plate. The permissible stresses are:

Concrete in compression:	5.3 N/mm²
Bolts in tension:	130 N/mm²

4 The stanchion base shown in Figure 9.30 is subjected to a direct load of 34 kN and a moment of 30 kN m. The entire moment on the

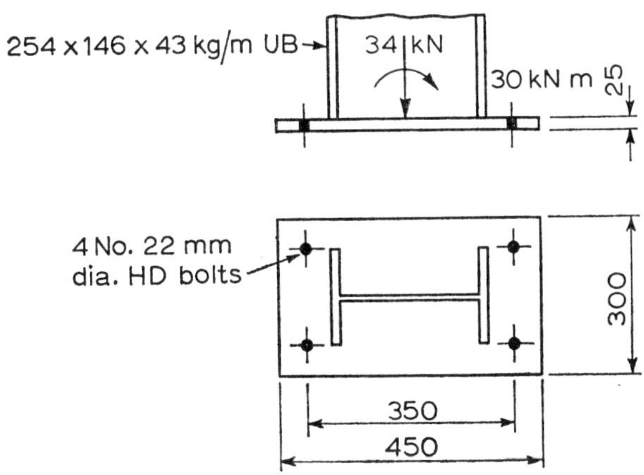

Figure 9.30

base is due to wind. The permissible stresses for the steel and concrete are:

Holding-down bolts	=	130 N/mm²
Concrete bearing stress	=	5.3 N/mm²
Base plate bending stress	=	185 N/mm²

These stresses may be increased by 25% as a wind allowance. The modular ratio m for the concrete is 15. The net area of the 22 mm diameter bolts is 303 mm^2. Check:

(a) The maximum bearing stress on the concrete
(b) The tensile stress in the holding-down bolts
(c) The base-plate thickness, using the bearing stress calculated in (a).

Design the weld between the stanchion and the base slab.

5 A stanchion base 600 mm×450 mm is subjected to:

Axial load $W = 1070$ kN

Moment $M_{XX} = 14$ kN m acting about the major principal axis

$M_{YY} = 7$ kN m acting about the minor principal axis

Determine the maximum bearing stress on the concrete.

6 Design a two-tier square grillage to carry an axial load from the stanchion of 1600 kN. The weight of the foundation is 150 kN. The safe bearing pressure on the ground is 180 kN/m^2. The stanchion is a 254×254×89 kg/m UC.

7 Determine the depth that the following sections must be embedded to develop the strength of the respective section in bending:

203×203×71 kg/m	UC
457×191×67 kg/m	UB
88.9 dia. × 4	CHS
88.9×88.9×3.6	RHS

In all cases the permissible stresses are:

p_s = bending stress in the steel = 165 N/mm^2

p_c = bending stress in compression in the concrete = 7 N/mm^2

10

Purlins and sheeting rails

Purlins support roof decking on flat or sloping roofs. Sheeting rails support the wall cladding on industrial buildings.

10.1 PURLINS

10.1.1 Design Considerations and Loading

Purlins are designed as beams carrying roof loads. They may be simply supported over one span or continuous over two or more spans. Clause 45 of BS 449 states that provisions regarding lateral stability and limiting deflections of beams do not apply to purlins. The decking restrains the purlin laterally and excessive deflection is generally not of importance.

Roof loads are due to dead weight of roof material and the purlins and the imposed loading. Many roofing materials are available, the most important being:

Steel or aluminium corrugated sheet

Steel trough decking

Asbestos corrugated sheet

Glazing—translucent sheeting of various types

These are used in conjunction with insulation board. Felt sheeting and insulation board are used on flat roofs. Weights vary from 30.6 to 102 kg/m^2, that is 0.3 to 1.0 kN/m^2. This includes the self weight of the purlins. Manufacturers' catalogues should be consulted.

The imposed loading is specified in CP3, Chapter V, Part 1, 1967, Clauses 6.2 and 6.3. These clauses state:

'6.2 Flat Roofs. On flat roofs and sloping roofs up to and including 10°, where access (in addition to that necessary for cleaning and repair) is provided to the roof, allowance shall be made for on imposed load, including snow, of 1.5 kN/m² measured on plan, or a load of 1.8 kN concentrated on a square with a 300 mm side.

'On flat roofs and sloping roofs up to and including 10°, where no access is provided to the roof (other than that necessary for cleaning and repair), allowance shall be made for an imposed load, including snow, of 0.75 kN/m² measured on plan, or a load of 0.9 kN concentrated on a square with a 300 mm side.

'6.3 Sloping Roofs. On roofs with a slope greater than 10° and with no access provided to the roof (other than that necessary for cleaning and repair), the following imposed loads including snow shall be allowed for:

'(1) for a roof-slope of 30° or less: 0.75 kN/m² measured on plan or a vertical load of 0.9 kN concentrated on a square with a 300 mm side.

'(2) for a roof-slope 75° or more: no allowance necessary. For roof slopes between 30° and 75°, the imposed load to be allowed for may be obtained by linear interpolation between 0.75 kN/m² for a 30° roof slope and nil for a 75° roof slope.'

10.1.2 Design of a Purlin for a Flat Roof

The roof consists of a steel deck with insulation board and bituminous felt and steel purlins of total weight 76.5 kg/m², that is 0.75 kN/m². The purlins span 4 m simply supported and are at 1.75 m centre to centre. The imposed load is 1.5 kN/m². The roof arrangement and loading are shown in Figure 10.1.

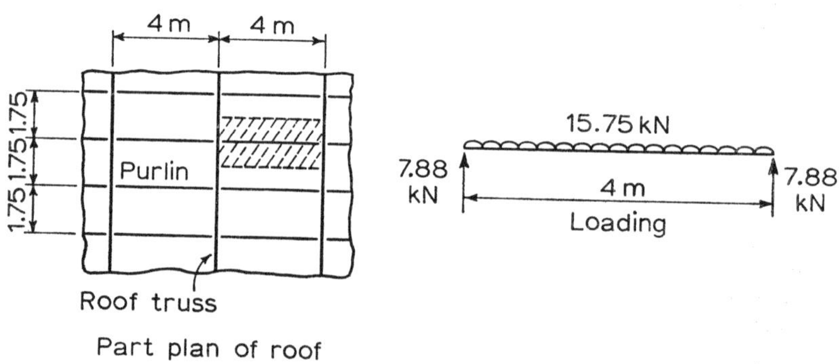

Figure 10.1

Dead load	$= 4\times1.75\times0.75$	$= 5.25$ kN
Imposed load	$= 4\times1.75\times1.5$	$= 10.5$ kN
Total		$= 15.75$ kN
Moment	$M = (15.75\times4)/8$	$= 7.87$ kN m
Modulus	$Z = (7.87\times10^6)/165$	$= 4.76\times10^4$ mm^3

The allowable stress is 165 N/mm^2 because the roof deck restrains the compression flange of the purlin against lateral buckling.

Purlin section: 305×165×42 kg/m UB.

$$Z = 559.6 \text{ cm}^3$$

This is the lightest section from steel tables with the required modulus.

10.2 PURLINS ON SLOPING ROOFS

10.2.1 Design Considerations

Consider a channel purlin on a sloping roof as shown in Figure 10.2. The load on an interior purlin is from a width of roof equal to the purlin spacing S. The load on this portion of roof is made up of dead and imposed load and acts vertically. This load is resolved normal and tangential to the roof slope to cause bending about the XX and YY axes of the purlin. The purlin is designed as a beam bending about both axes. The sheeting restrains the compression flange under normal

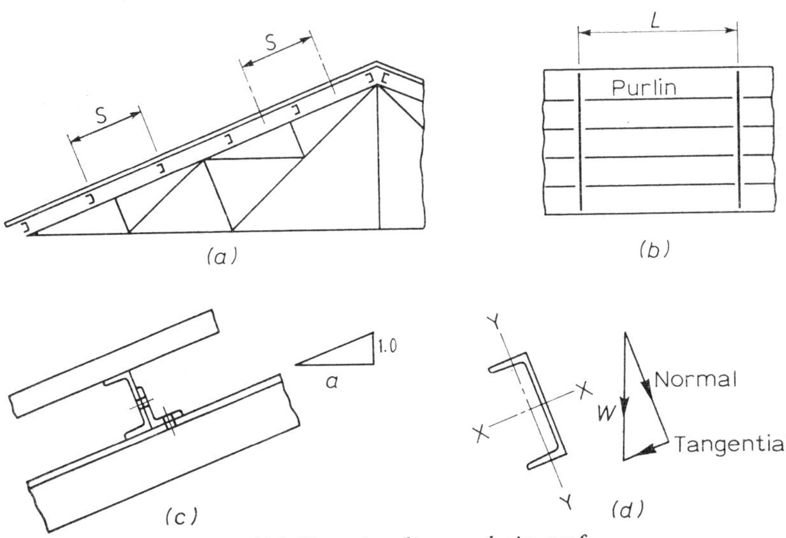

Figure 10.2 Channel purlin on a sloping roof

loading from lateral instability. So a maximum stress of 165 N/mm^2 for grade 43 steel in bending is permitted.

In the example shown in Figure 10.2 the purlins have the same effective span for bending about both axes. Sag rods can be introduced to reduce the span for bending about the weak axis caused by the tangential component of the load. This is shown in Figure 10.3.

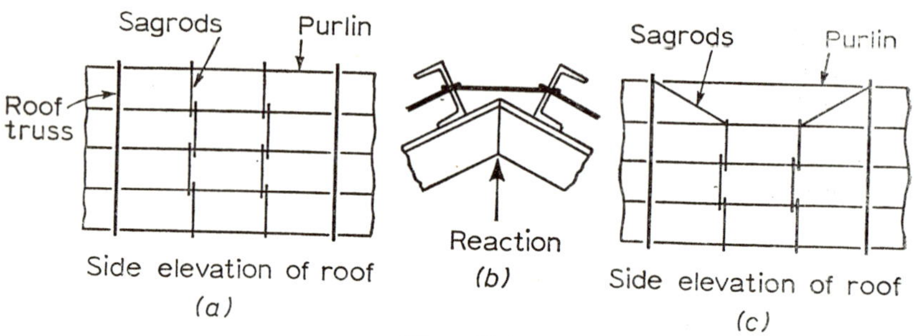

Figure 10.3

The purlins are tied across at the apex of the truss as shown in (b). In this case the top purlins must be designed for the reaction from the sag rods. Alternatively, the sag rods may be anchored on the roof trusses as shown in (c).

Sag rods are made from 16–20 mm diameter bar with screwed ends. Depending on the span, one, two or more lines of sag rods may be used as shown in (a). The purlin now acts as a continuous beam for bending about the weak axis. This can be analysed as a continuous beam by moment distribution or the coefficients for the moments in equal span continuous beams can be used. These are given in structural steel and design handbooks such as the 'Steel Designer's Manual'. Stresses due to the horizontal bending must be combined with those due to vertical bending as set out in section 32.4.

10.2.2 Design of a Purlin on a Sloping Roof

Design a channel purlin for a roof slope of 1 in 2.5. The purlin is simply supported and spans 6.0 m between roof trusses; the purlin spacing is 1.75 m. The total dead load is 30.6 kg/m^2, that is 0.3 kN/m^2 on the slope and the imposed load is 0.75 kN/m^2 on plan.

Dead load on slope	$= 0.3 \times 6 \times 1.75$	$= 3.15$ kN
Imposed loading on slope	$= 0.75 \times 6 \times 1.75 \times (2.5/2.69)$	$= 7.2$ kN
Total Vertical Loading		$= 10.35$ kN
Normal component	$= (2.5/2.69) \times 10.35$	$= 9.53$ kN
Tangential component	$= (1/2.68) \times 10.35$	$= 3.81$ kN

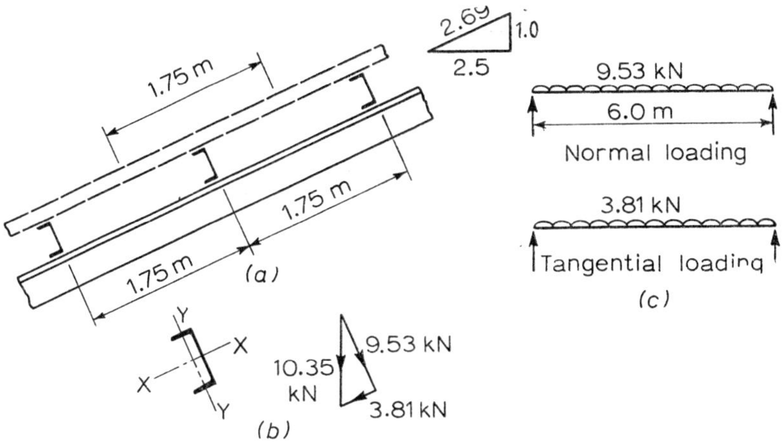

Figure 10.4

The arrangement of the purlins on the roof slope, the components of the load and the normal and tangential loading on the purlin are shown in Figure 10.4.

$$M_{XX} = (9.53\times6)/8 = 7.15 \text{ kN m}$$

$$M_{YY} = (3.81\times6)/8 = 2.86 \text{ kN m}$$

Try 203×76×23.82 kg/m channel

$$Z_{XX} = 192.0 \text{ cm}^3$$

$$Z_{YY} = 27.59 \text{ cm}^3$$

The stresses are:

$$(f_b)_{XX} = (7.15\times10^6)/(192\times10^3) = 37.2 \text{ N/mm}^2$$

$$(f_b)_{YY} = (2.86\times10^6)/(27.59\times10^3) = 104 \text{ N/mm}^2$$

$$\text{Combined stress} = 141.2 \text{ N/mm}^2$$

Section is satisfactory.

10.2.3 Design of a Purlin with One Sag Rod

Redesign the purlin in the example in section 10.2.2 if one sag rod is introduced at mid span. The purlins are 6 m span and spaced at 1.75 m. Refer to Figure 10.4. Calculate the size of sag rod required if the length of the roof slope is 10.77 m.

The loading on the purlin is taken from the example above and the normal and tangential loads on the purlin are shown in Figure 10.5(a) and (b) respectively. The sag rod provides the mid span reaction to

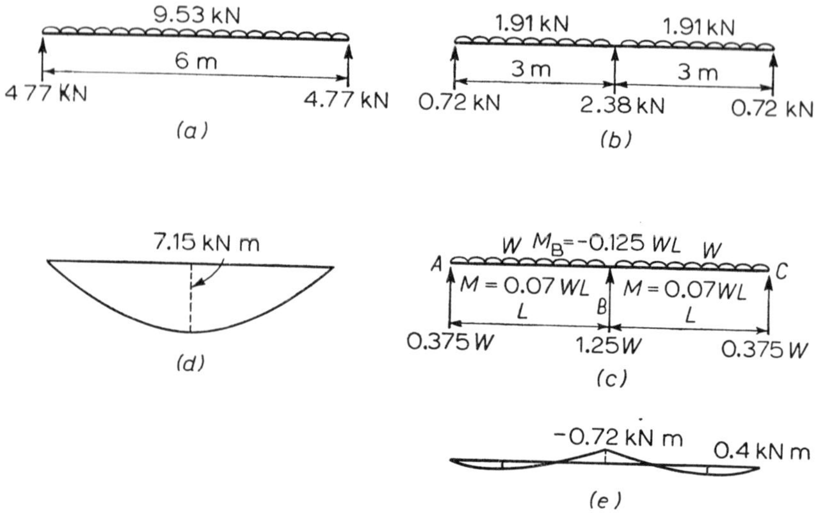

Figure 10.5

make the purlin into a two-span continuous beam to carry the tangential load.

The coefficients for moments and reactions for a two-span continuous beam are shown in Figure 10.5(c). These have been taken from the *Steel Designer's Manual*, Chapter 4, 'Continuous Beams'. The bending-moment diagrams for the normal and tangential loads are shown in (d) and (e) respectively. The size of the sag rod is determined first.

$$\text{Maximum sag rod load} = (10.77/1.75)\times 2.38 = 14.7 \text{ kN}$$

Require 16 mm diameter sag rod. Safe load in tension is 20.4 kN for this size of bar. Refer to Table 2 of the Appendix.

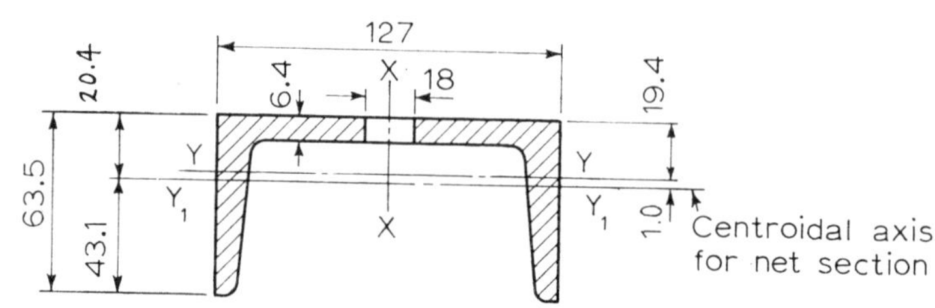

Figure 10.6

For the purlin, try 127×64×14.9 kg/m channel. The channel will be weakened by the 18 mm diameter hole at the point of maximum moment. The properties of the net section are calculated below. The channel section is shown in Figure 10.6. The hole is treated as a negative area. For the XX axis:

$$I_{XX} = 482.6\times10^4-(6.4\times18^3)/12 = 482.3\times10^4 \text{ mm}^4$$

$$Z_{XX} = (482.3\times10^4)/63.5 = 75.9\times10^3 \text{ mm}^4$$

The effect is negligible. For the Y_1Y_1 axis:

$$\bar{x} = [(18.98\times10^2\times19.4)-(18\times6.4\times3.2)]/[18.98\times10^2-(18\times6.4)] = 20.4$$

$$I_{Y_1Y_1} = (67.24\times10^4+18.98\times10^2\times1.0^2)-(18\times6.4\times17.2^2) -(18\times6.4^3)/12 = 63.83\times10^4 \text{ mm}^4$$

$$Z_{Y_1Y_1} = (63.83\times10^4)/43.1 = 14.8\times10^3 \text{ mm}^3$$

This is the minimum value of the modulus of section. The stresses are:

$$(f_b)_{XX} = (7.15\times10^6)/(75.9\times10^3) = 94.3 \text{ N/mm}^2$$

$$(f_b)_{YY} = (0.72\times10^6)/(14.8\times10^3) = 48.6 \text{ N/mm}^2$$

Combined stress $= 142.9$ N/mm²

This section is satisfactory.

10.3 PURLIN DESIGN IN ACCORDANCE WITH BS 449

10.3.1 Angle Purlins

Angle purlins can be designed in accordance with Clause 45 of BS 449, where rules are given for purlins on roofs not exceeding 30° pitch. The following conditions apply:

Minimum imposed load = 0.75 kN/m² (on plan)
Distance centre to centre of purlin supports, e.g. roof trusses $= L$ mm
Total distributed load on the purlin including dead load and snow, but excluding wind, both assumed to act normal to the roof $= W$ kN

Then the leg length in the plane of action of the maximum component of the load shall be not less than $L/45$; the length of the other leg shall not be less than $L/60$ and the section modulus in centimetre units shall not be less than $(WL/1.8)\times10^{-3}$.

Example 1 Design of an angle purlin.
Design an angle purlin to span 4 m. The purlin spacing is 1.75 m. The total dead load is 30.6 kg/m², that is, 0.3 kN/m² measured on the slope and the imposed load is 0.75 kN/m² on plan. The roof slope is 1 in 3.

Dead load	$= 0.3 \times 4 \times 1.75$	$= 2.1$ kN
Imposed load	$= 0.75 \times 4 \times 1.75 \times (3/3.16)$	$= 4.98$ kN
Total load		$= 7.08$ kN

Leg length in the plane of action of the maximum

component of load	$= 4000/45$	$= 89$ mm
Length of other leg	$= 4000/60$	$= 67$ mm
Section modulus	$= (7.08 \times 4000)/(1.8 \times 10^3)$	$= 15.8$ cm^3
Select $89 \times 76 \times 9.5$ angle;		$Z = 18.0$ cm^3

10.3.2 Tubular Purlins

Rules are given for tubular purlins for grade 43 steel for roof slopes up to and including 30°. The following is from Clause 45 of BS 449:

> 'The diameter of purlin shall not be less than $L/64$
> The section modulus in centimetre units shall be not less than $(WL/2) \times 10^{-3}$.'

All terms are as defined above for angle purlins.

Example 2 Design of a circular hollow section purlin.

Redesign the purlin of the previous example using a circular hollow section.

Diameter	$= 4000/64$	$= 62.5$ mm
Section modulus	$Z = (7.08 \times 4000)/(2 \times 10^3)$	$= 14.16$ cm^3
Select 76.1 O.D.$\times$4.5 CHS;		$Z = 17$ cm^3

10.4 SHEETING RAILS

10.4.1 Arrangement and Loading

Sheeting rails carry a portion of the wind load and the weight of the sheeting on the walls of buildings in addition to self weight. Angles, channels and circular hollow sections are usually used for sheeting rails.

Clause 46 of BS 449 requires that sheeting rails be designed for wind pressures and vertical loads if any. The requirements of the code regarding limiting deflection and lateral stability of beams do not apply.

Cladding materials are the same as used for roofs, that is, metal corrugated sheeting and asbestos sheeting, which may be used with insulation board, and various types of patent glazing.

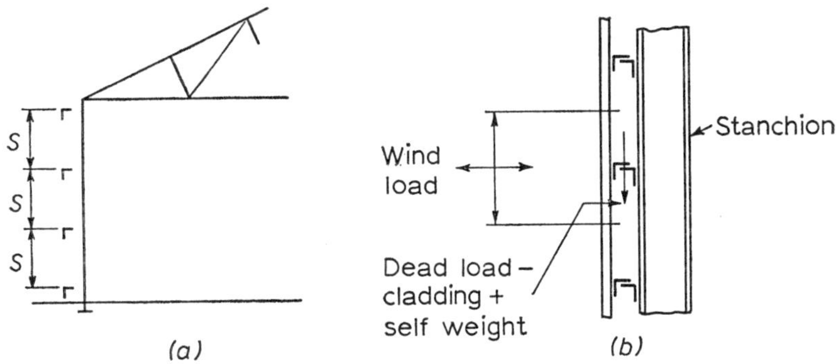

Figure 10.7

Wind loads are in accordance with CP3, Chapter V, Part 2, 1970. For design examples, suitable values for wind load will be assumed.

The arrangement of sheeting rails on the side of a building is shown in Figure 10.7(a). The loading on the rails is shown in (b) on the figure. The wind load may act in either direction due to pressure or suction on the building walls.

10.4.2 Design of Sheeting Rails

The design of sheeting rails is based on the assumption that bending takes place in the vertical and horizontal planes. Eccentricity of the vertical loading is not taken into account. The maximum stress depends on the section used. For angle purlins the maximum stress is the sum of the stresses due to vertical and horizontal bending, and stresses at various points will require investigation. For tubular purlins, the resultant bending moment is found and this is used to calculate the maximum stress.

The theory for beams bending about two axes was given in Chapter 3. The sheeting is taken to be stiff enough to make the sheeting rail bend in accordance with the assumptions above and also prevents any tendency to lateral instability. The design theory will be illustrated in the following examples. Sag rods may be used to support the dead load of the wall.

10.4.3 Design of an Angle Sheeting Rail

A simply supported sheeting rail spans 4 m. The rails are at 1.5 m centres. The total weight of cladding and self weight of the rail can be taken as 40.8 kg/m^2, that is, 0.4 kN/m^2. The wind loading on the wall will be taken as $\pm$0.5 kN/m^2. This would have to be carefully estimated

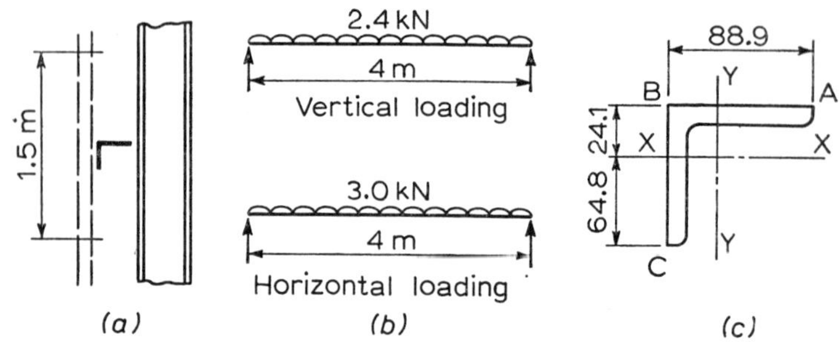

Figure 10.8

for a particular building and the maximum suction and pressure may be different. The sheeting rail arrangement is shown in Figure 10.8(a).

Vertical loading $= 1.5\times4\times0.4 = 2.4$ kN

Horizontal loading $= 1.5\times4\times0.5 = 3.0$ kN

The loading is shown in Figure 10.8(b)

Vertical moment $M_{XX} = (2.4\times4)/8 = 1.2$ kN m

Horizontal moment $M_{YY} = (3.0\times4)/8 = 1.5$ kN m

Try 89×89×6.3 angle. The section properties taken from steel tables are as follows. Refer to Figure 10.8(c).

$$I_{XX} = I_{YY} = 81 \text{ cm}^4$$

For bending about the XX axis:

$$Z_B = 81/2.41 = 33.6 \text{ cm}^3$$

$$Z_C = 81/6.48 = 12.5 \text{ cm}^3$$

For bending about the YY axis:

$$Z_A = 12.5 \text{ cm}^3$$

$$Z_B = 33.6 \text{ cm}^3$$

The stresses for bending about each axis are calculated. Vertical bending about XX axis:

$$f_A = f_B = -(1.2\times10^6)/(33.6\times10^3) = -35.7 \text{ N/mm}^2 \text{ compression}$$

$$f_C = (1.2\times10^6)/(12.5\times10^3) = 88 \text{ N/mm}^2 \text{ tension}$$

Horizontal bending about YY axis:

$$f_C = f_B = \pm(1.5\times10^6)/(33.6\times10^3) = \pm44.6 \text{ N/mm}^2 \text{ tension or compression}$$

$$f_A = \pm(1.5\times10^6)/(12.5\times10^3) = \pm120 \text{ N/mm}^2 \text{ tension or compression}$$

The maximum stresses are:

At A f_A max $= -(35.7+120) = -155.7$ N/mm^2

At C f_C max $= 88+44.6 = 132.6$ N/mm^2

The maximum allowable stress

$$p_{bc} = 165+25\% \text{ for wind} = 206 \text{ N/mm}^2$$

Thus it would be possible to select and check a lighter sheeting rail section.

10.4.4 Design of a Tubular Sheeting Rail

Redesign the sheeting rail in the previous example using a circular hollow section.

Vertical bending moment $M_{XX} = 1.2$ kN m—dead load

Horizontal bending moment $M_{YY} = 1.5$ kN m—wind momet

Resultant bending moment:

$$M_R = (1.2^2+1.5^2)^{1/2} = 1.9 \text{ kN m}$$

Permissible bending stress $p_{bc} = 206$ N/mm^2

Section modulus $Z = (1.9\times10^6)/206 = 9.22\times10^3$ mm^3

Require 60.3 dia. × 4 CHS, where $Z = 9.47$ cm^3

10.5 PROBLEMS

1 The purlins for a flat roof to which access is provided in addition to that necessary for cleaning and repair are at 2 m centres and span 4.5 m. The purlins are simply supported. The total dead load is 102 kg/m^2, that is 1 kN/m^2. Design the purlin.

2 Using Clause 45 of BS 449, design an angle purlin for a sloping roof. The roof slope is 1 in 2.5. The purlin spacing is 1.5 m, the span is 4.5 m and the purlin is simply supported. The total dead load is 40.8 kg/m^2, that is, 0.4 kN/m^2 on the slope length.

3 Redesign the purlin in Example 2 using a circular hollow section.

4 The sheeting rails on the side of a building are simply supported, span 3.5 m and are at 2 m centres. The total dead load is 51 kg/m^2, that is, 0.5 kN/m^2. The wind loading is 0.8 kN/m^2 and may act in either direction. Design the sheeting rail using an angle section.

5 Redesign the sheeting rail in Example 4 using a circular hollow section.

11

Trusses and lattice girders

Trusses and lattice girders are made up of small members acting in tension or compression. They are usually used to carry the roofs of buildings. They are more economical than beams and girders for this purpose.

11.1 TYPES AND LOADING

11.1.1 Types and Arrangement

In these structural elements, the loading in the plane of the member is resisted by axial forces in either tension or compression in the individual members. Some types of roof trusses and lattice girders are shown in Figure 11.1. These members generally carry roof loads. Lattice girders are also used to carry wind and floor loads. Trusses are used for spans above say 6 m and up to 50 m. Spacing between trusses is usually 5 to 10 m. The panel layout depends on decking or sheeting and purlin spacing adopted to suit this. The purlins need not be located at the node points of the truss. This introduces bending into the top chord and this will be considered later in the chapter.

Truss or lattice girder members are shown in Figure 11.2. These are double or single angles, tees, channels or the structural hollow sections. The choice of member is in the hands of the designer. Structural hollow sections are finding increasing use in modern construction. In very heavy trusses, such as bridge trusses, compound and built-up members are used. Consideration of these types is outside the scope of this book.

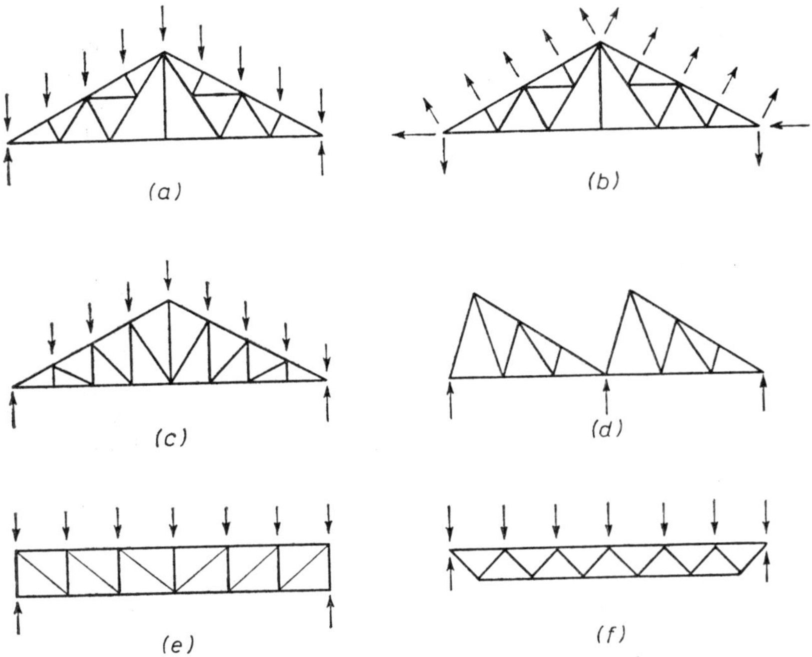

Figure 11.1 Some types of roof truss and lattice girder

Figure 11.2 Truss or lattice girder members

11.1.2 Dead and Imposed Loads

The dead load is due to sheeting or decking, insulation board, felt and weight of steel purlins and the truss itself. This load may range from 51 to 102 kg/m, that is, 0.5 to 1.0 kN/m^2 for average loads. The dead load should be estimated for any particular case. Typical values are given in the worked examples.

The imposed loading on roofs is taken from CP3, Chapter V, Part 1, 1967. This loading may be summarised as follows:

(1) Where there is only access to the roof for maintenance and repair—0.75 kN/m^2
(2) Where there is access in addition to that in (1)—1.5 kN/m^2

The full clause is set out in section 10.1.1 where the design of purlins is treated. The loading is divided in proportion to roof area over the panel point and is taken as a point load applied at the panel point. Loading between panel points is considered later. Refer to loading shown in Figure 11.1. In cases where lattice girders are used for floor beams, appropriate floor loading would be used and the deflection of the truss must be limited in this case. Trusses or lattice girders supporting roof loads will be considered here.

11.1.3 Wind Loads

These are important in light roofs where the wind suction can cause reversal of load in the members.

The wind blowing over the roof causes suction or pressure on the windward slope and a suction on the leeward slope. Refer to Figure 11.1(b). These loads act normal to the roof surface. The load is assumed to be applied through the purlins to the panel points of the truss.

The wind load is taken from CP3, Chapter V, Part 2, 1970. This wind load depends on the building dimensions and roof slope among other factors. Where the wind load is taken into account it is determined for the particular example.

In many cases, particularly for flat roofs with heavy decking, the wind load will not cause a reversal of stress. Here the uplift due to wind is less than the dead load and the wind load need not be considered in design. In this case the truss is designed only for dead plus imposed load. Wind is particularly important where it causes a reversal of stress. A light angle member is satisfactory when acting as a tie, but buckles readily when required to act as a strut.

11.2 ANALYSIS OF TRUSSES

11.2.1 Loads applied at Panel Points

For sloping roof trusses or girders with sloping chords, it is generally quicker to analyse by a force diagram. The truss is assumed to be pin jointed and the loads applied at the panel points. For a parallel-chord lattice girder as shown in Figure 11.1(e) or (f), where the web members are vertical or at a constant slope, the method of joint resolution is often quicker than a stress diagram. This can depend on the designer's personal choice. Finally, in some light wind trusses or in cases where the forces are required in only a few members, the method of sections can be used. This is very useful in a light lattice girder where only the force in the maximum loaded member would be found. The member is designed for this force and made uniform throughout.

These methods of analysis will be shown in the design examples that follow. For a full discussion on these methods the student should refer to textbooks on structural mechanics. Suitable references are given at the end of the book.

11.2.2 Purlin Loads Not at Panel Points

The case where the purlin loads are not applied at the panel points is shown in Figure 11.3(a). The analysis in this case is made in two parts. These are:

1 The roof loads are assumed divided between the panel points as shown in (b). The truss is analysed for this case and the axial loads in the members are found.
2 The top chord is now treated for the actual loads applied through the purlins. The normal components N of the vertical loads W (see Figure 11.3(c)) cause bending in the top chord. The top chord is treated as a fixed ended continuous beam loaded as shown in (d). The beam is analysed by moment distribution and the bending moment diagram can be drawn as shown in (f). The supports of the continuous beam are the internal members of the truss.

The top chord is then designed for axial load and bending moment. For the method of moment distribution and theory of continuous beams, the student is referred to textbooks on structural mechanics. As an approximation, the top chord may be treated as a series of

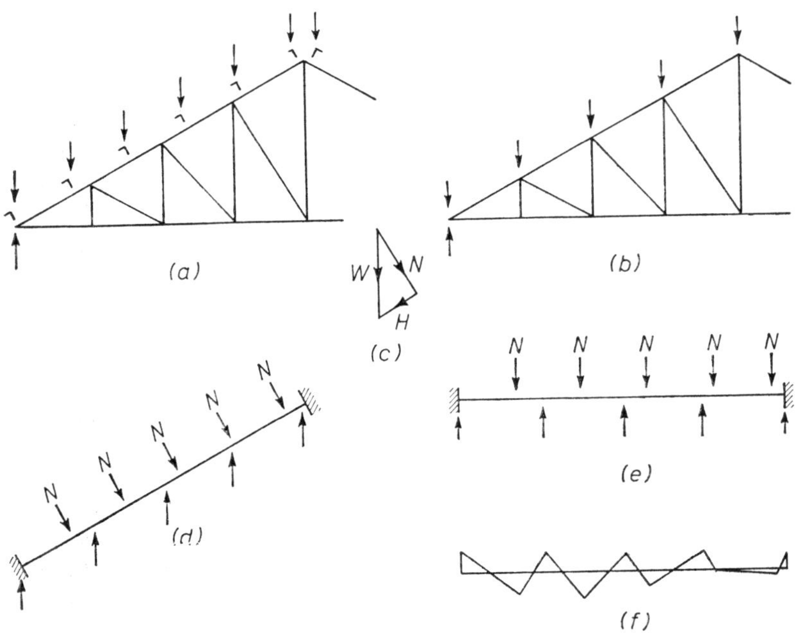

Figure 11.3

fixed-end beams between the supports provided by the internal truss members. These moments may be used instead of carrying out an analysis by moment distribution. This method of analysis will be shown in an example later in this section.

11.3 DESIGN OF TRUSS MEMBERS

11.3.1 Maximum Slenderness Ratio for Struts

For very lightly loaded members the maximum slenderness ratio permitted in BS 449 will often control the size of members.

The maximum slenderness ratio of struts is given in Clause 33. This states:

'The ratio of the effective length, or of the length centre to centre of connections, to the appropriate radius of gyration, shall not exceed the following values:

'1. For any member carrying loads resulting from dead weights with or without imposed loads and for single bolted or riveted single angle struts— 180

'2. For any member carrying loads resulting from wind forces only, and provided that the deformation of such member does not cause an increase of stress, in any part of the structure beyond the permissible stress— 250
(See also subclause 44a)

'Clause 44a. For any member normally acting as a tie in a roof truss but subject to reversal of stress resulting from the action of wind, the ratio of the effective length to the least radius of gyration shall not exceed— 350'

Note that there is no limitation on members that always act as ties. These may be flats or rounds.

11.3.2 Design of Angles as Struts

Rules for the design of angles as struts are given in Clause 30(c) of BS 449.

'(i) For single-angle discontinuous struts connected to gussets or to a section either by riveting or bolting by not less than two rivets or bolts in line along the angle at each end, or by their equivalent in welding. The eccentricity of the connection with respect to the centroid of the strut may be ignored and the strut designed as an axially loaded member provided the calculated average stress does not exceed the allowable stresses given in Table 17, where l is taken as 0.85 times the length of the strut centre to centre of intersections at each end, and r is the minimum radius of gyration.

'Single-angle struts with single, bolted or riveted connections shall be treated similarly but the calculated stress shall not exceed 80% of values given in Table 17, and the full length l centre to centre of intersections shall be used. In no case, however, shall the ratio of slenderness for such single angle struts exceed 180.

'(ii) For double-angle discontinuous struts, back-to-back connected to both sides of a gusset or section by not less than two bolts or rivets in line along the angles at each end or by the equivalent in welding, the load may be regarded as applied axially. The effective length l shall be taken as between 0.7 and 0.85 times the distance between intersections, depending on the degree of restraint, and the calculated average stress shall not exceed the values obtained from Table 17 for the ratio of slenderness based on the minimum radius of gyration about a rectangular axis of the strut. The angles shall be connected together in their length so as to satisfy the requirements of Clause 37.

(iii) Double-angle discontinuous struts back-to-back connected to one side of a gusset... shall be designed as for single angles....

(iv) The provisions of this clause are not intended to apply to continuous-angle struts such as those forming the rafters of trusses....'

Clause 37 states:

'Compression members composed of two angles, channels or tees, back to back in contact or separated by a small distance shall be connected together by riveting, bolting or welding so that the maximum ratio of slenderness l/r of each member between the connections is not greater than 40 or greater than 0.6 times the most unfavourable ratio of slenderness of the strut as a whole, whichever is the less....'

The full clause in the code should be consulted in this case.

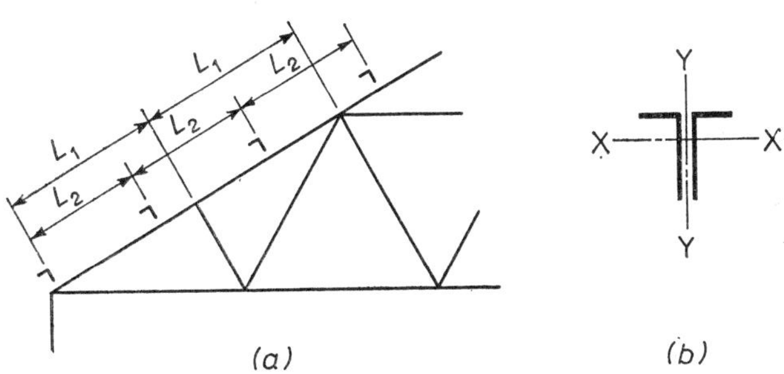

Figure 11.4

For a continuous strut forming the rafter of a roof truss, the effective lengths are as follows. (See Figure 11.4.)

XX axis: $l = 0.7\,L_1$

YY axis: $l = L_2$

where L_1 is the length between panel points on the top chord and L_2 is the spacing of the roof purlins.

It can be seen that buckling in the plane of the roof truss, that is, about the XX axis, is reduced by the presence of gussets and joints. However, for buckling about the YY axis, the purlins are pin connected to the top chord and the spacing of the purlins is used as the effective length.

11.3.3 Design of Angles and Tees as Ties

The design of tension members is controlled by Clause 41. An extract from this clause is:

> The direct stress in axial tension p_t on the net area of section shall not exceed the value given in Table 19. For grade 43 steel for rolled sections up to and including 40 mm in thickness
>
> $$p_t = 155\ \text{N/mm}^2$$

The design of angles and tees as ties is controlled by Clause 42 of BS 449. This states:

> '(a) Eccentric connections. When eccentricity of loading occurs in connections of angles and tees in tension, the net areas to be used in computing the mean tensile stress shall be as given by the following rules:
>
> '1 Single angle connected through one leg. To the net sectional area of the connected leg, add the sectional area of the unconnected leg multiplied by
>
> $$3a_1/(3a_1+a_2)$$
>
> where a_1 = the net sectional area of the connected leg
>
> a_2 = the sectional area of the unconnected leg
>
> '2 A pair of angles connected together along their length, as specified in sub-clause 51e or 54g and attached to the same side of a gusset or the equivalent by only one leg of each angle. To the net sectional area of the connected part, add the sectional area of the unconnected part multiplied by
>
> $$5a_1/(5a_1+a_2)$$
>
> where a_1 = the net sectional area of the connected part
>
> a_2 = the sectional area of the unconnected part

'3 A single tee attached by its table only.
As for 2 above.

'(b) Double angles or tees placed back to back and connected to each side of a gusset or to each side of part of a rolled section. For computing the mean tensile stress the net sectional area of the pair shall be taken, provided the membes are connected together along their length as specified in subclause 51e or 54g.'

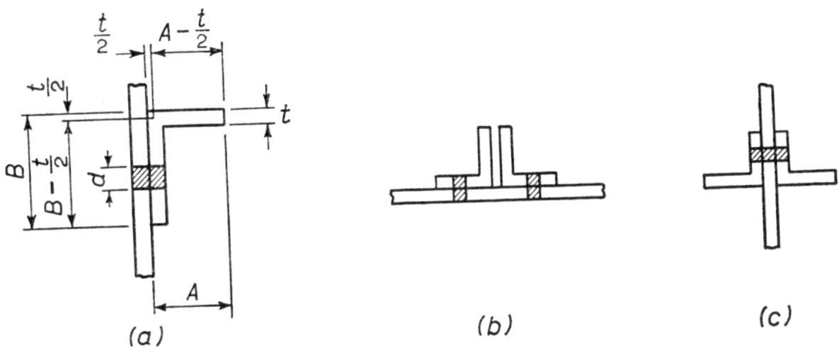

Figure 11.5

Figure 11.5(a) shows a single angle connected through one leg to a gusset plate by bolting.

Referring to the above:

$$a_1 = (B - t/2 - d)t$$

$$a_2 = (A - t/2)t$$

The dimensions of the angle are shown in the figure. A double-angle member connected to the same side of a gusset plate is shown in (b) in the figure and a member connected on each side of the gusset is shown in (c). If a single angle is welded to a gusset or another member on one leg, then the full sectional area of the connected leg is used in the formula above.

Clause 51(e) mentioned above concerns the spacing of tacking rivets for members composed of two rolled sections.

Clause 54(g) concerns intermittent welding of tension members composed of, for example, two angles back-to-back together or separated by a distance piece. The pitch of welded connections or distance pieces shall not exceed 1000 mm. The full clauses should be consulted.

11.3.4 Safe-Load Tables

Safe-load tables are of great assistance in the design of truss members. These are given in the:

Handbook on Structural Steelwork published jointly by the British Constructional Steelwork Association Ltd and The Constructional Steel Research and Development Organisation.

Some of the tables given are for:

1 Struts: Equal angles / Unequal angles — With two or more bolts or rivets in line or welded at ends

2 Compound struts composed of:
Two equal angles back to back with gusset between angles.
Two unequal angles back to back with gusset between angles.

3 Ties: Equal and unequal angles, based on the net section with various diameters and numbers of holes deducted.

These tables are based on theory set out above.

11.4 TRUSS DESIGN

11.4.1 Design Points

Not every member of the truss need be designed for the force in it. This would result in a different section for every member. The designer may use his judgment and group members together making them all the size required for the maximum loaded member in the group. For example, it is common practice to make the top and bottom chords of trusses uniform throughout and to use continuous members.

The connections in a truss should be arranged so that the centres of gravity of the members meeting at the joint are concurrent. If this is not the case, there is an out-of-balance moment at the joint and this should be taken into account in design.

The design of joints has been dealt with in Chapter 2. Figure 11.6 shows some typical column cap and end connections, internal truss joints in welded construction and bolted and welded field splices. Joints are shown for construction in angles, tees and structural hollow sections.

Internal joints in bolted and riveted construction using gusset plates are not given as these have been almost completely superseded by welded connections.

11.4.2 Construction Details

The truss is set out on the centroidal axes of the members and these should be arranged so that the axes meet at a point at the joint. Where bolted splices are used, the bolts are set out on the gauge lines of the

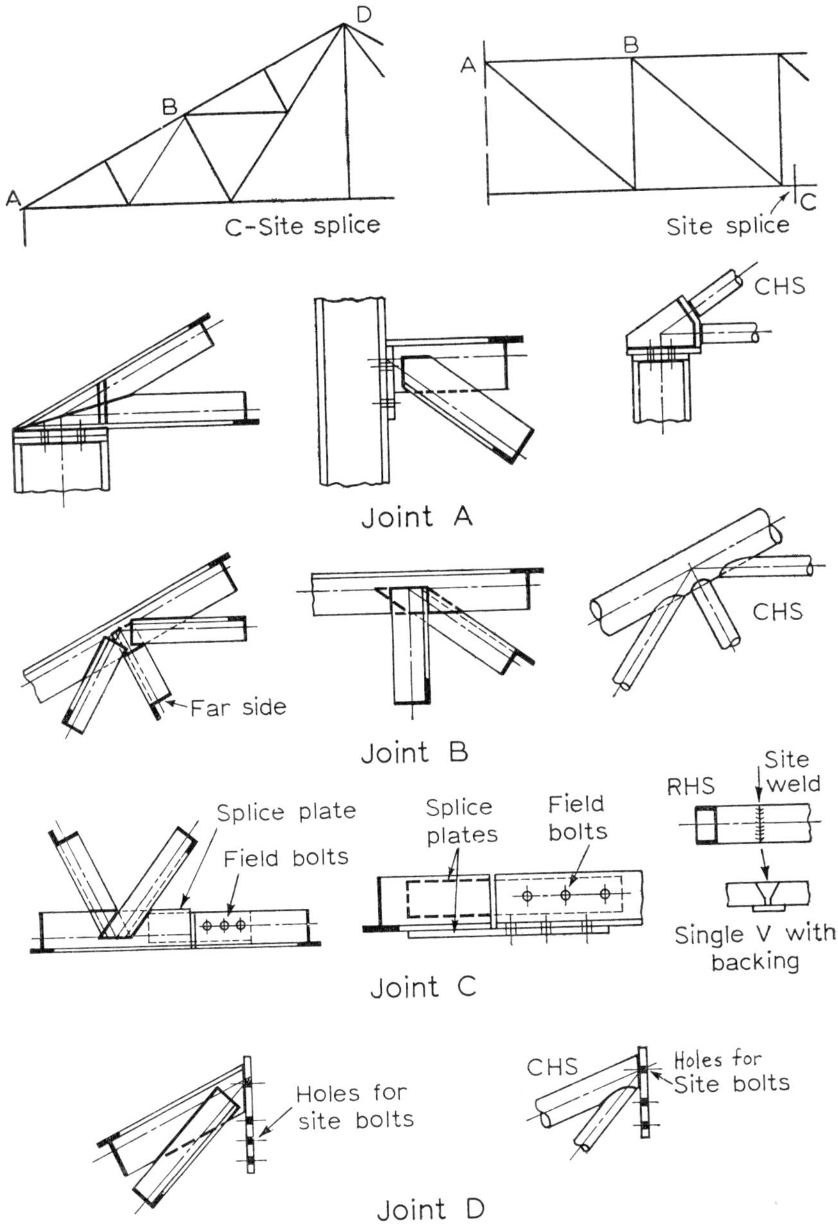
D
A
B
C-Site splice
B
A
C
Site splice
CHS
Joint A
Far side
CHS
Joint B
Splice plate
Field bolts
Splice
plates
Field
bolts
RHS
Site
weld
Single V with
backing
Joint C
Holes for
site bolts
CHS
Holes for
Site bolts
Joint D

Figure 11.6

members. The position of gauge lines for drilling and the spacing of holes are given in the *Handbook on Structural Steelwork*.

In the welded trusses in rolled sections shown in Figure 11.6, it is convenient to use tee sections for the chords and to weld the internal members directly to either one or both sides of the stalks of the tees. Angles can be used for the chords of light trusses. With structural hollow sections, one member can be welded directly to the other after the end has been mitred or shaped.

The welding of joints is carried out manually. Balanced welding may be used as set out in section 2.3.3 for joints in angles and tees. However, it is common practice to weld round both sides of the joint as this provides a seal for the edges in contact and prevents corrosion between the surfaces. Joints in structural hollow sections must be completely sealed by welding to prevent internal corrosion.

11.5 DESIGN OF A FLAT-ROOF TRUSS

The section through a flat-roof building of 20 m span is shown in Figure 11.7. The roof trusses are at 5 m centres longitudinally and the purlins are at 2 m centres. The total roof load including the imposed

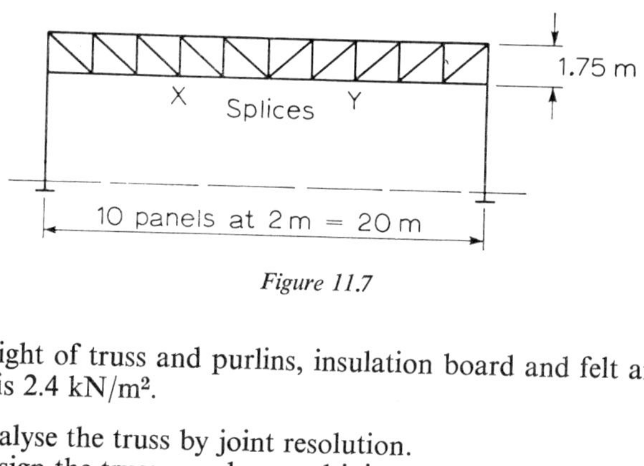

Figure 11.7

load, weight of truss and purlins, insulation board and felt and steel decking is 2.4 kN/m².

(a) Analyse the truss by joint resolution.
(b) Design the truss members and joints.

The truss is to be of welded construction with bolted field joints of black bolts at splices X and Y.

(a) Analysis of truss

Panel point load $= 2.4 \times 2 \times 5 = 24$ kN

The joint loads and the forces in the members of one half of the truss, found by joint resolution, are shown in Figure 11.8.

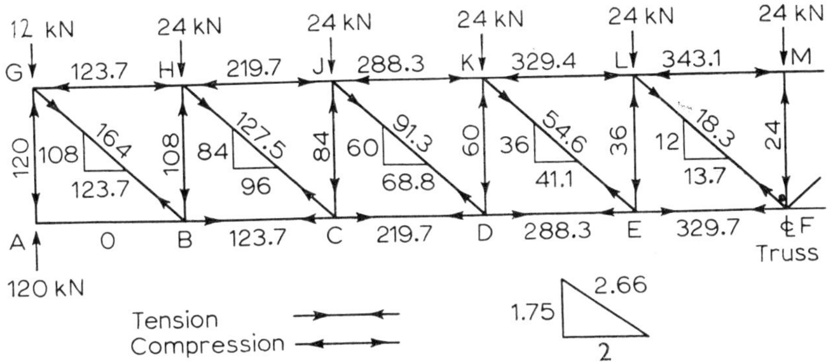

Figure 11.8

(b) Design of Truss

1 *Top chord* K–M Maximum force = 343.1 kN compression. Try 146×127×22 kg/m Struct. Tee. The tee section is shown in Figure 11.9 and the properties from the *Handbook on Structural*

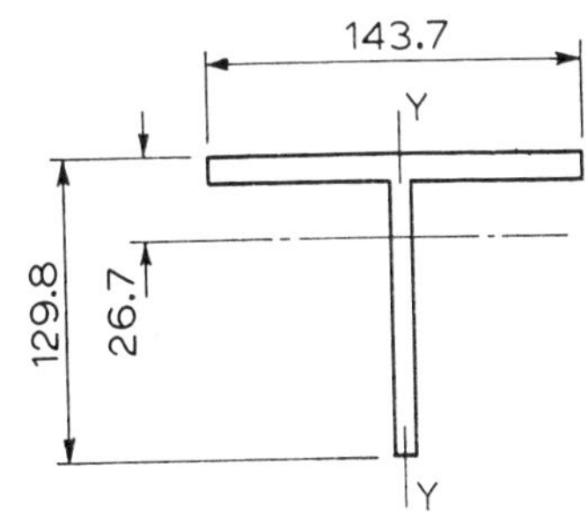

Figure 11.9

Steelwork are:

$A = 27.5 \text{ cm}^2$

$r_{XX} = 3.56 \text{ cm}$

$r_{YY} = 3.39 \text{ cm}$

The slenderness ratios are:

XX axis $l/r_{XX} = (0.7\times2000)/35.6 = 39.4$

YY axis $l/r_{YY} = 2000/33.9 = 59$

$p_c = 126 \text{ N/mm}^2$ from Table 17

$f_c = (343.1\times10^3)/(27.5\times10^2) = 125 \text{ N/mm}^2$

Member: 146×127×22 kg/m Struct. Tee

2 *Top Chord* G–K Maximum force = 288.3 kN compression. Try 146×127×19 kg/m Struct. Tee. The properties of this section are:

$A = 23.7\ \text{cm}^2$

$r_{XX} = 3.54\ \text{cm}$

$r_{YY} = 3.34\ \text{cm}$

$l/r_{YY} = 2000/33.4 = 60$

$p_c = 126\ \text{N/mm}^2$

$f_c = (288.3\times10^3)/(23.7\times10^2) = 122\ \text{N/mm}^2$

Member 146×127×19 kg/m Struct. Tee

3 *Bottom Chord* D–F

There is a splice at D, so the member must be able to resist the following forces in tension:

Net section at the splice = 288.3 kN

Gross section at the centre = 329.7 kN

Try 146×127×19 kg/m Struct. Tee with a gross area of 23.7 cm². The arrangement of the joint is shown in Figure 11.10.

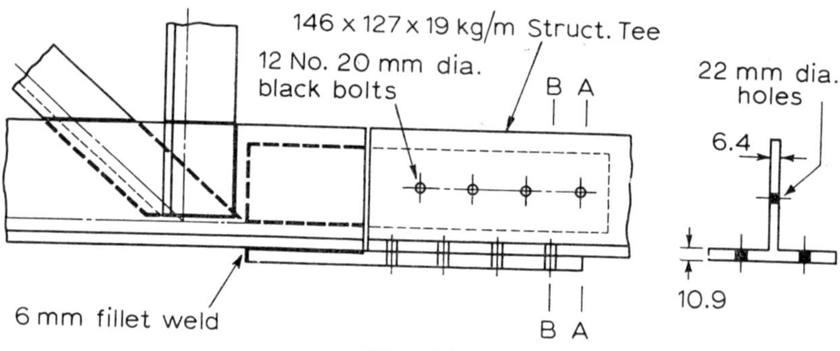

Figure 11.10

Use 20 mm diameter black bolts for the splice.

Single-shear value per bolt = 25.1 kN

No. of bolts = 288.3/25.1 = 11.5 Use 12 bolts.
Tensile stress on the gross section

$f_t = (329.7\times10^3)/(23.7\times10^2) = 139\ \text{N/mm}^2$

Tensile stress on the net section at AA

Net area $= 2370-(22\times6.4) = 2229\ \text{mm}^2$

$f_t = (288.3\times10^3)/2229 = 129\ \text{N/mm}^2$

Tensile stress on the net section at BB

Force $= 288.3 - 25.1 = 263.2$ kN

Net area $= 2370 - (2 \times 22 \times 10.9) = 1890$ mm^2

$f_t = (263.2 \times 10^3)/1890 = 139$ N/mm^2

Member 146×127×19 kg/m Struct. Tee

4 *Bottom Chord* A–D
Maximum tension on the gross section is 219.7 kN.

Area required $= (219.7 \times 10^3)/155 = 1415$ mm^2

Member: 102×127×11 kg/m Struct. Tee. Gross area = 142 cm^2.

5 *Struts* HB, JC Maximum force = 108 kN compression.

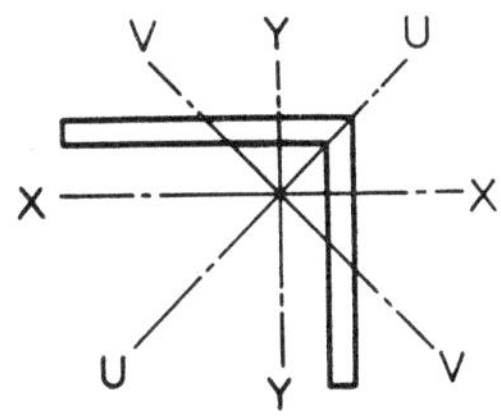

Figure 11.11

Try 89×89×7.9 Angle. Referring to Figure 11.11, the properties from the steel handbook are:

$A = 13.47$ cm^2

$r_{VV} = 1.74$ cm, the minimum radius of gyration

$l/r_{VV} = (1750 \times 0.85)/1.74 = 85.5$

$p_c = 96.5$ N/mm^2

$f_c = (108 \times 10^3)/(13.47 \times 10^2) = 80.5$ N/mm^2

Member 89×89×7.9 Angle

6 *Struts* KD, LE, MF Maximum force = 60 kN compression. Try 76×76×6.2 Angle. The properties are:

$A = 9.12$ cm^2

$r_{VV} = 1.49$ cm

$l/r_{VV} = (0.85 \times 1750)/14.9 = 99.8$

$p_c = 79$ N/mm^2

$f_c = (60 \times 10^3)/(9.12 \times 10^2) = 66$ N/mm^2

Member: 76×76×6.2 Angle

7 *Ties* GB, HC Maximum force = 164 kN tension.
Try 89×89×7.9 Angle connected through one leg as shown on Figure 11.12.

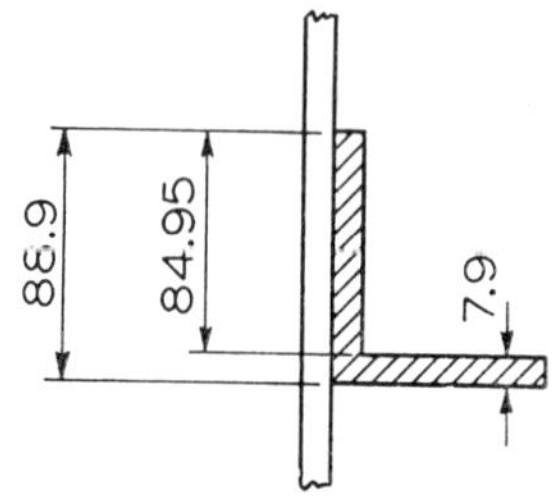

Figure 11.12

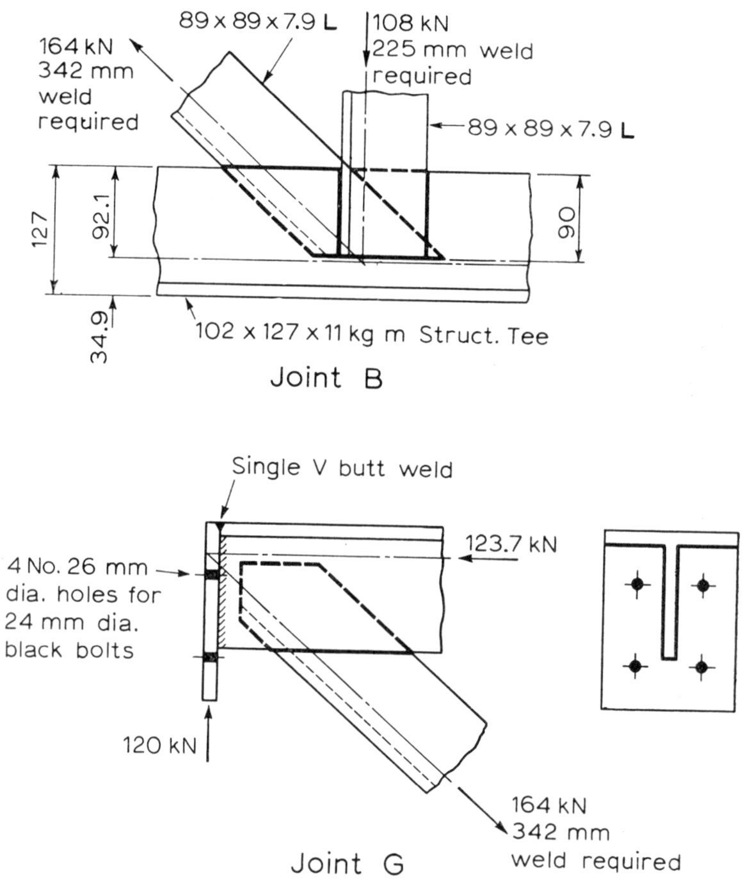

Figure 11.13

Area of connected leg = area of unconnected leg.

$a_1 = a_2 = 7.9 \times 84.95 = 670 \text{ mm}^2$

$$\text{Effective area} = 670 + [(3 \times 670)/\{(3 \times 670) + 670\}]670 = 1172 \text{ mm}^2$$

$$f_t = (164 \times 10^3)/1172 = 140 \text{ N/mm}^2$$

Member: 89×89×7.9 Angle

8 *Ties* JD, KE, LF   Maximum force = 91.3 kN.

Try 64×64×6.2 Angle connected through one leg.

$$a_1 = a_2 = (63.5 - 3.1)6.2 = 374 \text{ mm}^2$$

$$\text{Effective area} = 374 + \tfrac{3}{4} \times 374 = 654 \text{ mm}^2$$

$$f_t = (91.3 \times 10^3)/654 = 140 \text{ N/mm}^2$$

Member: 64×64×6.2 Angle

9 *Welds*

Fillet weld of 6 mm leg length and strength 483 N/mm will be used. The members are to be continuously welded all round, where possible, to seal the joints. Enlarged details of some of the joints in the truss are shown in Figure 11.13. The effective weld lengths required at the

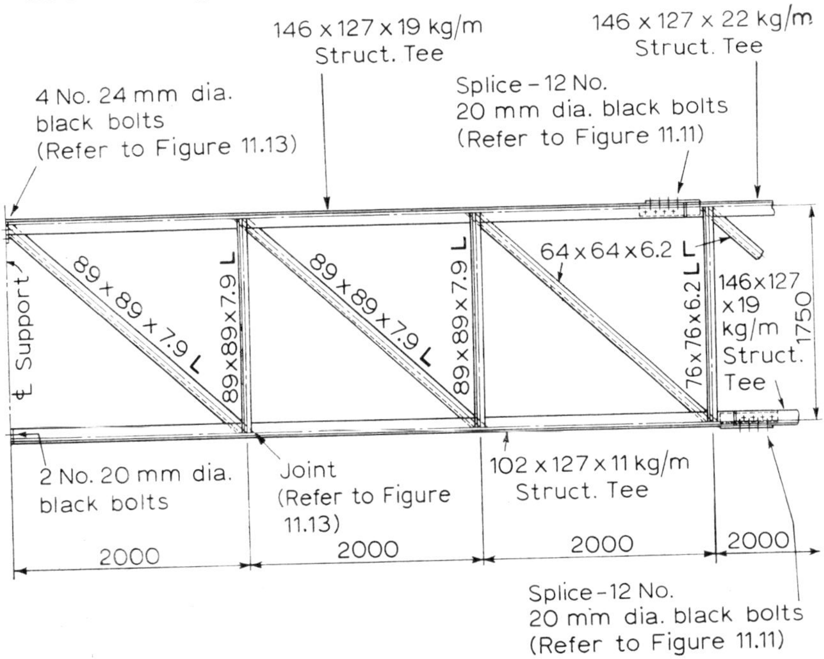

Figure 11.14 Part of roof truss. See Figures 11.1 and 11.13 for enlarged details of some joints

ends of the members are also given. The lengths shown are adequate to resist the member forces.

The splice is shown in Figure 11.10 and the same detail is used for both top and bottom chords. Design for the splice plates and welds is not given. A drawing of part of the truss is shown in Figure 11.14.

11.6 DESIGN OF A LATTICE GIRDER

The plan of a flat roof for a factory is shown in Figure 11.15. The roof is to be carried by lattice girders of 20 m span at 4 m centres. The roofing is a pattern of square plastic pyramid roof lights and flat roof

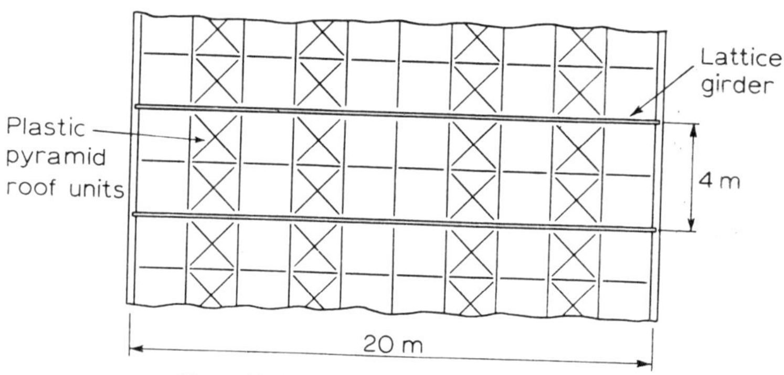

Figure 11.15 Plan of a flat roof for a factory

units in panels 2 m square. This allows for fixing and drainage. The panels are supported on a secondary beam system between the lattice girders. Roof ventilators are installed on some of the roof deck units.

The roof loading is as follows:

Dead load including an estimate for the weight of the lattice girder = 61 kg/m^2 = 0.6 kN/m^2
Imposed load = 0.75 kN/m^2

The wind load need not be taken into account in the design of the lattice girders as the building height is 5 m and the ventilators will reduce uplift.

Design the lattice girder using rectangular hollow sections and welded construction. The girder will be fabricated in one piece in the shop. If required the girder could be shop fabricated in two pieces and connected together on site with a bolted or welded splice. The depth of the girder centre to centre of chords is to be 1200 mm. Appearance of the girder is important as the steelwork in the roof will be exposed to view.

The British Steel Corporation have standardised production of structural hollow sections on two grades of steel:

Grade 43 C—mild steel

Grade 50 C—high yield strength steel

Grade 43 C steel is to be used for the girder. All section properties are taken from the publication of the British Steel Corporation, Tubes Division, 'Structural Hollow Sections, Technical Data'.

(a) Loading and analysis

The lattice girder is shown in Figure 11.16(a). The panel point loading is

$$4\times2\times1.35 = 10.8 \text{ kN}$$

The analysis by joint resolution is shown in (b) on the figure. Check the force in the centre bottom chord member LM by the method of sections by taking moments about F:

$$1.2\ F_{LM} = (54-5.4)10-10.8(2+4+6+8)$$

$$F_{LM} = 225 \text{ kN}$$

This checks the value calculated above.

(b) Design of lattice girder

In order to save weight, the chord sections will be reduced at D in the top chord and at K in the bottom chord. The positions of the shop splices are shown in Figure 11.16(a).

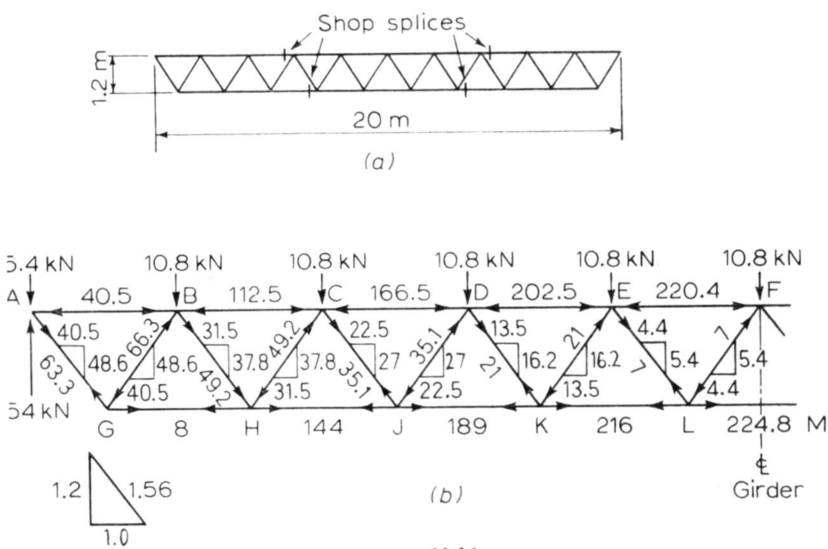

Figure 11.16

1 *Top chord* DEF Maximum force = 220.4 kN compression. Try 101.6×101.6×4.9 RHS. The properties of this section are (see Figure 11.17):

$A = 18.8 \text{ cm}^2$

$r_{XX} = r_{YY} = 3.94 \text{ cm}$

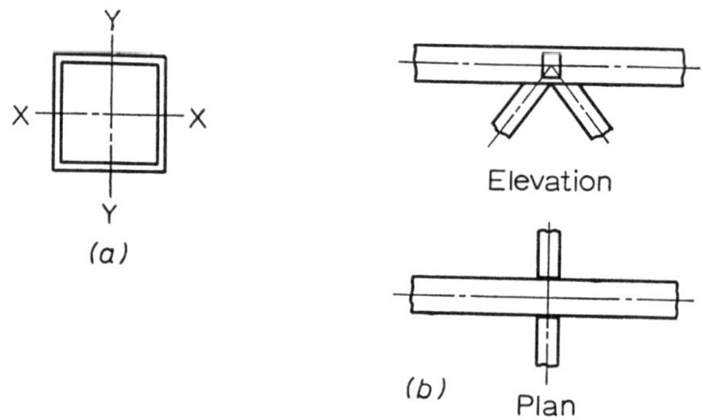

Figure 11.17

Referring to Figure 11.17(b), the top chord is supported at the panel points by the internal members and the secondary roof beams. The effective lengths are:

XX axis— $l = 0.85 \times 2000 = 1700$ mm

YY axis— $l = 2000$

$l/r = 2000/39.4 = 50.8$

$p_c = 133 \text{ N/mm}^2$

$f_c = (220.4 \times 10^3)/1880 = 117.6 \text{ N/mm}^2$

Member: 101.6×101.6×4.9 RHS

2 *Top chord* ABCD Maximum force = 166.5 kN compression. Try 101.6×101.6×4 RHS. The properties of this section are:

$A = \quad = 15.8 \text{ cm}^2$

$r_{XX} = r_{YY} \quad = 3.99 \text{ cm}$

$l/r = 2000/39.9 = 50.2$

$p_c = 133 \text{ N/mm}^2$

$f_c = (166.5 \times 10^3)/1580 = 105.6 \text{ N/mm}^2$

A lighter section could be used here, but 101.6×101.6 section will be adopted for uniformity.

Member: 101.6×101.6×4 RHS

3 *Bottom chord* KLM Maximum force = 224.8 kN tension

Net area = $(224.8\times10^3)/155$ = 1450 mm^2

Member: 88.9×88.9×4.9 RHS, A = 16.3 cm^2

4 *Bottom chord* GHJK Maximum force = 189 kN tension

Net area = $(189\times10^3)/155$ = 1220 mm^2

Member: 88.9×88.9×3.6 RHS, A = 12.4 cm^2

5 *Compression diagonal* BG Force = 63.3 kN compression

Try 50.8×50.8×3.2 RHS. The properties are:

$$A = \quad = 6.14 \text{ cm}^2$$

$$r_{XX} = r_{YY} \quad = 1.94 \text{ cm}$$

$$l/r = (1560\times0.85)/19.4 = 68.4$$

$$p_c = 117.5 \text{ N/mm}^2$$

$$f_c = (63.3\times10^3)/614 = 103.5 \text{ N/mm}^2$$

Member: 50.8×50.8×3.2 RHS

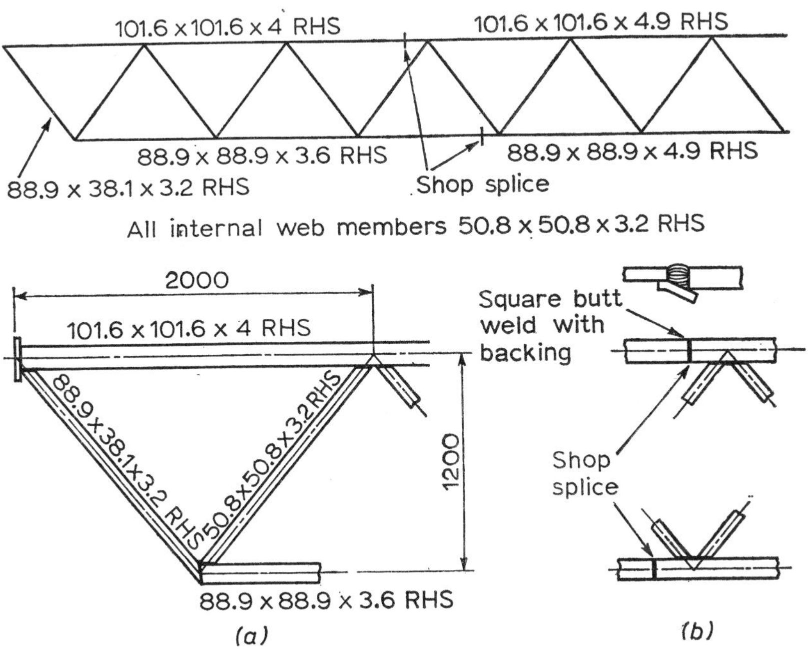

Figure 11.18

6 *End Diagonal* AG Force = 63.3 kN tension

Net area = $(63.3\times10^3)/155 = 411\ mm^2$

For uniformity adopt 88.9×38.1×3.2 RHS, $A = 7.81\ cm^2$

All other internal members will be made 50.8×50.8×3.2 RHS for uniformity. The joints can be designed as set out in Chapter 2. The member sizes are shown in Figure 11.18(a). Details for the end panels and at the splices are shown on (b) and (c) respectively on the figure. The welds in the chords can be square butt welds and backing members should be used as shown. The backing member requires special fitting in this case.

The chords of the truss could be designed in Grade 50 C structural hollow sections. This would show that a considerable saving in weight can be achieved by using the higher grade steel.

11.7 DESIGN OF A ROOF TRUSS FOR AN INDUSTRIAL BUILDING

A section through an industrial building is shown in Figure 11.19(a). The frames are at 5 m centres and the length of the building is 50 m.

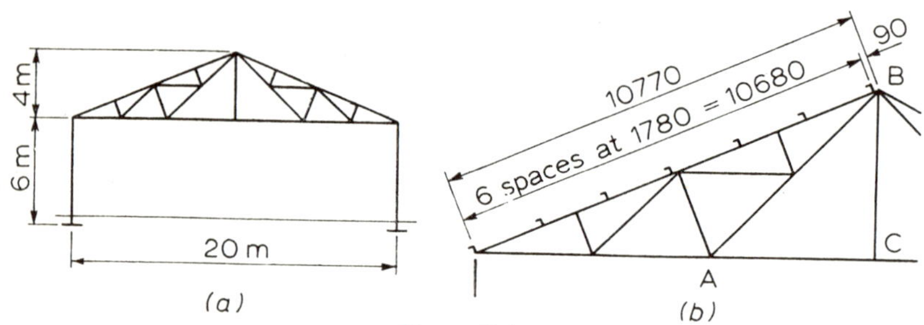

Figure 11.19

The purlin spacing on the roof is shown in (b) in the figure. The loading on the roof is as follows:

1 Dead load—measured on the slope length

Sheeting and insulation board	= 25.6 kg/m²	= 0.25 kN/m²	
Purlins	= 10.2 kg/m²	= 0.1 kN/m²	
Truss	= 10.2 kg/m²	= 0.1 kN/m²	
Total dead load	= 46 kg/m²	= 0.45 kN/m²	

2 Imposed load—measured on plan = $0.75\ kN/m^2$

measured on slope = $0.7\ kN/m^2$

3 Wind loading. This is taken from CP3, Chapter V, Part 2, 1970. Derivation of the wind loading is set out in full below, where all necessary data regarding the building are given. The code should be consulted for the method, tables of coefficients, etc.

Design the roof truss using angles and tees for welded construction. The truss is to be fabricated in two parts in the shop for transport to site. Field joints are to be bolted. These are at A, B and C as shown in Figure 11.19(b).

(a) Dead and imposed loads

Because of symmetry only half of the truss is considered.
Dead loads:

End panel point $= \frac{1}{8}\times 0.45\times 10.77\times 5 = 3.03$ kN

Internal panel points $= \frac{1}{4}\times 0.45\times 10.77\times 5 = 6.06$ kN

Imposed loads:

End panel points $= 3.03\times(0.7/0.45) = 4.71$ kN

Internal panel points $= 6.06\times(0.7/0.45) = 9.42$ kN

The dead loads are shown in Figure 11.22(a).

(b) Wind load

Location—North East England

Basic wind speed $= 45$ m/s

Topography factor $S_1 = 1$

Building is sited on the outskirts of a city with obstructions up to 10 m high

Ground roughness $= 3$

Building size Class B

Height to the top of the structure or part on which the load is required— $H = 10$ m

Factor from Table 3 $S_2 = 0.74$

Statistical factor $S_3 = 1$

Design wind speed $V_s = 0.74\times 45 = 33.3$ m/s

Dynamic pressure from Table 4 $q = 682$ N/m^2

The external pressure coefficients C_{pe} are shown in Figure 11.20. These are taken from Table 8. The internal pressure coefficients C_{pi} are taken from Appendix E. Here, the case is taken where there is only a negligible probability of a dominant opening occurring during a severe storm. C_{pi} is then taken as the more onerous of the values $+0.2$ and -0.3.

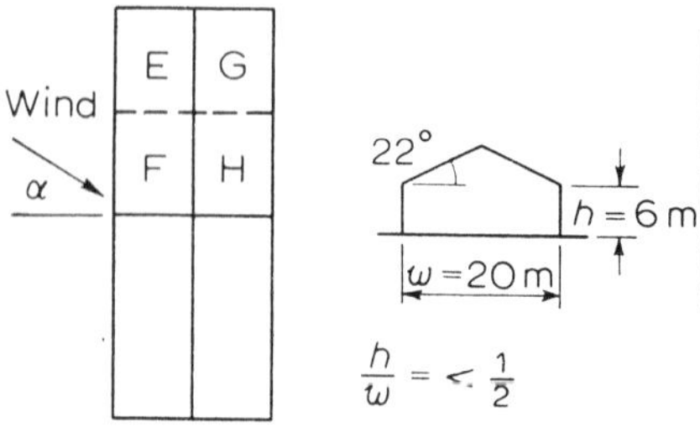

Roof angle	Wind angle 0°		Wind angle 90°	
	EF	GH	EG	FH
22°	-0.32	-0.4	-0.6	-0.7

Figure 11.20

For design of the roof truss, the condition of maximum uplift is the only one that need be investigated. Thus, a truss is selected from section FH of the roof shown in Figure 11.20 where C_{pe} is a maximum and C_{pi} is taken as +0.2, the case of internal pressure. The wind loading normal to the roof slope is then:

$$(682/10^3)(C_{pe}-C_{pi})$$

The wind loads on the roof are shown in Figure 11.21 for the two cases of the wind transverse and longitudinal to the building.

The wind loads at the panel points normal to the top chord for the case of wind longitudinal to the building are:

End panel points $= \frac{1}{8}\times 0.613\times 10.77\times 5 = 4.11$ kN

Internal panel points $= 8.22$ kN

The wind loads are shown in Figure 11.22(b).

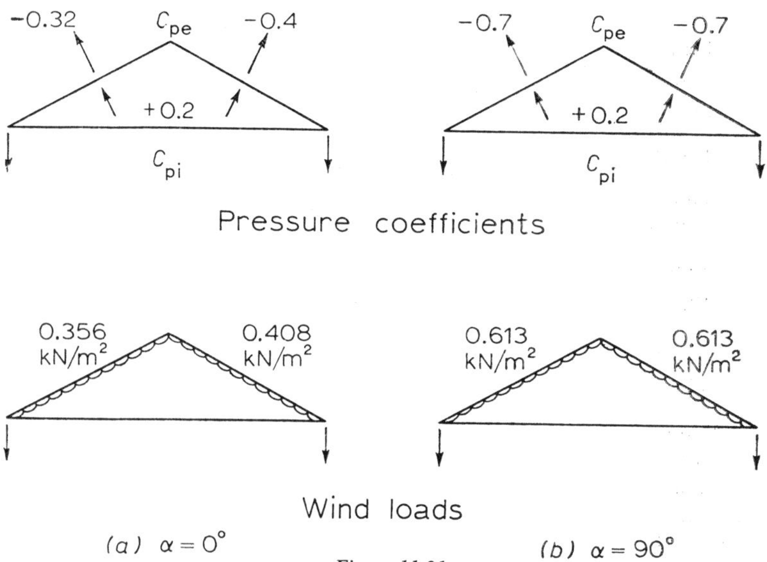

Figure 11.21

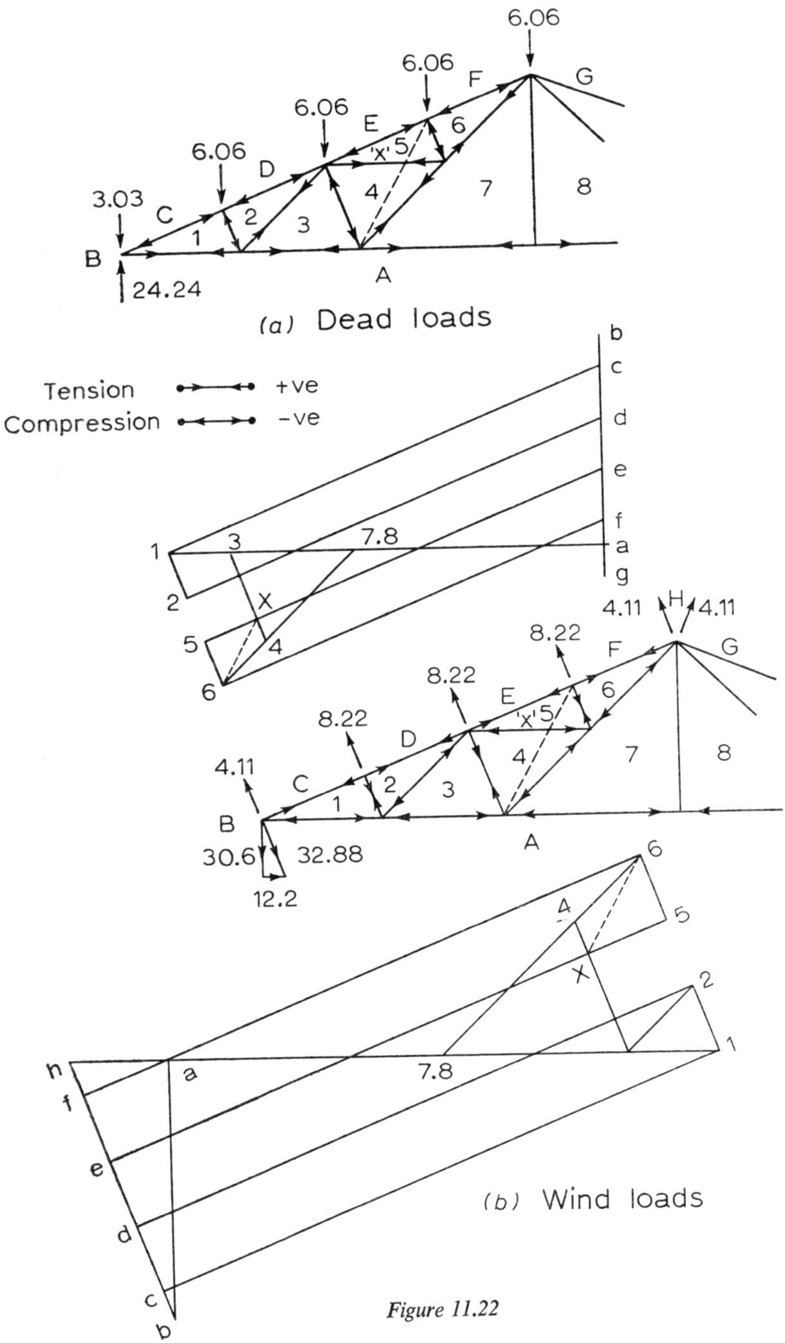

(a) Dead loads

(b) Wind loads

Figure 11.22

(c) Analysis

Because of the symmetry of loading in each case, only one half of the truss is considered. The truss is analysed by the force diagram method and the analyses are shown in Figure 11.22. The student should refer to textbooks on structural mechanics for this method. In this case members 4–5 and 5–6 must be replaced by the fictitious member 6–X to locate point 6 on the force diagram. Then point 6 is used to locate points 4 and 5. Refer to the force diagrams in Figure 11.22.

The dead load case is analysed and the forces in the members due to the imposed loads are found by proportion. Only the case of maximum uplift is considered in the wind load analysis. This occurs when the wind is blowing longitudinally along the building and causes the greatest number of reversals of force from tension to compression or vice versa in the truss members.

The forces in the members of the truss are tabulated for the various load cases. The following two combined load cases are given:

1 Dead and imposed load. This gives the maximum forces in the members.
2 Dead and wind load. This gives the possible force reversals.

The forces and combined load cases are given in *Table* 11.1.

Table 11.1 Forces in members of roof truss (kN)

	Member	*Dead load*	*Imposed load*	*Wind load*	*Dead+ imposed*	*Dead+ wind*
C–1	Top chord	−56.6	−94.4	+74.5	−151.0	+17.9
D–2		−54.4	−90.8	+74.5	−145.2	+20.1
E–5		−52.5	−87.5	−73.7	−140.0	+21.2
F–6		−50.2	−83.6	−73.7	−133.8	+23.5
A–1	Bottom chord	+52.5	+87.5	−67.9	+140.0	−15.4
A–3		+45.0	+75.0	−56.5	+120.0	−11.5
A–7		+29.3	+65.5	−34.0	+104.8	− 4.7
1–2	Struts	− 5.5	− 9.2	+ 8.4	− 14.7	+ 2.9
3–4		−11.4	−19.0	+16.8	− 30.4	+ 5.4
5–6		− 5.5	− 9.2	+ 8.4	− 14.7	+ 2.9
2–3	Ties.	+ 8.0	+13.3	−11.4	+ 21.3	− 3.4
4–5		+ 8.0	+13.3	−11.4	+ 21.3	− 3.4
4–7		+15.8	+26.4	−11.2	+ 42.2	+ 4.6
6–7		+23.5	+39.2	−33.0	+ 62.7	− 9.5
7–8		0	0	0	0	0

Convention + tension
− compression

(d) Bending in the top chord

The top chord is analysed as a continuous beam for bending caused by the normal components of the purlin loads. The method of moment distribution is used. The student should refer to textbooks on structural mechanics for this method.

Purlin load	$= 1.78\times5\times1.05$	$=$	10.2 kN
Normal component	$= (10.2\times10)/10.77$	$=$	9.5 kN
End purlin at L	$= 9.5\times(0.98/1.78)$	$=$	5.23 kN

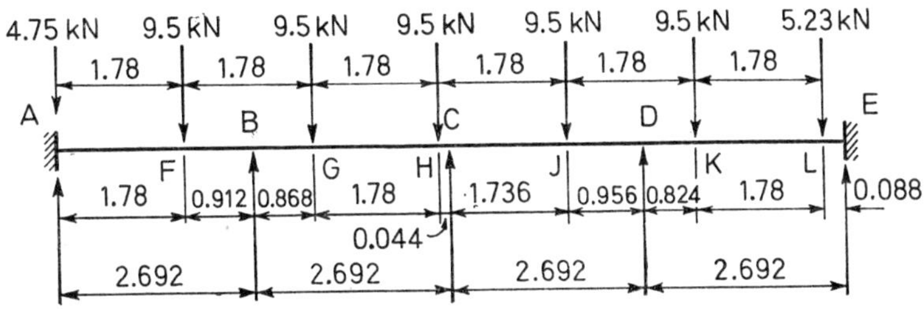

Figure 11.23

The top chord loading is shown in Figure 11.23. The fixed end moments are calculated below:

Span AB $M_{AB} = (9.5\times0.912^2\times1.78)/2.692^2 = 1.94$ kN m

$M_{BA} = (9.5\times0.912\times1.78^2)/2.692^2 = 3.82$ kN m

Span BC $M_{BC} = (9.5\times0.868\times1.824^2)/2.692^2 + (9.5\times2.648\times0.044^2)/2.692^2 = 3.78$ kN m

$M_{CB} = (9.5\times1.824\times0.868^2)/2.692^2 + (9.5\times2.648^2\times0.044)/2.692^2 = 2.22$ kN m

Span CD $M_{CD} = (9.5\times1.736\times0.956^2)/2.692^2 = 2.08$ kN m

$M_{DC} = (9.5\times0.956\times1.736^2)/2.692^2 = 3.78$ kN m

Span DE $M_{DE} = (9.5\times0.824\times1.858^2)/2.692^2 + (5.23\times2.604\times0.088^2)/2.692^2 = 3.8$ kN m

$M_{ED} = (9.5\times1.858\times0.824^2)/2.692^2 + (5.23\times0.088\times2.604^2)/2.692^2 = 2.1$ kN m

Because the spans are equal and the chord section is uniform, the distribution factors are 0.5 each side of the joints B, C and D. The moment distribution is shown in Figure 11.24(a). The results after two cycles show that the fixed end moments could be used in design without sacrificing too much accuracy.

A	0.5 B	B 0.5	0.5 C	C 0.5	0.5 D	D 0.5	E
1.94	−3.82	3.78	−2.22	2.08	−3.78	3.8	−2.1
0.00	0.02	0.02	0.07	0.07	−0.01	−0.01	0.00
0.01	0.00	0.04	0.01	−0.01	0.04	0.00	−0.0
0.00	−0.02	−0.02	0.00	0.00	−0.02	−0.02	0.00
1.95	−3.82	3.82	−2.14	2.14	−3.77	3.77	−2.11

(a) Moment distribution

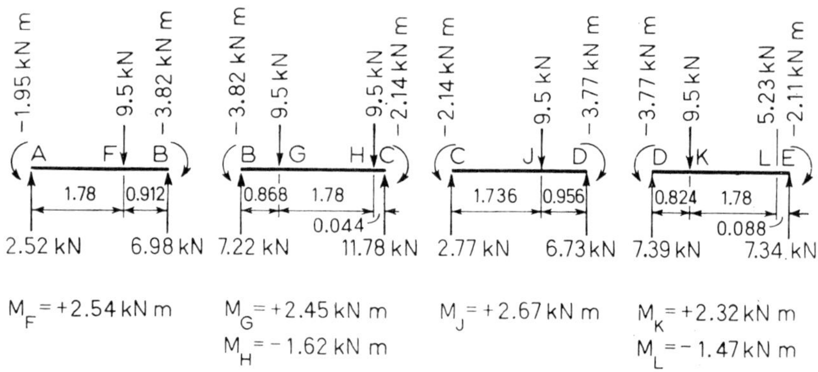

(b) Separate spans. Reactions and Internal moments

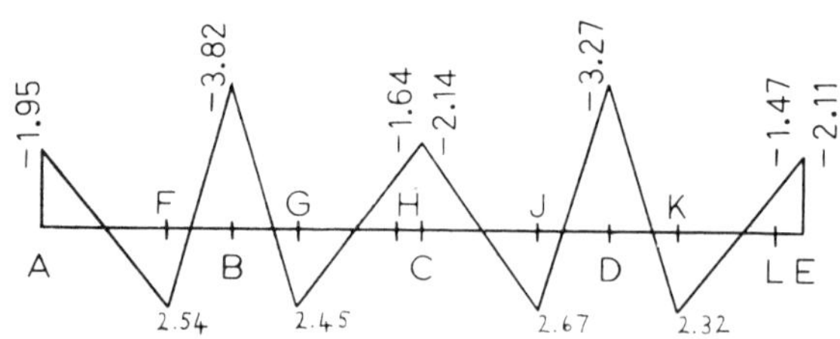

(c) Bending moment diagram-Moments (kN m)

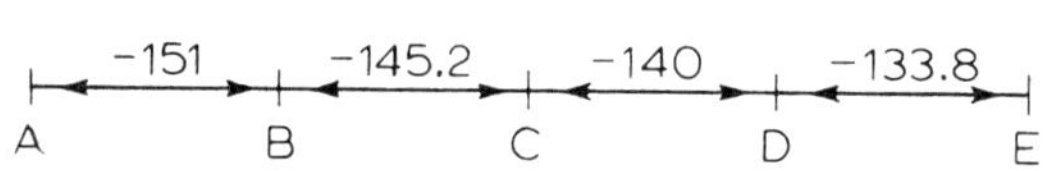

(d) Axial loads (kN)

Figure 11.24

Each span of the beam is treated separately to find the reactions and internal moments. This information is shown in Figure 11.24(b). The bending moment diagram for the complete top chord is shown in (c) in the figure and the axial loads from the force diagram are shown in (d).

(e) Design of truss members

Top Chord

Design data Axial load $= -151$ kN

Moments $M_B = -3.82$ kN m

$M_F = 2.54$ kN m

Refer to Figure 11.24(c) and (d) for the positions of the load and moments. For the case of dead and wind load, there is a maximum tension in the top chord of 23.5 kN. However, the conditions above of maximum moment and compressive load will control the design.

Try 165×152×23 kg/m Struct. Tee. The properties from the *Handbook on Structural Steelwork* are:

$A = 29.4$ cm^2

$Z_{XX} = 44.1$ cm^3 at the bottom of the stalk

$= 174.7$ cm^3 at the table of the tee

$r_{XX} = 4.29$ cm

$r_{YY} = 3.74$ cm

The section of the tee is shown in Figure 11.25.

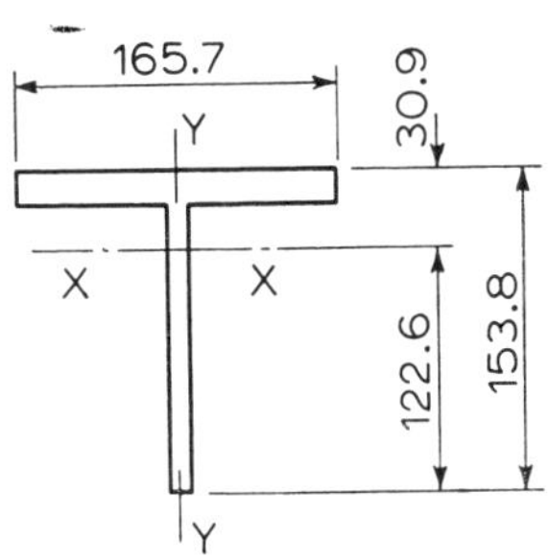

Figure 11.25

The permissible stress in axial compression is determined as follows.

$$l/r_{XX} = (0.7\times2692)/42.9 = 44$$

$$l/r_{YY} = 1780/37.4 = 47.6$$

$$p_c = 135.6 \text{ N/mm}^2 \text{ from Table 17.}$$

17

The permissible stress in bending will be taken as 165 N/mm^2 as the section at B, where the stalk of the tee is in compression, is supported by the internal truss member against lateral buckling. Reference should be made to Clause 20(c) of BS 449 to determine the allowable stress in compression in bending if buckling is likely to occur in the stalk of the tee.

The maximum stress condition occurs at B and the stresses here are:

$f_c = (151 \times 10^3)/2940 \qquad = 51.3 \text{ N/mm}^2$

$f_{bc} = (3.82 \times 10^6)/(44.1 \times 10^3) = 86.5 \text{ N/mm}^2$

Combined:

$$(51.3/135.6)+(86.5/165) = 0.38+0.525 = 0.905$$

165×152×23 kg/m Struct. Tee is satisfactory.

Bottom Chord Member A–1. Refer to Figure 11.22

Design loads = 140 kN—tension

= −15.4 kN—compression

Net area required = $(140 \times 10^3)/155 = 905$ mm^2

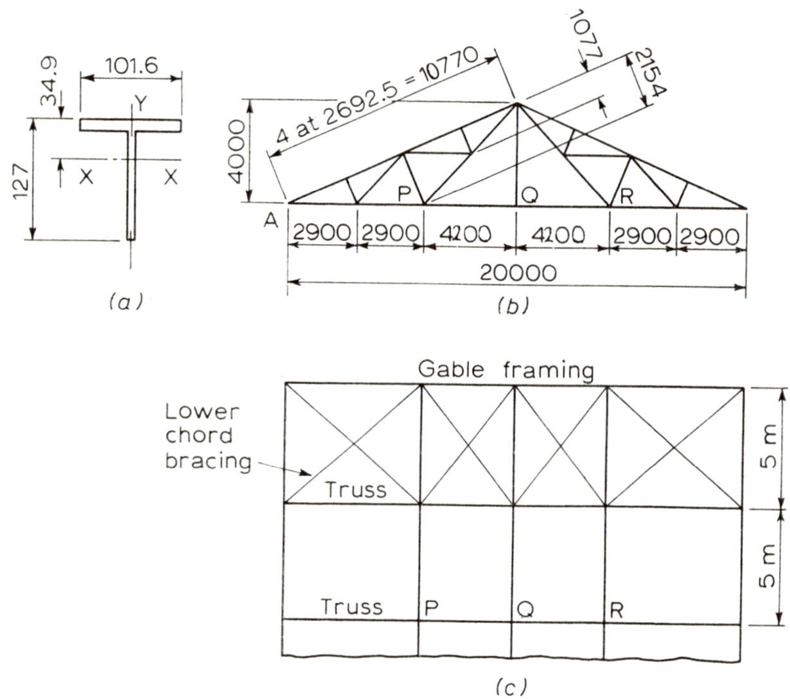

Figure 11.26

Try 102×127×11 kg/m Struct. Tee. The properties from steel tables are:

$$A = 14.2 \quad \text{cm}^2$$

$$r_{YY} = \quad 2.02 \text{ cm}$$

This section is shown in Figure 11.26(a).

When the bottom chord of the truss is in compression under maximum uplift from wind load, lateral supports will be provided at points P, Q and R by the lower chord bracing shown on Figure 11.26(b) and (c). The lengths of the truss members have been calculated and these are shown in (b). Referring to this figure, the effective length for buckling about the YY axis for member AP is 5800 mm.

The permissible compressive stress is given by

$$l/r_{YY} = 5800/20.2 = 287$$

For a member normally acting as a tie in a roof truss but subjected to reversal of stress resulting from wind load, the maximum l/r ratio shall not exceed 350.

$p_c = 12 \text{ N/mm}^2$ from Table 17

$f_c = (15.4\times10^3)/(14.2\times10^2) = 10.9 \text{ N/mm}^2$

102×127×11 kg/m Struct. Tee is satisfactory.

Bottom Chord *Member* A–7

Design loads $= 104.8$ kN—tension

$= -4.7$ kN—compression

Net area $= (104.8\times10^3)/155 = 675 \text{ mm}^2$

Try 102×127×11 kg/m Struct. Tee.

A bolted field joint will be made at joint P, Figure 11.26(b). This is shown in Figure 11.27. Use 20 mm diameter black bolts, the shear value of the joint shown

$$= 6\times25.1 = 150.6 \text{ kN}$$

The holes for the bolts are 22 mm diameter, so the net area of the tee:

$$= 1420-(22\times5.8)-(2\times22\times7.8) = 949.5 \text{ mm}^2$$

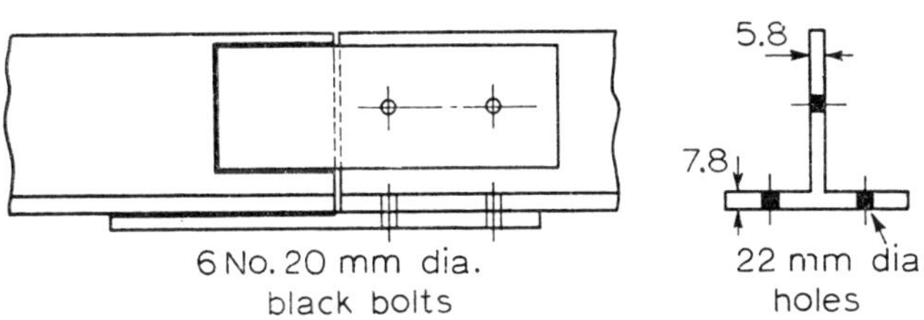

Figure 11.27

The tee is satisfactory as a tension member and is now checked as a strut with a lateral restraint at mid span from the lower chord bracing.

$l/r_{YY} = 4200/20.2 = 208$

$p_c = 22$ N/mm^2 from Table 17

$f_c = (4.7\times10^3)/1420 = 3.3$ N/mm^2

102×127×11 kg/m Struct. Tee is satisfactory.

Internal Struts. Members 1–2 *and* 5–6

Design loads = −14.7 kN—compression

= 2.9 kN—tension

Length = 1077 mm

Try 51×51×6.3 Angle. The properties are:

$A = 6.08$ cm^2

$r_{min} = 0.99$ cm

The effective length when the angle is connected by welding at the ends is 0.85 of the length between intersections.

$l/r_{min} = (0.85\times1077)/9.9 = 92.5$

$p_c = 88$ N/mm^2

Safe load $= (88\times608)/10^3 = 53.5$ kN

The safe load in tension taken from the safe load tables for the angle connected through one leg is 81 kN. A smaller size angle could be used here, but for practical reasons the size selected will be adopted.

51×51×6.3 Angle is satisfactory.

Internal Strut Member 3–4

Design loads = −30.4 kN—compression

= 5.4 kN—tension

Length = 2154 mm

Try 64×64×6.2 Angle. The properties are:

$A = 7.59$ cm^2

$r_{min} = 1.25$ cm

$l/r_{min} = (0.85\times2154)/12.5 = 146$

$p_c = 42$ N/mm^2

Safe load $= (42\times759)/10^3 = 31.8$ kN

Safe load in tension = 102 kN

64×64×6.3 Angle is satisfactory.

Internal Ties *Members* 2–3, 4–5

Design loads = 21.3 kN—tension
= −3.4 kN—compression

Length = 2900 mm

Try 51×51×6.3 Angle.
Safe load in tension = 81 kN

$l/r_{min} = (0.85 \times 2900)/9.9 = 250$

$p_c = 15\ \text{N/mm}^2$

Safe load in compression = −9.11 kN
51×51×6.3 Angle is satisfactory.

Internal ties *Members* 4–7, 6–7
Design for the maximum load in member 6–7.

Design loads = 62.7 kN—tension
= −9.5 kN—compression

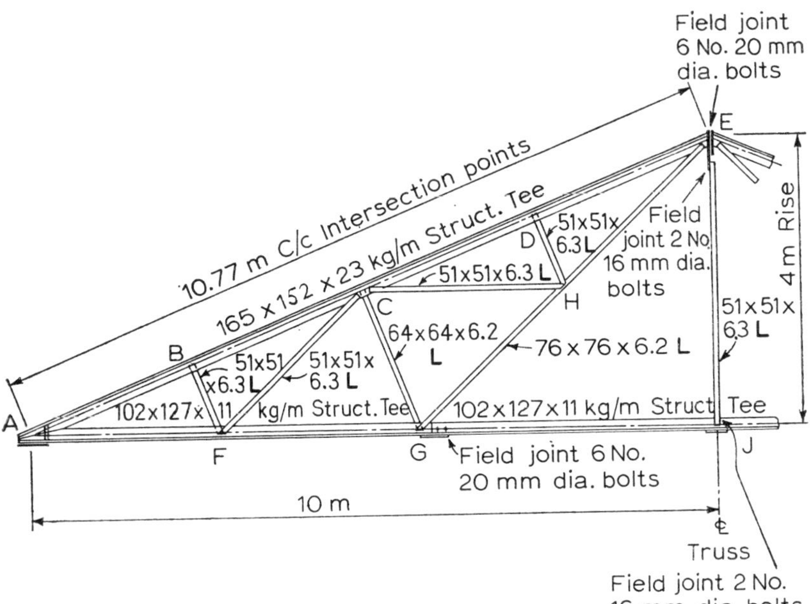

Figure 11.28 (Joint details are shown in Figure 11.29(a) and (b). All bolts are black bolts to diameters shown. Purlin cleats are not shown.)

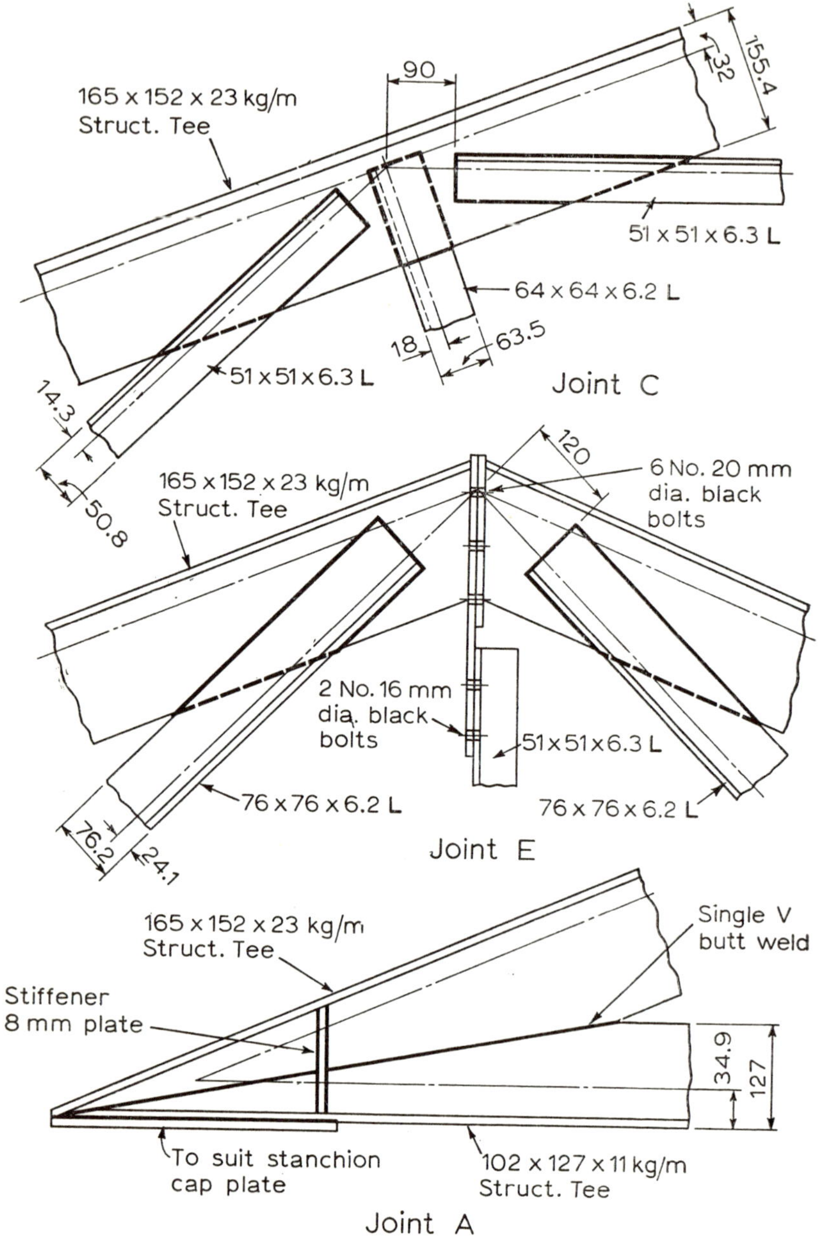

(*a*)

Figure 11.29

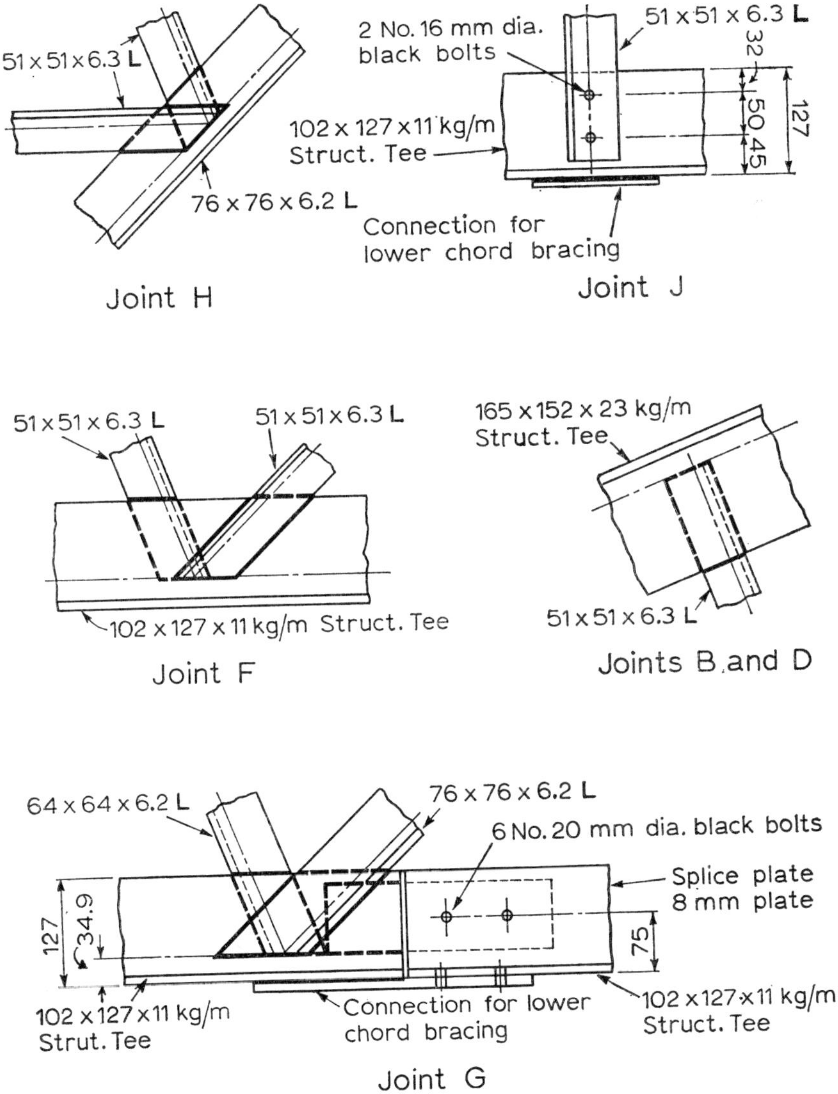

(b)

Figure 11.29

Length = 2900 mm—buckling of member 6–7 about its weakest axis

= 5800 mm—the two members 4–7 and 6–7 buckling laterally to the plane of the truss

Try 76×76×6.2 Angle. The properties are:

$A = 9.12\ \text{cm}^2$

$r_{min} = 1.49\ \text{cm}$

$r_{XX} = r_{YY} = 2.33\ \text{cm}$

For member 6–7, buckling about its weakest axis:

$$l/r_{min} = (0.85 \times 2900)/14.9 = 165$$

For members 6–7, 4–7, buckling laterally:

$l/r_{XX} = (0.85 \times 5800)/23.3 = 212$

$p_c = 21\ \text{N/mm}^2$

Safe load in compression = 19.1 kN

Safe load in tension = 123 kN

76×76×6.3 Angle is satisfactory.

Internal hanger *Member* 7–8 force is zero.
Use 51×51×6.3 Angle.
Length = 4000 mm

$l/r_{min} = (4000 \times 0.85)/9.9 = 342$

(f) Welding

Use 5 mm fillet weld, strength 400 N/mm. The welding on each joint is to be continuous to seal the two members. The welding on all internal joints, as shown in Figure 11.29, is more than sufficient to transmit the loads in the members.

(g) Sketch

A sketch of the truss is shown in Figure 11.28 and joint details are shown in Figure 11.29(a) and (b).

11.8 PROBLEMS

1 A flat roof building of 18 m span has 1.5 m deep trusses at 4 m centres. The trusses carry purlins at 1.5 m centres. The total roof load including imposed load, decking and steel is 1.5 kN/m².

(a) Analyse the truss by joint resolution.
(b) Design the truss using welded connections and bolted field splices.

2 Floor girders are 15 m span and at 3 m centres. The total floor load is 7.5 kN/m² which includes the imposed and dead loads. Design the floor girders using lattice girder construction in rectangular hollow sections.

3 A roof truss is shown in Figure 11.30. The trusses are at 6 m centres and the height to the eaves is 5 m. The length of the building is 35 m. The roof loading is:

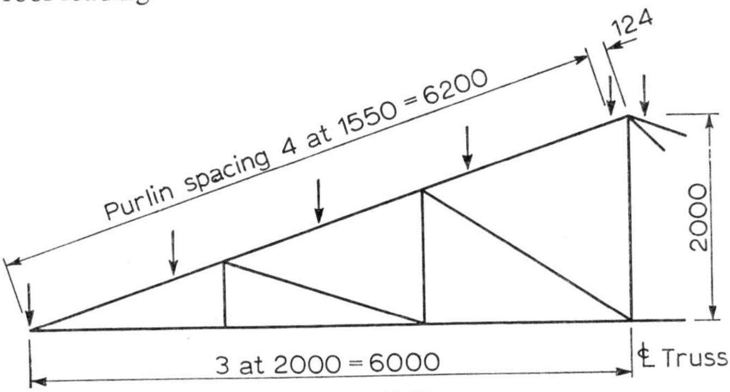

Figure 11.30

Dead load = 61 kg/m² that is 0.6 kN/m² measured on the slope
Imposed load = 0.75 kN/m² measured on plan

The wind load is to be in accordance with CP3, Chapter V, Part 2, 1970. The building in situated in the North East of England on the outskirts of a city.

(a) Analyse the truss for the roof loads.
(b) Analyse the top chord for the loading due to the purlin spacing shown. The total load from roof and purlins is 1.2 kN/m²
(c) Design the top chord.

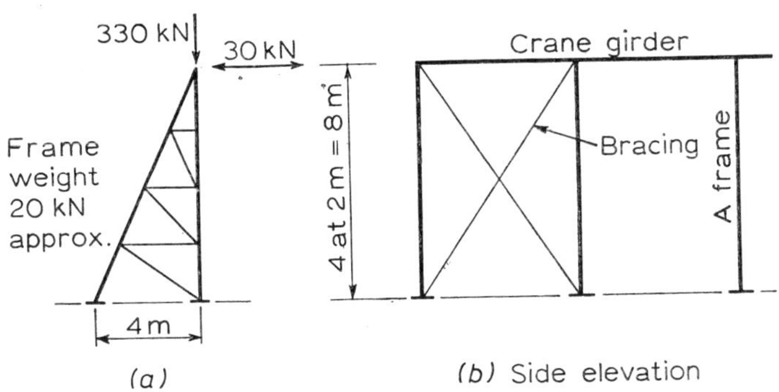

Figure 11.31 (a) 'A' frame for an open crane runway. (b) Side elevation of runway

4 An A-frame for an open crane runway is shown in Figure 11.31(a). The frame is restrained longitudinally at B. A side elevation of the runway is shown in (b) on the figure where the bracing in the end bay which provides longitudinal restraint is shown. The crane loading applied to the A frame is shown on (a) on the figure. Design members AB and BC of the frame.
What is the purpose of the internal members?

5 A cantilever truss and loading is shown in Figure 11.32.

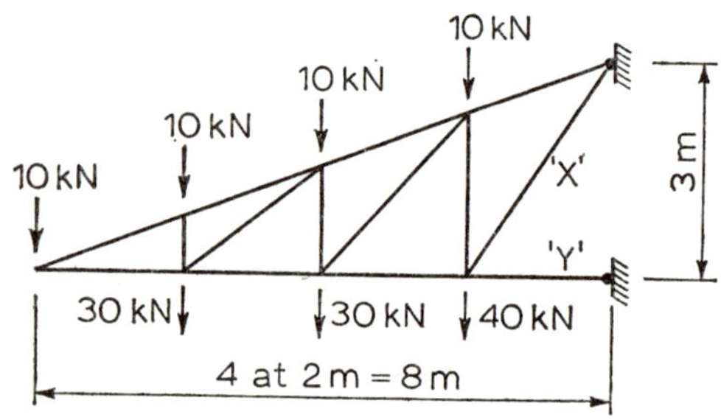

Figure 11.32 Contilever truss and loading

The bottom chord is supported laterally at the node points. Wind loading need not be considered.

(a) Find the forces in members 'X' and 'Y'.
(b) Check that the following sections are satisfactory.
Member X—89×76×7.8 Angle
Member Y—2 No. 76×76×7.8 Angles.

6 The section through a building is shown in Figure 11.33. The roof trusses are supported on stanchions A and B and cantilever out to the front of the building. The front has roller doors that run on tracks on

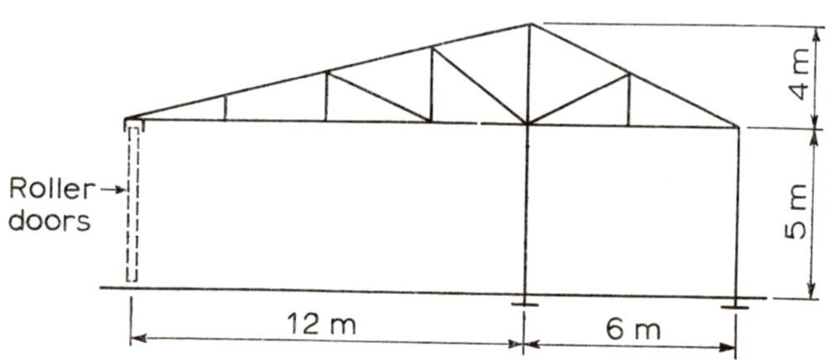

Figure 11.33

the floor. The frames are at 6 m centres and the length of the building is 48 m. The roof load is:

Total dead load = 81.5 kg/m^2 that is, 0.8 kN/m^2

Imposed load = 0.75 kN/m^2

These loads are measured on plan.
The wind load is to be in accordance with CP3, Chapter V, Part 2, 1970. The basic wind speed is 45 m/s and the location is in the suburbs of a city. The structure should be analysed for the conditions of door opened and closed. Analyse and design the truss shown for the worst loading condition.

12

Bracing

Bracing is generally required to resist horizontal loading in pin-jointed buildings. The buildings would be unstable and liable to collapse if the bracing was not provided. The bracing and building members form trusses to carry the loads.

12.1 ARRANGEMENT AND LOADING

12.1.1 Bracing for Single-Storey Industrial Buildings

The bracing for a single-storey building is shown in (a) in Figure 12.1. The internal frames resist transverse wind load by bending in the cantilever columns. However, the gable frame can be braced to resist this load as shown. The wind blowing longitudinally along the building causes pressure and suction forces on the windward and leeward gables respectively, and wind drag on the roof and walls. These forces are resisted by the roof and wall bracing shown.

If the building contains a crane, an additional load due to the longitudinal crane surge has to be taken on the wall bracing. Details for this loading are given in Clause 7 of BS 449. The force is 5% of the static wheel loads. A suitable bracing system is shown in Figure 12.1(b). Note that a longitudinal bracing member is required in the end bay separate to the crane girder. In any case the crane girder is inside the stanchion line, while the plane of the bracing is near the outside flange of the stanchion.

12.1.2 Bracing for a Multi-Storey Building

The bracing for a multi-storey building is shown in Figure 12.2. Vertical bracing is required on all elevations to stabilise the building. The wind loading is applied at floor level. It is usually assumed that

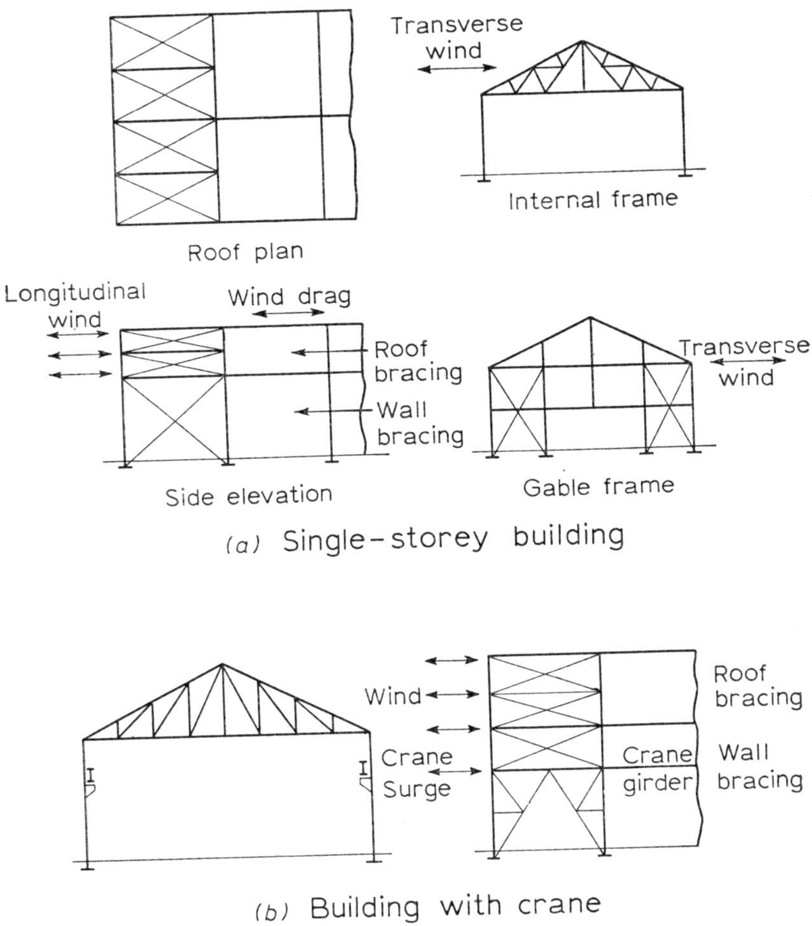

Figure 12.1

the floor slabs will transmit the load on the internal stanchions to the trusses at the ends of the building. If the building frame and cladding is erected before the floors are constructed, floor bracing must be supplied as shown in (c). This can be removed when the floors are in place. Floor bracing is also required if the floor slabs are not continuous.

In rigid jointed buildings the wind loading is resisted by bending in the beams and stanchions. Treatment of these frames is outside the scope of this book. However, if the joints are made with high-strength friction-grip bolts, then temporary bracing may be required to stabilise the frame before the joints have been completed.

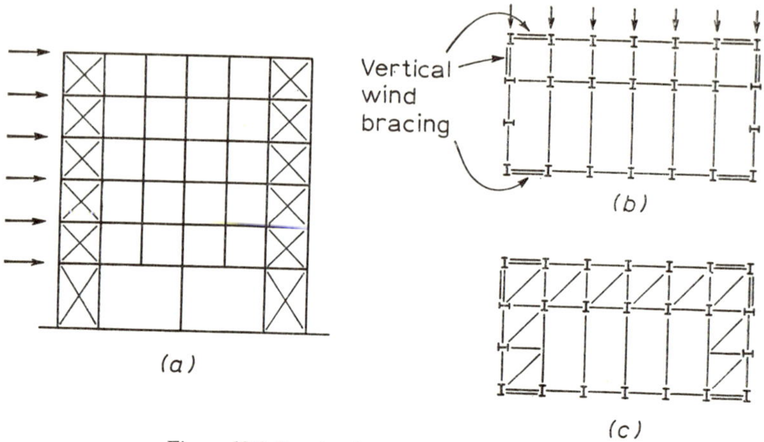

Figure 12.2 Bracing for a multi-storey building

12.2 DESIGN OF BRACING

The form of bracing can be single diagonal members or cross-bracing. The loading is generally due to wind or crane surge and is reversible. Single bracing must be designed to carry loads in tension and compression. With cross-bracing, only the member in tension is assumed to be effective. The diagonal member that would be in compression is ignored. A cross-bracing system, taking account of all the members, is redundant, but if one set of diagonal members is ignored, the remaining truss is statically determinate.

The bracing members are the diagonal web members of the truss formed with the main building members, stanchions, purlins, ridge or eaves struts, etc. The forces in the bracing members are found in the same manner as those in a truss. The members are designed as ties or struts as set out in Chapter 11. The design of simple bracing systems are set out in the examples that follow. Secondary stresses due to bending are introduced where sheeting rails, etc., apply wind loads to stanchions or other frame members at points other than the node points of the bracing truss. These effects are usually ignored. Bracing members are often very lightly loaded and minimum-size angles or tees are chosen for practical reasons.

12.3 DESIGN OF BRACING FOR A SINGLE-STOREY BUILDING

The end frame and bracing in the end bay of a single-storey building are shown in Figure 12.3. The end bay at the other end of the building

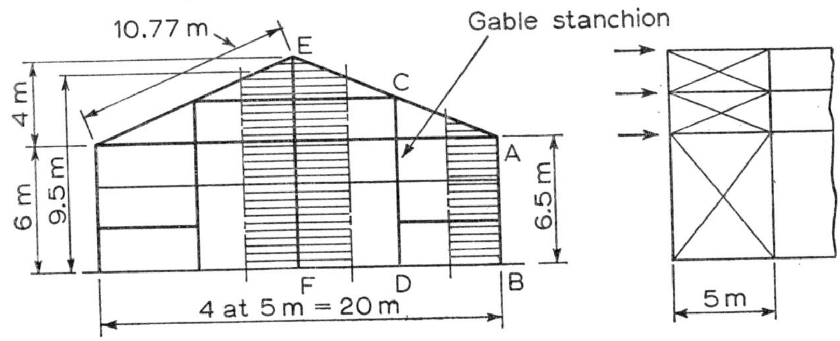

Figure 12.3 End frame and bracing in end bay of a single-storey building

is also braced. The length of the building is 50 m and the truss and stanchion frames are at 5 m centres. The building dimensions are shown on the figure. Particulars of the wind loading are set out below. Design the roof and wall bracing to resist the longitudinal wind loading.

(a) Wind loading Refer to CP3, Chapter V, Part 2, 1970.

Basic wind speed		= 45 m/s
Topography factor	S_1	= 1
Ground roughness		= 3
Building size—Class B		
Height to ridge	H	= 10 m
Factor	S_2	= 0.74—Table 3
Factor	S_3	= 1.0
Design wind speed		= 45×0.74
		= 33.4 m/s
Dynamic pressure	q	= 682 N/m² (from Table 4)

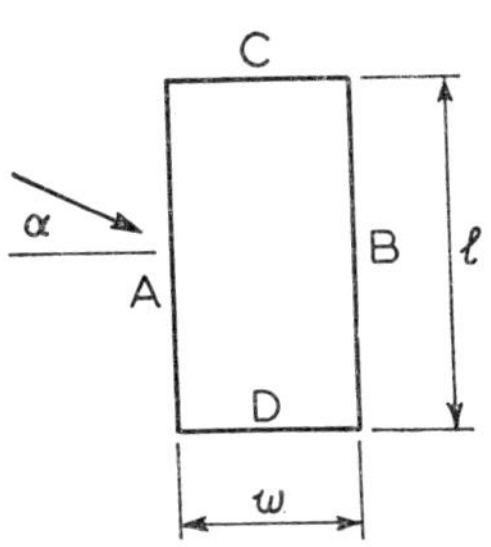

Wind angle α	C_{pe} for surface	
	C	D
0°	−0.6	−0.6
90°	+0.7	−0.1

Figure 12.4

The pressure coefficients C_{pe} on the end walls are from Table 7.

$$h/w = 6/20 = 0.3 < \tfrac{1}{2}$$
$$l/w = 50/20 = 2.5; \quad \tfrac{3}{2} < l/w < 4$$

The coefficients are set out in Figure 12.4.

The overall pressure coefficient is 0.8. This is not affected by variation in internal pressure. Alternatively, the force coefficient C_f given in Clause 7.3 could be used to find the pressure on the building as a whole.

$$\text{Wind pressure} = (0.8\times682)/1000 = 0.53 \text{ kN/m}^2$$

The method for calculating the frictional drag is given in Clause 7.4.

Cladding—corrugated plastic coated steel sheet.

Factor $C_{f'} = 0.02$ for surfaces with corrugations across the wind direction.

Refer to Figure 12.5

$$d/h = 50/6 = 8.3 > 4$$
$$d/b = 50/20 = 2.5 < 4$$

The ratio d/h is greater than 4, so the drag must be evaluated.

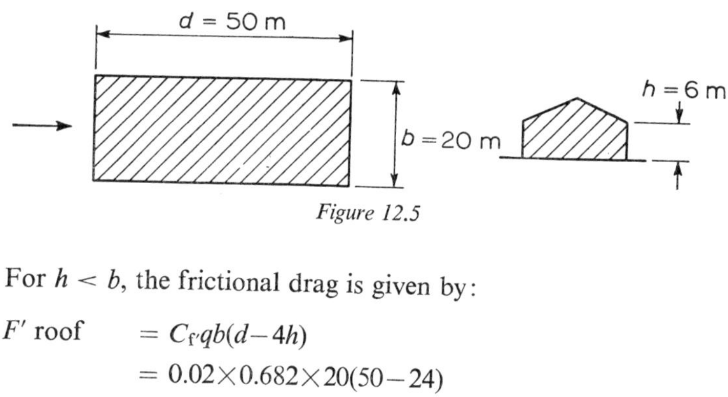

Figure 12.5

For $h < b$, the frictional drag is given by:

$$\begin{aligned} F' \text{ roof} &= C_{f'}qb(d-4h) \\ &= 0.02\times0.682\times20(50-24) \\ &= 7.2 \text{ kN} \\ F' \text{ walls} &= C_{f'}q2h(d-4h) \\ &= 0.02\times0.682\times12\,(50-24) \\ &= 4.26 \text{ kN} \end{aligned}$$

The total loading is the sum of the wind load on the gables and the frictional drag. This load is divided equally between the bracing at each end of the building.

(b) Loads on the bracing Refer to Figure 12.3

Point E

Load from the wind on the end wall gable stanchion EF

$$= 9.5\times5\times\tfrac{1}{2}\times0.53 = 12.6 \text{ kN}$$

Reaction at E, the top of the stanchion = 6.3 kN

The load at E from wind drag on the roof

$= \frac{1}{2}\times7.2\times\frac{1}{4}$ = 0.9 kN

Total load at E = 7.2 kN

Point C

Load from the wind on the end wall gable stanchion CD

$= 8\times5\times\frac{1}{2}\times0.53$ = 10.6 kN

Reaction at C, the top of the stanchion = 5.3 kN

Load at C from wind drag on the roof = 0.9 kN

Total load at C = 6.2 kN

Point A

Load from the wind on building stanchion AB

$= 2.5\times6.5\times\frac{1}{2}\times0.53$ = 4.3 kN

Reaction at A, the top of the stanchion = 2.15 kN

Load at A from wind drag on the roof and wall

$= (\frac{1}{8}\times7.2\times\frac{1}{2})+(\frac{1}{8}\times4.26)$ = 0.98 kN

Total load at A = 3.13 kN

(c) Roof bracing

The loading on the truss and the forces in the bracing members are shown in Figure 12.6.

Note that the members of the cross-bracing in compression have not been shown.

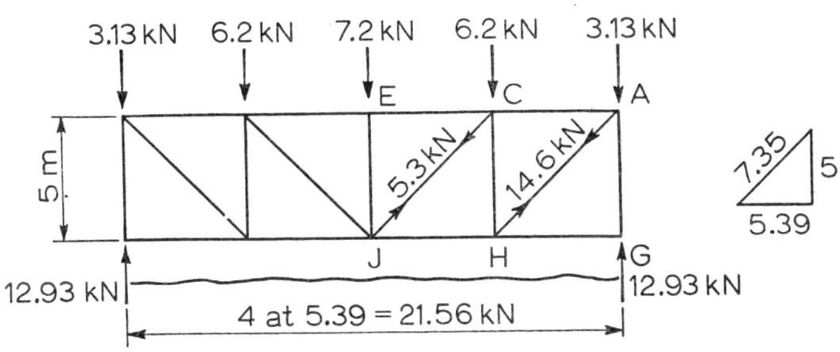

Figure 12.6

The maximum load in member AH = 14.6 kN
Use 51×51×6.3 Angle
The safe load for this member from safe load tables = 59 kN
This allows for 1 No. 18 mm diameter hole in the angle leg.
Make all members the same section.

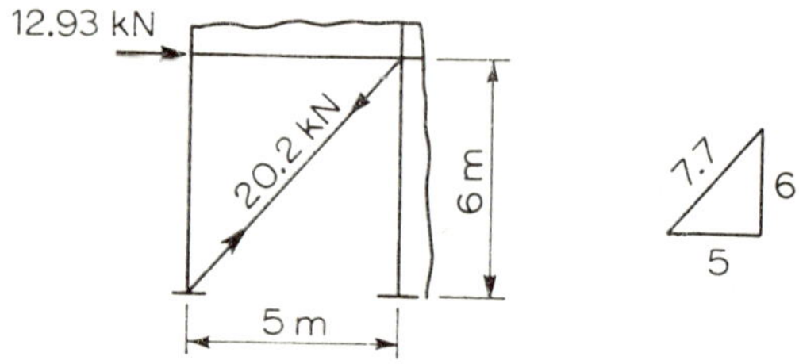

Figure 12.7

(d) Wall bracing

The loading on the truss and the force in the bracing is shown in Figure 12.7.

Use 51×51×6.3 Angle for these members.

12.4 DESIGN OF BRACING FOR AN OFFICE BUILDING

The framing plans for an office building are shown in Figure 12.8. The floors and roof are cast *in situ* in the form of reinforced concrete slabs. It is assumed that the slabs will transmit the wind load from the internal stanchions to the end bracing. Horizontal bracing at each floor level will not be used.

Design the wind bracing.

(a) Wind Loading

The data for the wind loading are:

Basic wind speed		$V = 45$ m/s
Topography factor		$S_1 = 1.0$
Ground roughness		$= 4$
Class	—B	
Factor S_2		
Roof	$H = 13$ m	$S_2 = 0.66$
2nd floor	$H = 11$ m	$S_2 = 0.63$
Ist floor	$H = 7$ m	$S_2 = 0.58$

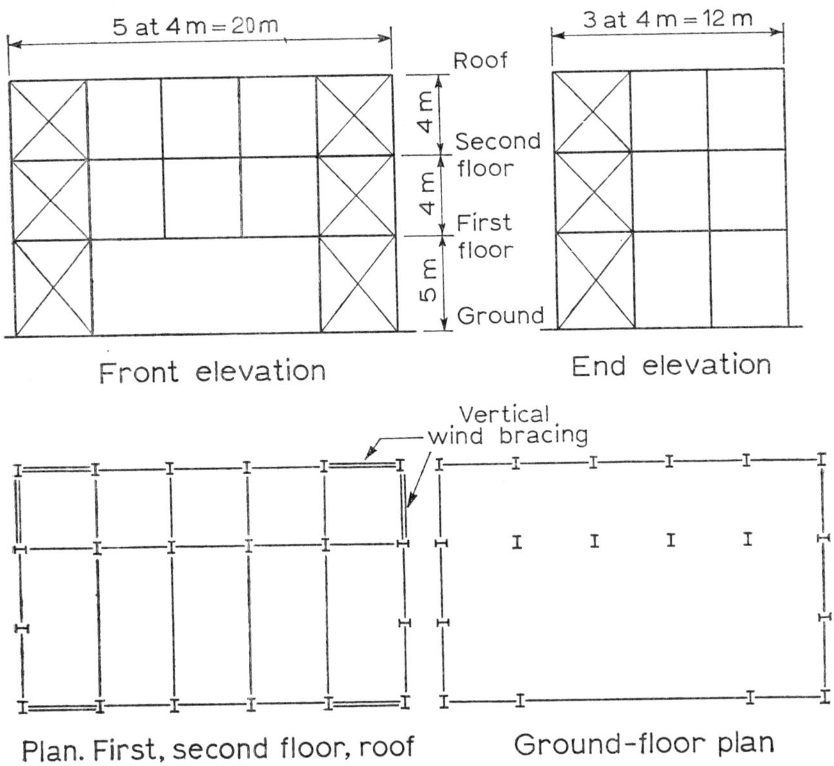

Figure 12.8 Framing plans for an office building

The height of the building is divided into parts for the measurement of H, as follows:

Roof—to the top of the building
Second floor—mid way between the second floor and roof
First floor—mid way between the first and second floors
Refer to Figure 12.10

Statistical factor $S_3 = 1$

Design wind speeds and dynamic pressures are:

Roof	$V_s = 45\times0.66 = 29.6$ m/s;	$q = 536$ N/m^2
Second floor	$V_s = 45\times0.63 = 28.4$ m/s;	$q = 496$ N/m^2
First floor	$V_s = 45\times0.58 = 26.1$ m/s;	$q = 418$ N/m^2

The force coefficients for wind on the building as a whole are taken from Table 10 of CP3, Chapter V, Part 2, 1970. These are shown in Figure 12.9 for wind transverse and longitudinal.

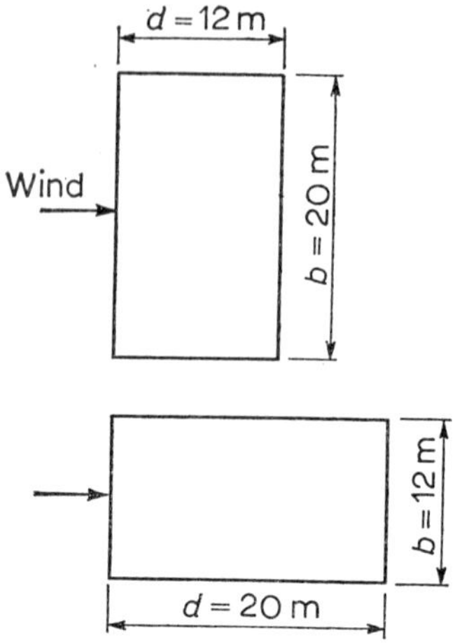

ℓ = length of building = 20 m
w = width of building = 12 m
h = height of building = 13 m

Wind transverse		
ℓ/w	b/d	C_f for $h/b = 0.65$
1.66	1.66	0.981

Wind longitudinal		
ℓ/w	b/d	C_f for $h/b = 1.08$
1.66	0.6	0.816

Figure 12.9

$$\text{Force} = C_f q A_e$$

where A_e is the effective frontal area of the structure. Wind drag on the roof or walls need not be taken into account as neither the ratio d/h nor d/b is greater than 4.

(b) Transverse bracing

The loads at the floor levels are:

$$P = 0.981 \times 536 \times 2 \times 10 \times 1/10^3 = 10.6 \text{ kN}$$
$$Q = 0.981 \times 496 \times 4 \times 10 \times 1/10^3 = 19.5 \text{ kN}$$
$$R = 0.981 \times 418 \times 4.5 \times 10 \times 1/10^3 = 18.5 \text{ kN}$$

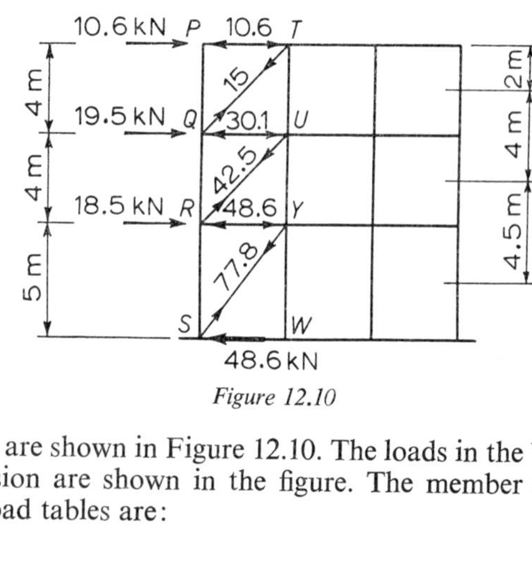

Figure 12.10

These loads are shown in Figure 12.10. The loads in the bracing members in tension are shown in the figure. The member sizes selected from safe-load tables are:

Members *QT*, *RU*. Load 42.5 kN
51×51×6.3 Angle. Safe load = 59 kN
Allows for 1 No. 18 mm diameter hole.
Member *SU*. Load 77.8 kN
64×64×7.9 Angle. Safe load = 94 kN
Allows for 1 No. 22 mm diameter hole

(c) Longitudinal bracing
The loads at the floor levels are:

$D = 0.816 \times 536 \times 2 \times 6 \times 1/10^3 = 5.26$ kN
$E = 0.816 \times 496 \times 4 \times 6 \times 1/10^3 = 9.7$ kN
$F = 0.816 \times 418 \times 4.5 \times 6 \times 1/10^3 = 9.2$ kN

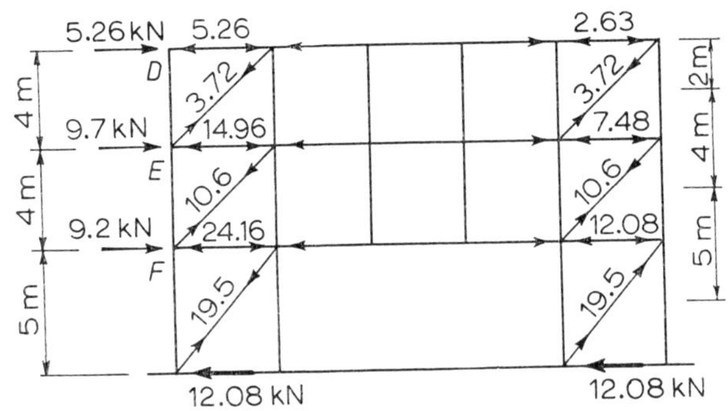

Figure 12.11

These loads are shown in Figure 12.11. The loads are divided betwen the bracing at each end of the building. The loads in the bracing members in tension are shown on the figure. The loads in these members are small. The following section will be used for all members:

51×51×6.3 Angle. Safe load = 59 kN
Allows for 1 No. 18 mm diameter hole in the angle leg.

12.5 PROBLEMS

1 The end framing and bracing for a single-storey building are shown in Figure 12.12. The location of the building is on an industrial estate

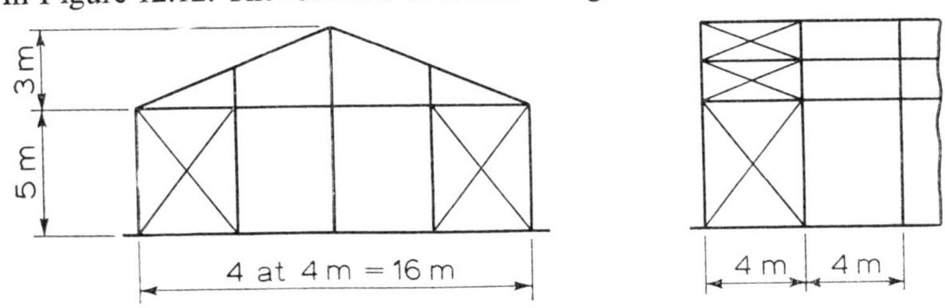

Figure 12.12

on the outskirts of a city in the north east of England. The length of the building is 32 m. Design the bracing in the end frame, roof and side wall. The wind loads are to be in accordance with CP3, Chapter V, Part 2, 1970.

2 The framing for a square tower building is shown in Figure 12.13. The bracing is similar on all four faces. The building is located in a city centre in an area where the basic wind speed is 50 m/s. Design the bracing.

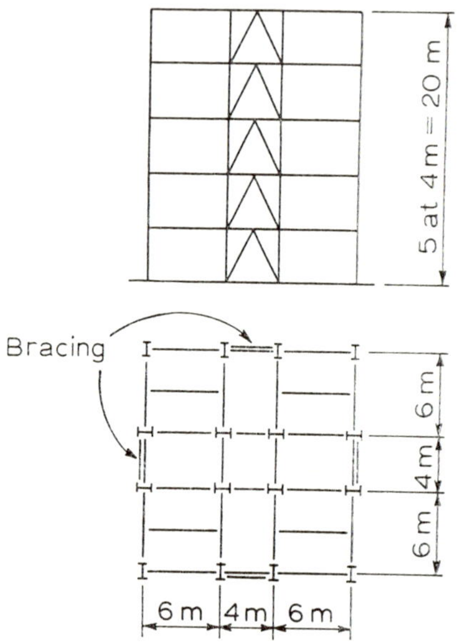

Figure 12.13 Framing for a square tower building

3 The side wall bracing for wind and crane surge is shown in Figure 12.14. The loading from the wind and crane surge at eaves and crane girder level is shown on the figure. Design the bracing.

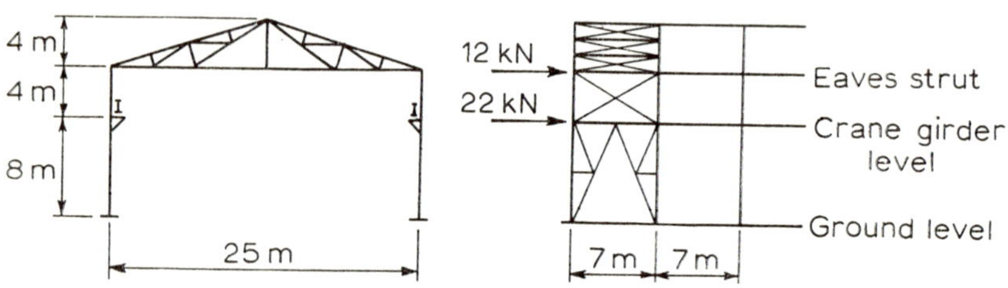

Figure 12.14

13

Detailing

Drawings are the means by which the requirements of architects and engineers are communicated to the fabricators and erectors. Drawings must be presented in an acceptable manner. Detailing is given for selected structural elements.

13.1 DRAWINGS

Drawings are needed to show the layout and to describe and specify the requirements of a building. They show the location, general arrangement and details for fabrication and erection. Drawings are also used for estimating quantities and cost and for making material lists for ordering materials.

Sufficient information must be given on the designers' sketches for the draughtsman to make up the arrangement and detail drawings.

A classification of drawings is set out below:

Site or location plans These show the location of the building in relation to other buildings, site boundaries, streets, roads, etc.
General arrangement This consists of plans, elevations and sections to set out the function of the building. These show locations and leading dimensions for offices, rooms, work areas, machinery, cranes, doors, services, etc. Materials and finishes are specified.
Marking plans These are the framing plans for the steel-frame building showing the location and mark numbers for all steel members in the roof, floors and various elevations.
Foundation plans These show the setting out for the stanchion bases and holding-down bolts and should be read in conjunction with detail drawings of the foundations.

Sheeting plans These show the arrangement of sheeting and cladding on buildings.
Key plan If the work is set out on various drawings, a key plan may be provided to show the portion of work covered by the particular drawing.
Detail drawings. These show the details of structural members and give all information regarding materials, sizes, welding, drilling, etc. for fabrication. The mark number of the detail refers to the number on the marking plan.

Detail drawings and marking plans will be dealt with here.

13.2 GENERAL RECOMMENDATIONS

13.2.1 Scales, Drawing Sizes and Title Blocks

The following scales are recommended:

Site, location, key plans	1 : 500 1 : 200

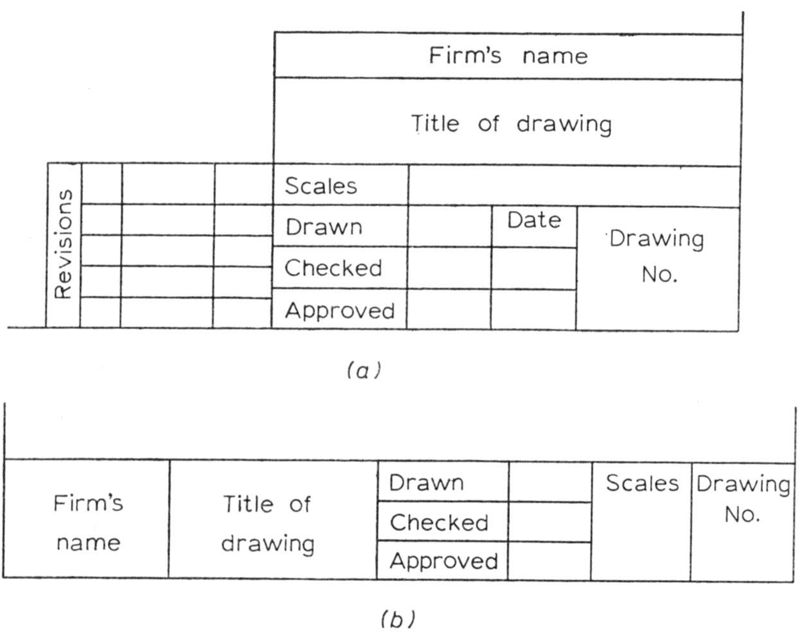

Figure 13.1 Typical title blocks

General arrangement	1 : 200 1 : 100 1 : 50
Marking plans	1 : 200 1 : 100
Detail drawings	1 : 25 1 : 10 1 : 5
Enlarged details	1 : 5 1 : 2 1 : 1

The following drawing sizes are used:

A4 210×297 —Sketches

A3 297×420 —Details

A2 420×594 —General arrangement, details

A1 594×841 —General arrangement, details

A0 841×1189—General arrangement, details

Title blocks on drawings vary to suit the requirements of individual firms and authorities. Typical title blocks are shown in Figure 13.1.

Material lists can either be shown on the drawing or on separate A4 size sheets. These generally give the following information:

Item or mark number
Description
Material
Number off
Weight
Etc.

13.2.2 Lines, Sections, Dimensions and Lettering

Recommendations regarding lines, sections and dimensions are shown in Figure 13.2. With regard to lettering, capitals should be used for titles of views and lower case for notes.

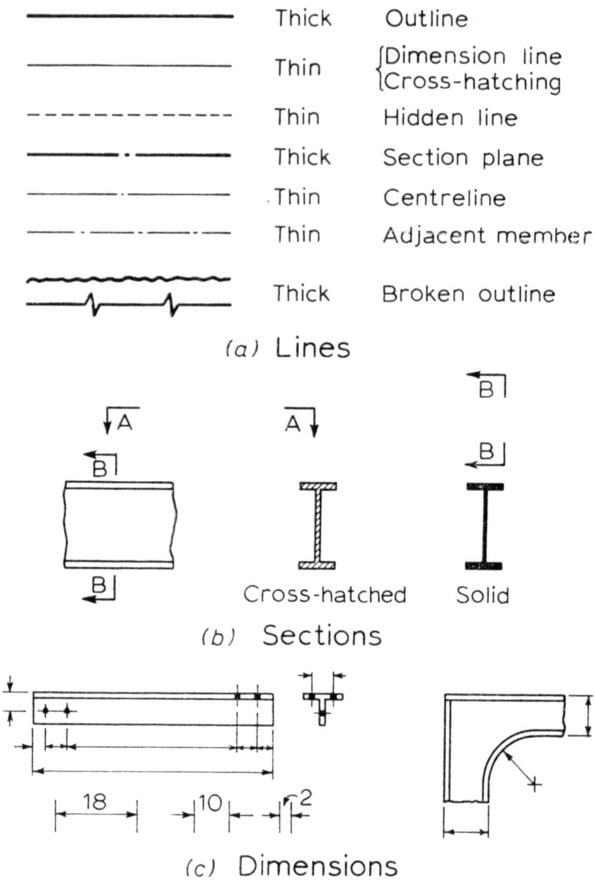

Figure 13.2 Recommendations for lines, sections and dimensions

13.3 STEEL SECTIONS

Rolled and formed steel sections are represented on steelwork drawings as set out in Figure 13.3.

The first two figures indicate the size of section, for example, depth and breadth. The last figure indicates the weight in kg/m for beams, columns, channels and tees. For angles and hollow sections the last figure gives the thickness of steel. With channels or angles the name may be written or the section symbols used as shown. For circular hollow sections, the outside diameter and wall thickness are given.

Section		Reference	Example
niversal beam	I	UB	610 x 178 x 91 kg/m UB
niversal column	I	UC	203 x 203 x 89 kg/m UC
oist	I	RSJ Joist	203 x 102 x 25.33 kg/m Joist
hannel	[	Channel	254 x 76 x 28.29 kg/m [
ngle	L	Angle	152 x 76 x 12.6 L
ee	T	Tee	178 x 305 x 46 kg/m Struct. Tee
ectangular hollow section	□	RHS	152.4 x 101.6 x 6.35 RHS
ircular hollow section	○	CHS	76.2 O.D. x 6.35 CHS

Figure 13.3 Representation of rolled and formed steel sections

Built-up sections can be shown by either

(a) true section for large scale views

or

(b) diagrammatically by heavy lines with the separate plates and sections separated for clarity for small scale views. Here only the depth and breadth of the section may be true to scale.

These two cases are shown in Figure 13.4. The section is often shown in the middle of a member inside the break lines in the length as shown in (c) on the figure. This saves having to draw a separate section.

Beams may be represented by lines, and stanchions by small-scale

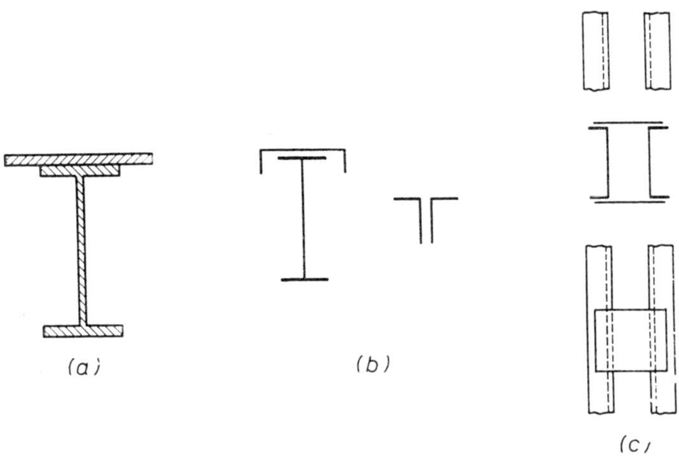

Figure 13.4 Representation of built-up sections

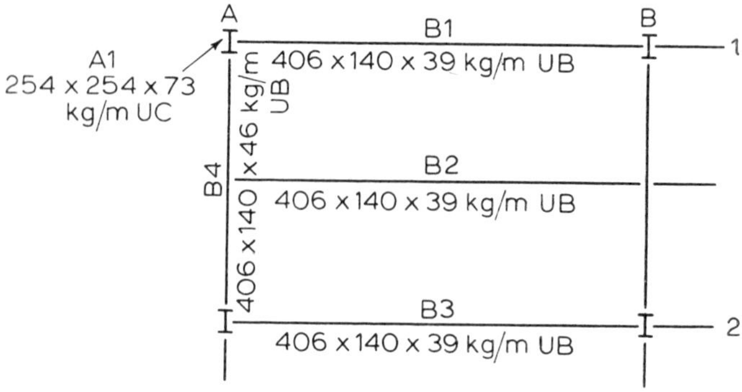

Figure 13.5 Representation of beams and stanchions

sections in heavy lines as shown in Figure 13.5. The mark numbers and sizes are written on the respective members. This system is used for marking plans.

13.4 GRIDS AND MARKING PLANS

Marking plans for single-storey buildings present no difficulty. Members are marked in sequences as follows:

Stanchions A1, A2, ... See grid referencing below.
Trusses T1, T2, ...
Crane girders CG1, CG2, ...
Purlins P1, P2, ...
Sheeting rails SR1, SR2, ...
Bracing B1, B2, ...
Gable stanchions GS1, GS2, ...
Etc.

Various numbering systems are used to locate beams and stanchions in multi-storey buildings. Two schemes an outlined below:

1 In plan, the stanchion grid is marked A, B, C... in one direction and 1, 2, 3... in the direction at right angles. Stanchions are located A1, C2, ...
Floors are numbered A, B, C... for ground, first, second... respectively.
Floor beams, for example, on the second floor are numbered B1, B2, ...
Stanchion lengths are identified, for example, A4–B is the stanchion on grid intersection A4, length between second and third floors.

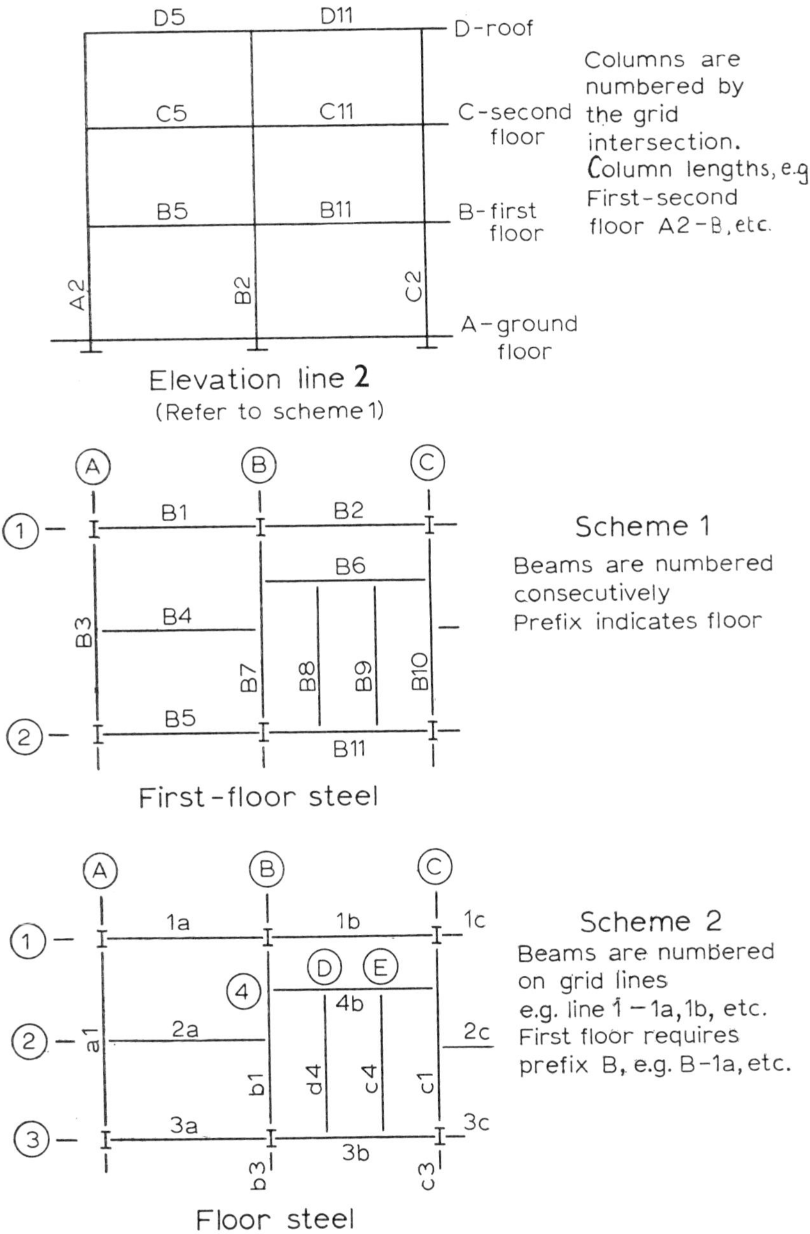
D5
D11
D-roof
C5
C11
C-second floor
B5
B11
B-first floor
A2
B2
C2
A-ground floor
Columns are numbered by the grid intersection.
Column lengths, e.g. First-second floor A2-B, etc.
Elevation line 2
(Refer to scheme 1)
A
B
C
1
2
B1
B2
B3
B4
B5
B6
B7
B8
B9
B10
B11
First-floor steel
Scheme 1
Beams are numbered consecutively
Prefix indicates floor
A
B
C
D
E
1
2
3
4
1a
1b
1c
2a
2c
3a
3b
3c
4b
a1
b1
c1
d4
c4
b3
c3
Floor steel
Scheme 2
Beams are numbered on grid lines
e.g. line 1 – 1a, 1b, etc.
First floor requires prefix B, e.g. B-1a, etc.

Figure 13.6

2 A grid line is required for each beam.
The stanchions are numbered by grid intersections as above. The beams are numbered on the grid lines with a prefix letter to give the floor if required.
For example:

Second floor—grid line 1—C-1a, C-1b,—

Second floor—grid line B—C-b1, C-b2,—

The systems are shown in Figure 13.6. The section size may be written on the marking plan.

13.5 BOLTS

13.5.1 Specification

The types of bolts in their order of usage are:

Black bolts
High-strength friction-grip bolts
Close-tolerance and turned bolts

The British Standards covering bolts are:

BS 4190: 1967—ISO Metric Black Hexagon Bolts and Nuts
BS 4395: Part 1: 1969—High-Strength Friction-Grip Bolts

The strength grade designation should be specified. BS 449 gives stresses for strength grade designation 4.6 for black bolts and close-tolerance bolts for normal usage. Bolts of other strength grade designations are available.

The nominal diameter is given in millimetres. Bolts are designated as M12, M16, M20, M22, M24, M27, M30, etc., where 12, 16 etc., is the diameter in millimetres. The length under the head in millimetres should also be given.

Examples in specifying bolts are as follows:

4 No. 16 mm dia. (or M16) black hex. hd. (hexagon head) bolts, strength grade 4.6×40 mm length
20 No. 24 mm dia. high-strength friction-grip bolts ×75 mm length

The high-strength friction-grip bolts may be abbreviated HSFG. The majority of bolts may be covered by a blanket note. For example:

All bolts M20 black hex. hd.
All bolts 24 mm dia. HSFG unless otherwise noted.

13.5.2 Drilling

The following tolerances for drilling are given in BS 449 Clause 59:

Black bolts—holes to have a maximum of 2 mm clearance for bolt diameters up to 24 mm and 3 mm for bolts over 24 mm diameter.

Close tolerance and barrel bolts—holes shall be drilled to a tolerance of +0.15 mm and −0 mm.

For high-strength friction-grip bolts, holes are drilled the same as for black bolts.

Drilling may be specified by notes as follows on the drawing:

All holes drilled 22 mm dia. for 20 mm dia. black bolts

All holes drilled 26 mm dia., unless otherwise noted.

13.5.3 Designating and Dimensioning

The representation for bolts and holes in plan and elevation on steelwork drawings is shown in Figure 13.7(a). Some firms adopt different

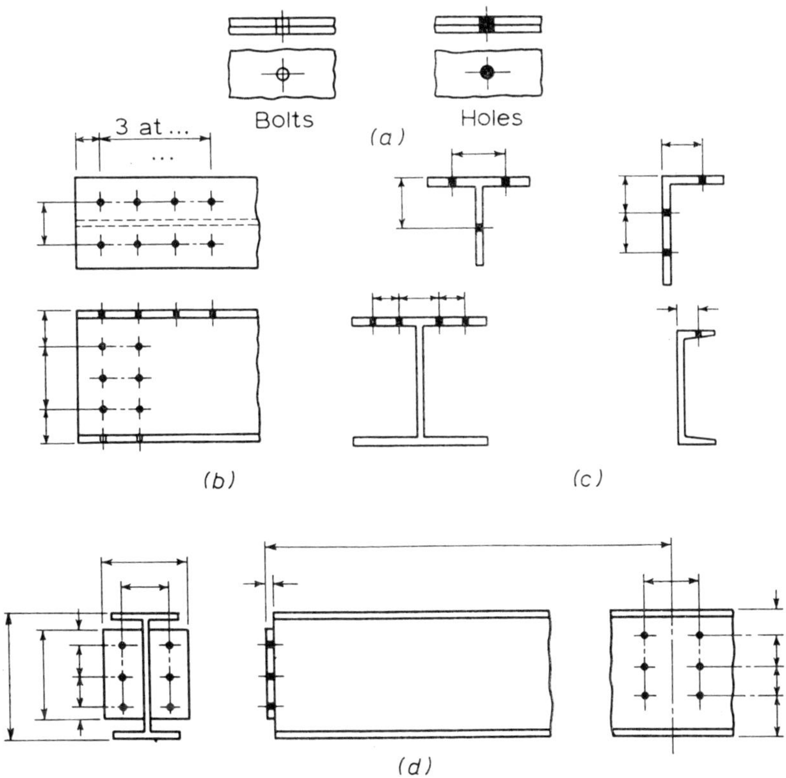

(d)

Figure 13.7

symbols for showing different types of bolts and to differentiate between shop and field bolts. If this system is used a key to the symbols must be given on the drawing.

Gauge lines for drilling for rolled sections are given in the Handbook on Structural Steelwork. Dimensions are given for various sizes of section as shown in (c) in Figure 13.7. Edge distances for drilling for various types of edge, for example, for machined or flame-cut edges are given in Table 21 of BS 449. This table is given in Chapter 2 of this book.

Details must show all dimensions for drilling as shown in Figure 13.7(b) and (d). The holes must be dimensioned off a finished edge of a plate or the back or end of a member. Holes are placed equally about centrelines. Sufficient end views and sections as well as plans and elevations of the member or joint must be given to show the location of all holes, gussets and plates.

13.6 WELDS

As set out in Chapter 2, the two types of weld are butt and fillet welds.

13.6.1 Butt Welds

The types of butt weld are shown in Figure 13.8 with the plate edge preparation and the fit-up for making the weld. The following terms are defined:

T = thickness of plate

g = gap between the plates

R = root face

α = minimum angle

Values of the gap and root face vary with the plate thickness but are of the order of 1 to 4 mm. The minimum angle between prepared faces is generally 50° to 60° for V preparation and 30° to 40° for U preparation. For thicker plates the U preparation gives a considerable saving in the amount of weld metal required. The student should consult the complete details given in:

1 Handbook for Welding Design—Volume 1 ‘Fundamentals’ (published for the Institute of Welding by Pitman)
2 BS 499: 1965: Part 2, ‘Welding Terms and Symbols’.

Welds may be indicated on drawings by symbols from Table 1 of BS 499. These are shown in Figure 13.9. Using these symbols, butt

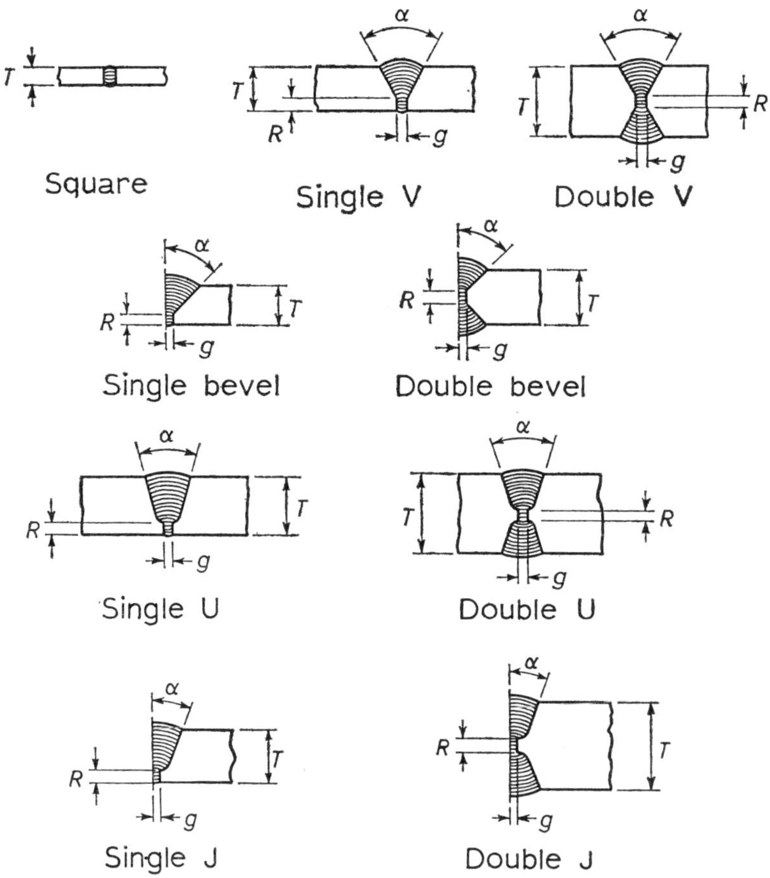

Figure 13.8 Butt welds

welds are indicated on a drawing as shown in Figure 13.10(a). Reference should be made to BS 499 for a complete set of examples using these symbols.

The weld name may be abbreviated, for example, DVBW for double V butt weld. An example of this is shown in (b) on the figure. Finally, the weld may be listed by its full description and an enlarged detail given to show the edge preparation and fit up for the plates. This method is shown in (c) on the figure. Enlarged details should be given in cases where complicated welding is required. Here, detailed instructions from a welding engineer may be required and these should be noted on the drawing.

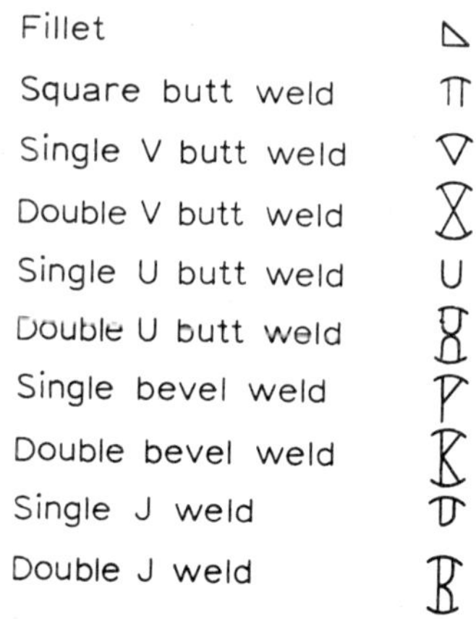

Figure 13.9 Symbols for welds

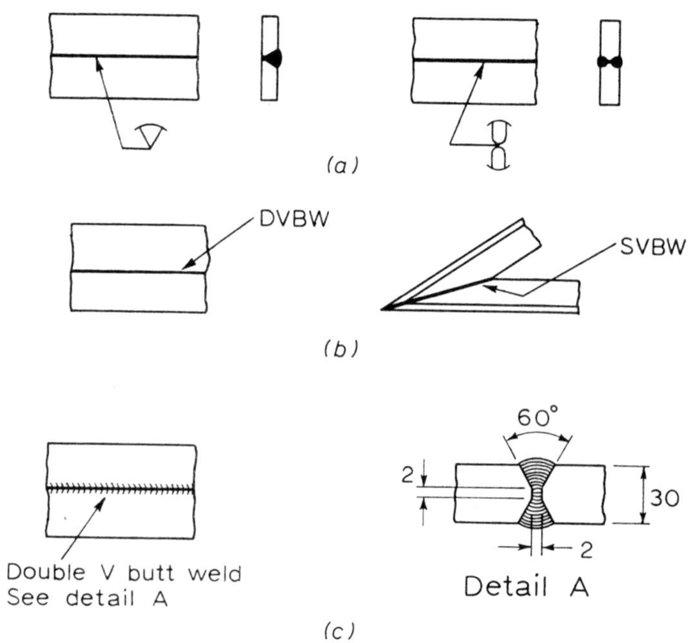

Figure 13.10

13.6.2 Fillett Welds

These welds are triangular in shape. As set out in Chapter 2, the size of the weld is specified by the leg length. Welds may be indicated symbolically as shown in Figure 13.11(a). BS 499 should be consulted for further examples. The weld size and type may be written out in full or

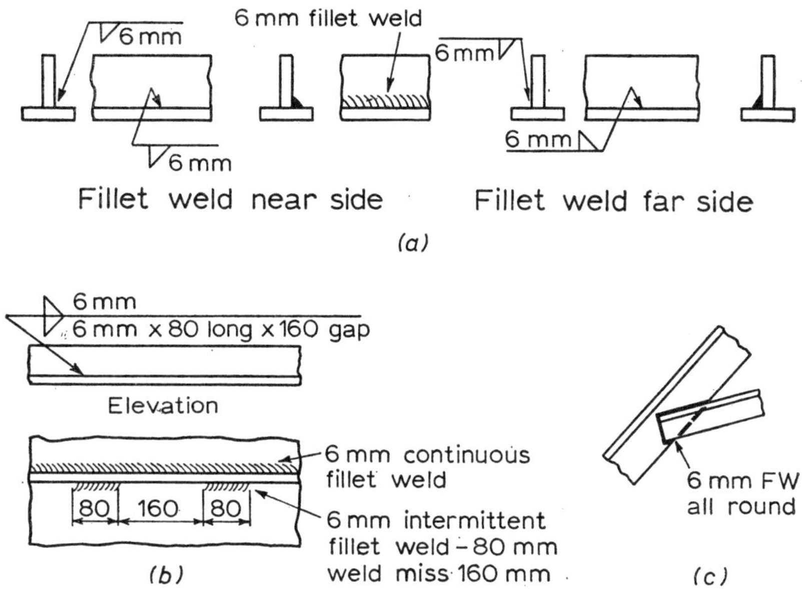

Figure 13.11 Representation of fillet welds

the words fillet weld abbreviated, for example, 6 mm FW. If the weld is of limited length its exact location should be shown and dimensioned. Intermittent weld can be shown by writing the weld size, then two figures which indicate length and space between welds. These methods are shown in (b) on the figure. The welds can also be shown and specified by notes as on the plan view in (b). Finally, a common method of showing fillet welds is given in (c) where thickened times are used to show the weld.

13.6.3 Welds for Tubular Structures

Some information on these welds was given in Chapter 2. Further recommendations for welding and jointing are given in literature provided by the British Steel Corporation, Tubes Division. Detail of joints for pipe lines are given in the *Handbook for Welding Design* mentioned previously. The student should refer to these publications for detailed information as this is beyond the scope of this book.

13.7 BEAMS

Detailing of beams, purlins and sheeting rails is largely concerned with showing the length, end joints, welding and drilling required. A typical example is shown in Figure 13.12. Sometimes it is necessary to show the connecting member and this may be shown by chain dash lines as shown in (b) in the figure.

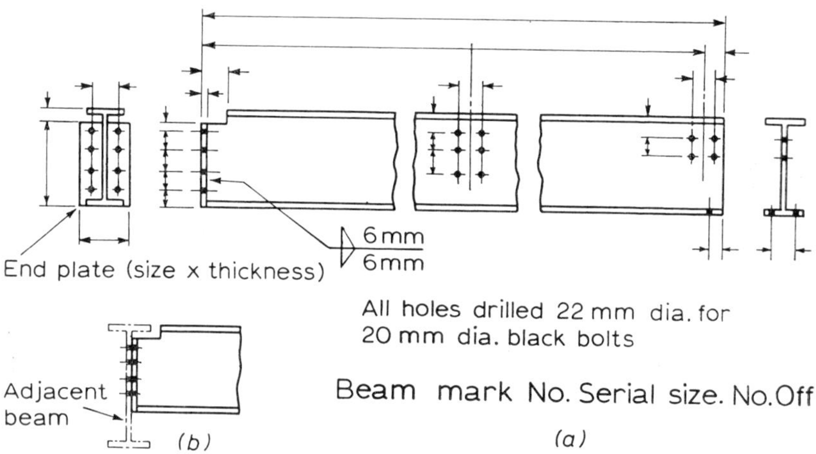

Figure 13.12

13.8 PLATE GIRDERS

The detail drawing of a plate girder shows the girder dimensions, flange and web plate sizes, sizes of stiffeners and end plates, their location and the details for drilling and welding. Any special instructions regarding fabrication should be given on the drawing. For example, preheating may be required when welding thick plates or 2 mm may require machining off flame-cut edges to reduce the likelihood of failure by brittle fracture or fatigue on high yield strength steels.

Generally all information may be shown on an elevation of the girder together with sufficient sections to show all types of end plates, intermediate and load bearing stiffeners. The elevation would show the location of stiffeners, brackets and location of holes and the sections

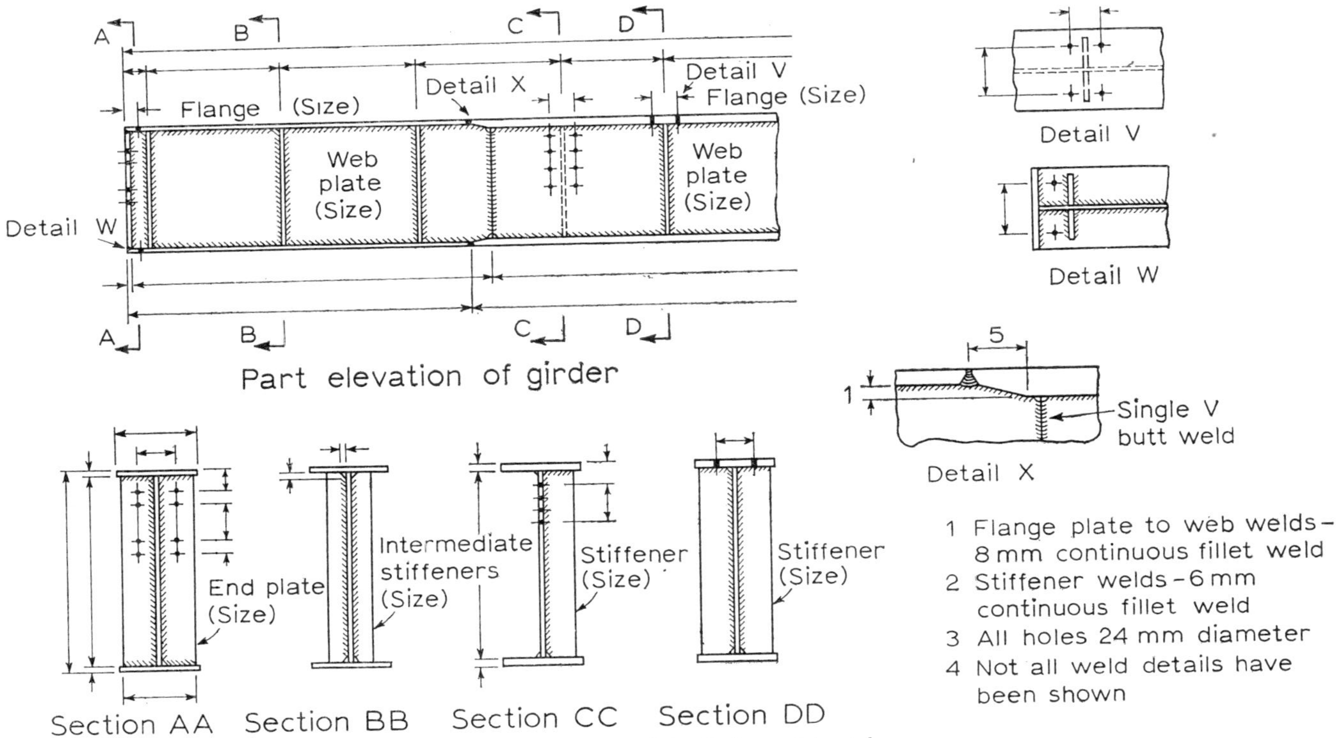

Figure 13.13 Typical details for a plate girder

complete this information. Plan views on the top and bottom flange are used if there is a lot of drilling or other features best shown on such a view. The draughtsman decides whether such views are necessary.

Part sectional plans are often used to show stiffeners in plan view. Enlarged details are frequently made to give plate weld edge preparation for flange and web plates, splices and for load-bearing stiffeners that require full-strength welds.

Notes are added to cover drilling, welding and special fabrication procedures as stated above. Finally, the drawing may contain a material list giving all plate sizes required in the girder. Typical details for a plate girder are shown in Figure 13.13.

13.9 STANCHIONS AND BASES

A typical detail of a stanchion for a multi-storey building is shown in Figure 13.14. The bottom stanchion length with base slab, drilling for floor beams and splice details is shown.

A compound crane stanchion for a single-storey industrial building is shown in Figure 13.15. The crane stanchion is a built-up section and the roof portion is a universal beam. Details at the stanchion cap, crane girder level and base are shown.

13.10 TRUSSES AND LATTICE GIRDERS

The rolled sections used in trusses are small in relation to the length of the members. Several methods are adopted to show the details at the joints. These are:

(a) If the truss can be drawn to scale of 1 in 10, then all major details can be shown on the drawing of the truss.
(b) The truss is drawn to a small scale, 1 in 25, and then separate enlarged details are drawn for the joints to a scale of 1 in 10 or 1 in 5.
(c) The truss centrelines of members are drawn to a small scale, say 1 in 25, then the joints are drawn to a large scale on this reference. The members are shown broken in length. Thus the truss shape and details are shown on one drawing.

Members should be designated by size and length, for example, where all dimensions are in mm:

$$102\times89\times9.5 \text{ Angle}\times2312 \text{ long}$$

$$101.6\times50.8\times4.9 \text{ RHS}\times1310 \text{ long}$$

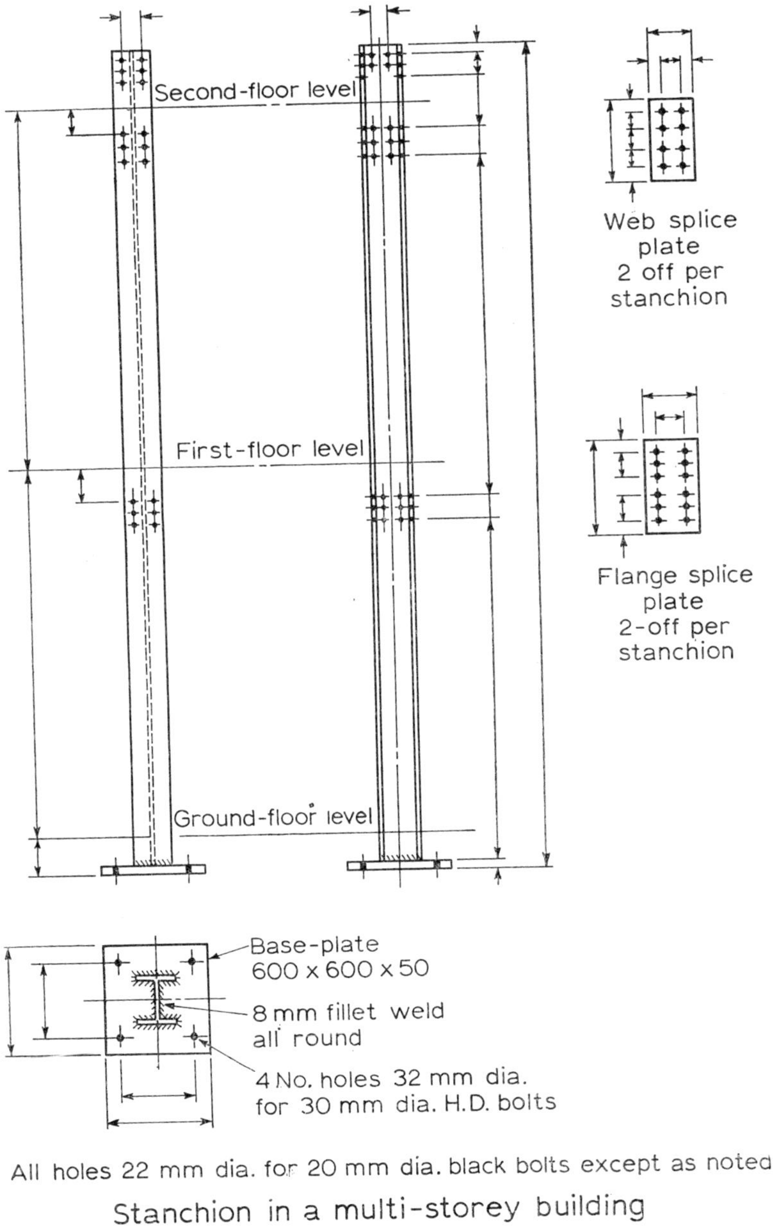

Figure 13.14 Stanchion in a multi-storey building

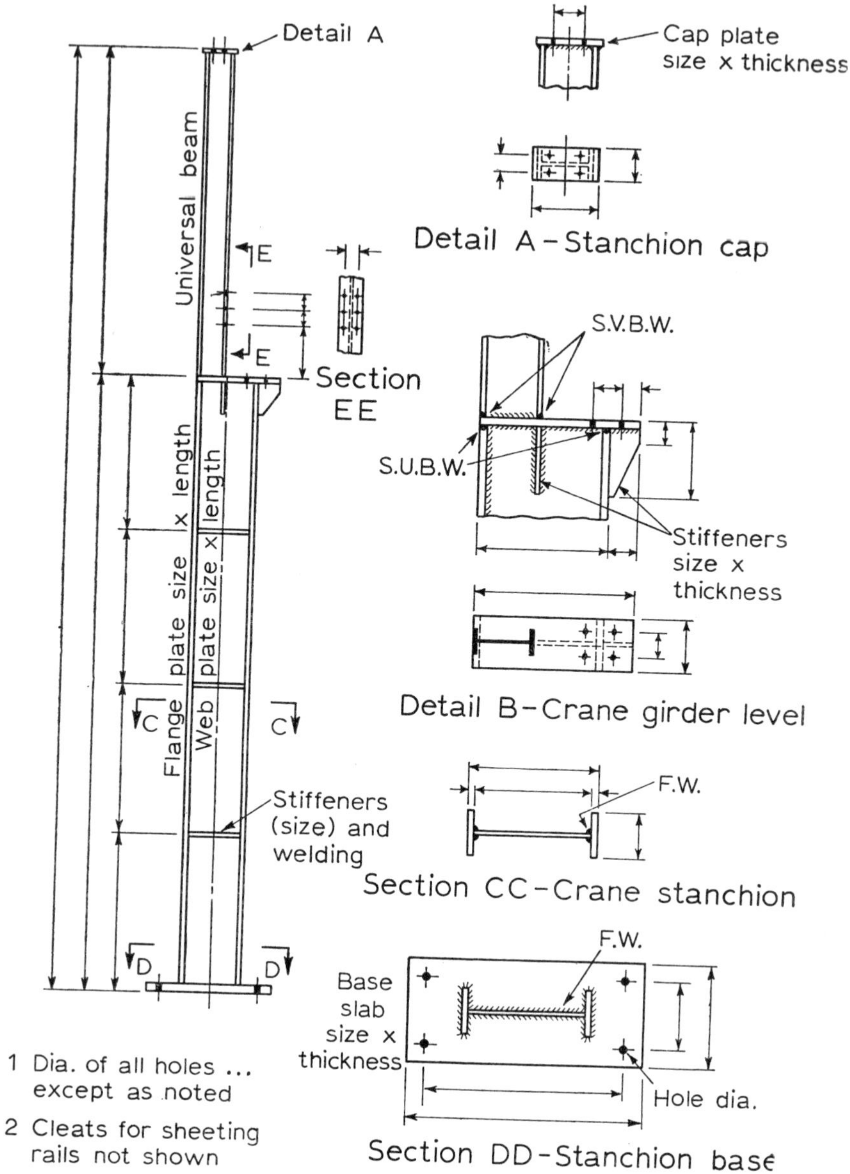

Figure 13.15 Compound crane stanchion for a single-storey industrial building

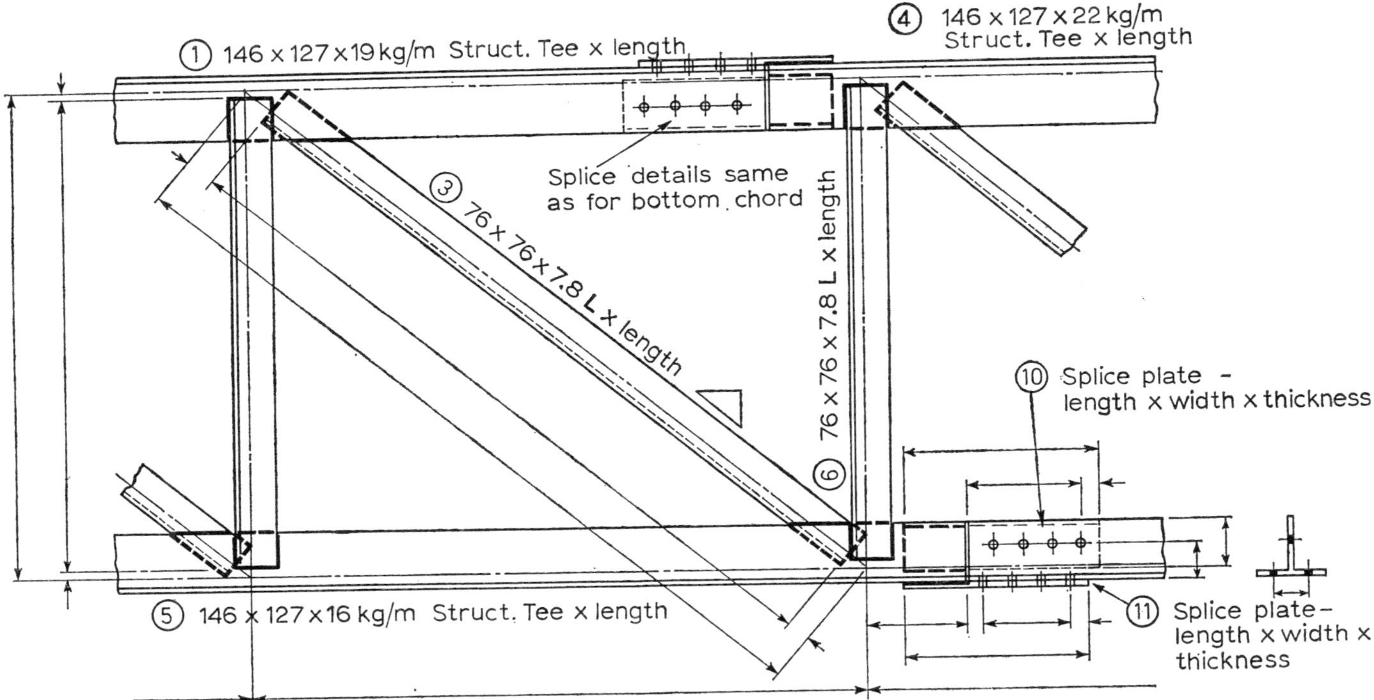

Notes: 1 All holes 22 mm dia. for 20 mm dia. black bolts
2 All weld 6 mm continuous fillet weld

All major details are shown on the truss elevation
All parts of the truss are given item numbers for listing

Figure 13.16 Portion of a flat roof truss

On sloping members it is of assistance in fabrication to show the slope of the member from the vertical and horizontal by a small triangle adjacent to the member.

The centroidal axes of the members are used to set out the frame and the members should be arranged so that these axes are coincident at the nodes of the truss. If this is not the case, the eccentricity causes secondary stresses in the truss.

Full details are required for bolted or welded splices, end plates, column caps, etc. The positions of gauge lines for drilling holes are given in the *Handbook on Structural Steelwork*. Dimensions should be given for edge distances, spacing and distance of holes from the adjacent node of the truss. Splice plates may be detailed separately.

Sometimes the individual members of a truss are itemised. In these cases, the separate members, splice plates, cap plates, etc., are given an item number. These numbers are used to identify the member or part on a material list.

The following figures are given to show typical truss detailing:

Figure 13.16 shows a portion of a flat roof truss with all the major details shown on the elevation of the truss. Each part is given an item number for listing.

Figure 13.17 shows a roof truss drawn to a small scale where the truss dimensions, member sizes and lengths are shown. Enlarged details are given of some of the joints. Reference should also be made to Figures 11.28 and 11.29.

Figure 13.18 shows enlarged joints drawn on a smaller scale layout of the truss.

Details of a truss in rectangular hollow sections are shown in Figure 11.18. Literature on tubular structures from the British Steel Corporation, Tubes Division, should also be consulted.

13.11 PROBLEMS

1 Four connections are shown in Figure 13.19. Draw the following views:

1 Eaves joint—plan and elevation. Scale: 1 in 5
2 Bottom chord joint—elevation and section. Scale: 1 in 5
3 Gusseted stanchion base—plan and elevation. Scale: 1 in 10
4 Crane girder bracket—plan and elevation. Scale: 1 in 10

2 The plan of a welded plate girder is shown in Figure 13.20. The particulars of the girder are:

Plate girder	1500×500
Flange plates	500× 45
Web plate	1410× 12

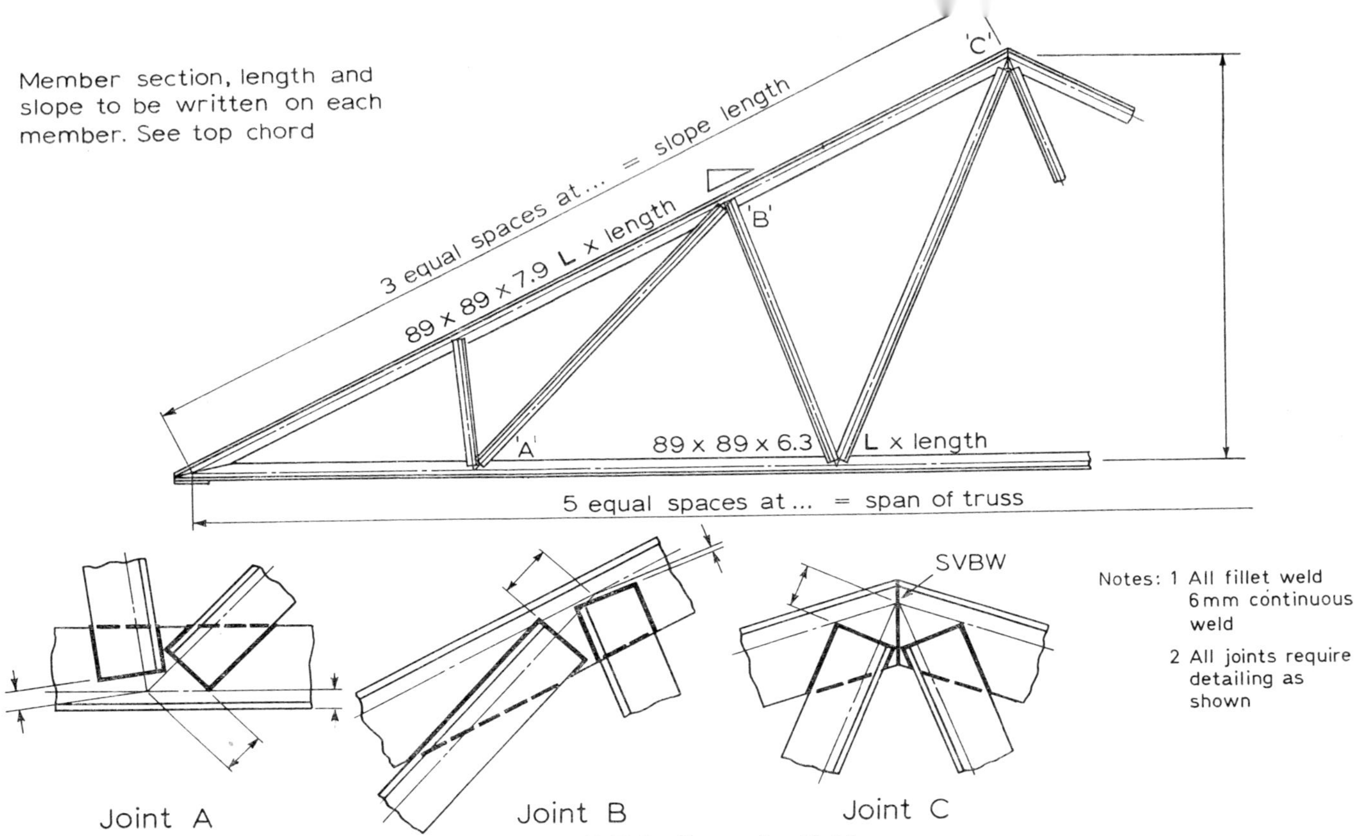

Figure 13.17 Small span all-welded truss

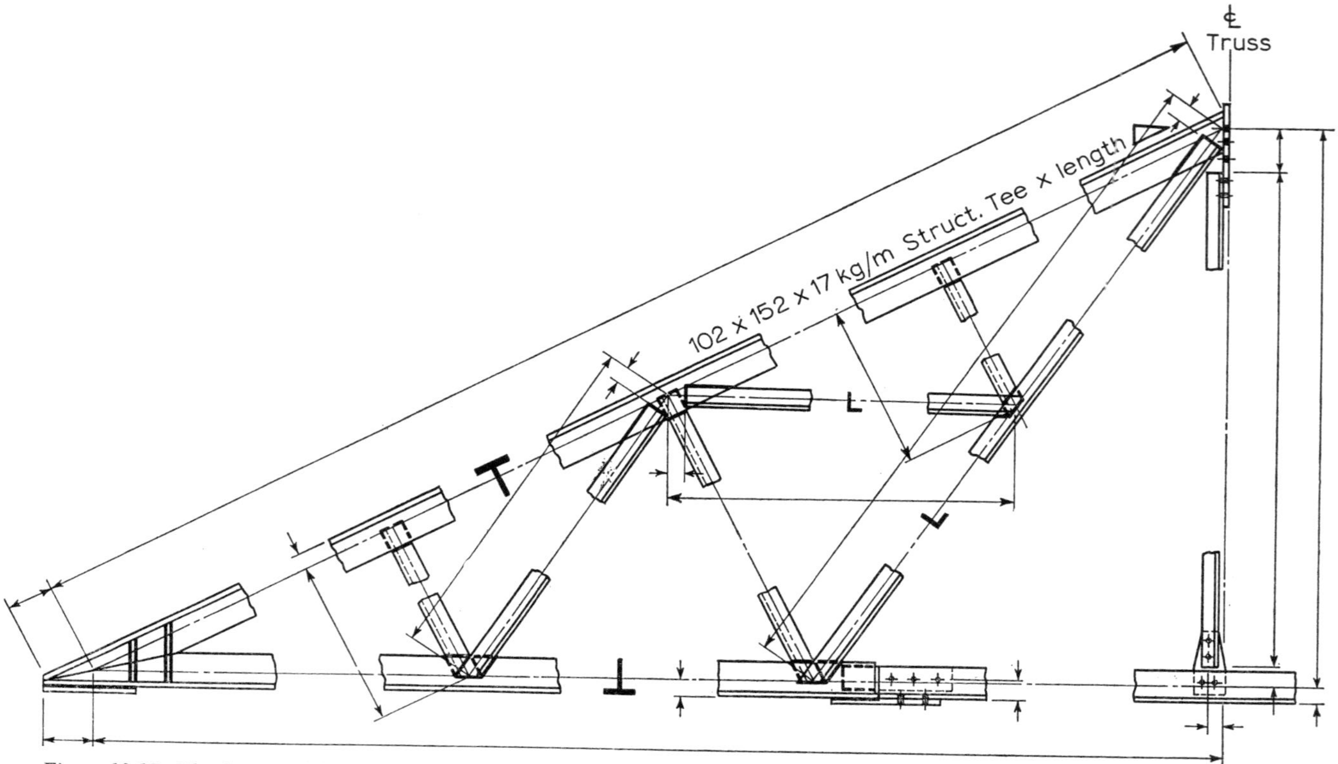

Figure 13.18. The layont of the truss is made to a small scale. The joints are superimposed on this to a larger scale. Member lengths and welding are shown

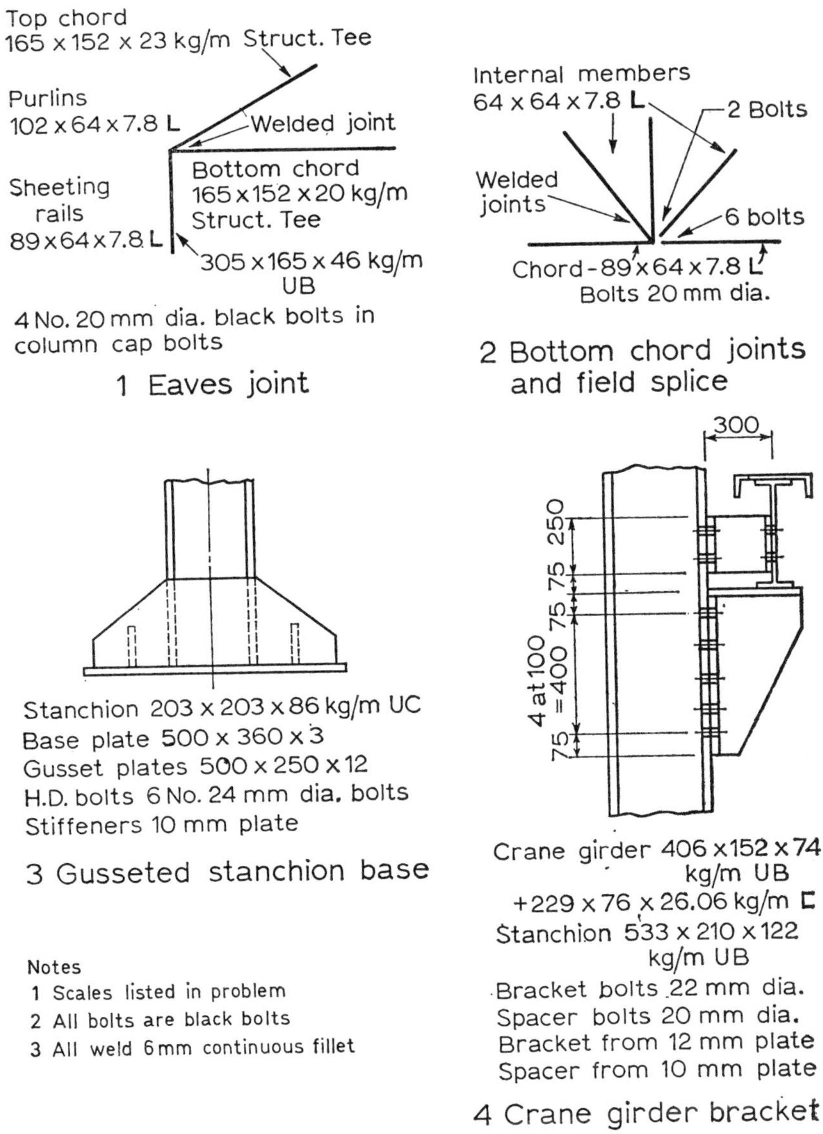
Top chord
165 x 152 x 23 kg/m Struct. Tee
Purlins
102 x 64 x 7.8 L
Welded joint
Bottom chord
165 x 152 x 20 kg/m
Struct. Tee
Sheeting rails
89 x 64 x 7.8 L
305 x 165 x 46 kg/m
UB
4 No. 20 mm dia. black bolts in column cap bolts
1 Eaves joint
Internal members
64 x 64 x 7.8 L
2 Bolts
Welded joints
6 bolts
Chord - 89 x 64 x 7.8 L
Bolts 20 mm dia.
2 Bottom chord joints and field splice
Stanchion 203 x 203 x 86 kg/m UC
Base plate 500 x 360 x 3
Gusset plates 500 x 250 x 12
H.D. bolts 6 No. 24 mm dia. bolts
Stiffeners 10 mm plate
3 Gusseted stanchion base
300
250
75
75
4 at 100 = 400
75
Crane girder 406 x 152 x 74 kg/m UB
+ 229 x 76 x 26.06 kg/m C
Stanchion 533 x 210 x 122 kg/m UB
Bracket bolts 22 mm dia.
Spacer bolts 20 mm dia.
Bracket from 12 mm plate
Spacer from 10 mm plate
4 Crane girder bracket
Notes
1 Scales listed in problem
2 All bolts are black bolts
3 All weld 6 mm continuous fillet

Figure 13.19

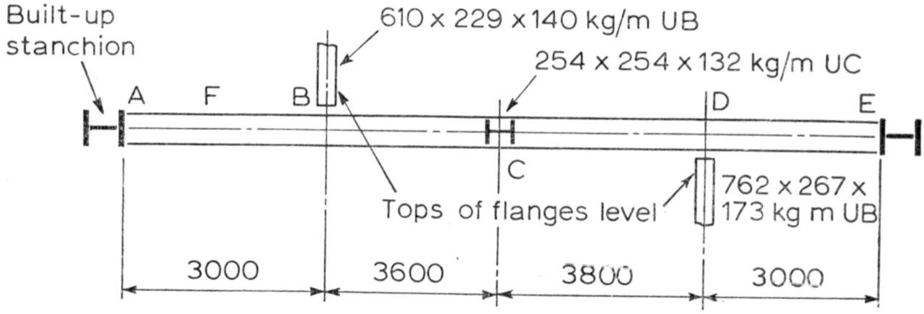

Figure 13.20 Plan of a welded plate girder

Load-bearing stiffeners	200× 20
Intermediate stiffeners	120× 10
End stiffener	500× 15

Welds:	Web to flange	8 mm continuous fillet weld
	Stiffeners	6 mm continuous fillet weld
Joints:	B—shear connection	10 No. 22 mm dia. close tolerance bolts
	D—shear connection	16 No. 22 mm. dia. close tolerance bolts
	A, E—girder supports	—supported on brackets on the building stanchions
	C—stanchion base plate bolted to the top flange of the girder	

(a) Draw the girder elevation. Scale: 1 in 25

(b) Draw sections at B, C, D and F. Scale: 1 in 25

(c) List the material in the plate girder.

3 Make a detail drawing of the cantilever truss shown in Figure 13.21. The truss is to be in welded construction. Scale: 1 in 10

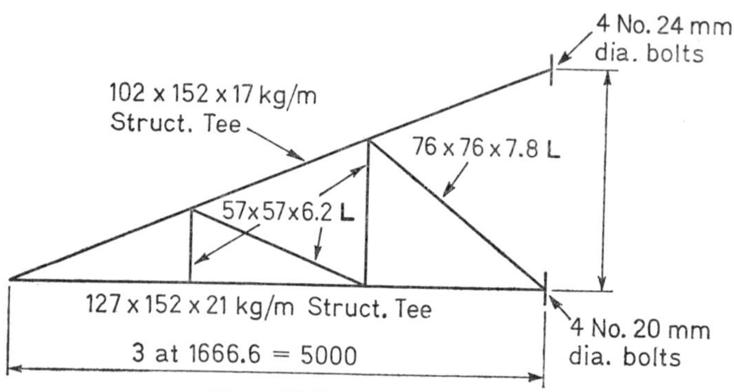

Figure 13.21 A cantilever truss

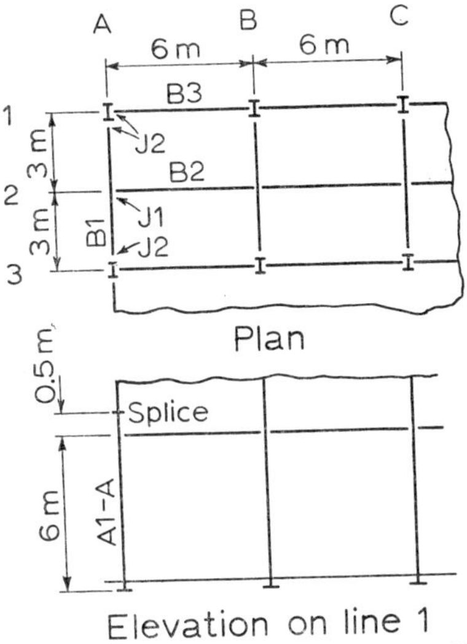

Beams
B1 – 406 × 140 × 46 kg/m UB
B2. B3 – 356 × 171 × 45 kg/m UB

Joints
J1 – 6 No. 20 mm dia. black bolts
J2 – 6 No. 20 mm dia. black bolts

Stanchion
A1 – 203 × 203 × 46 kg/m UC

Base slab
550 × 550 × 38

Figure 13.22

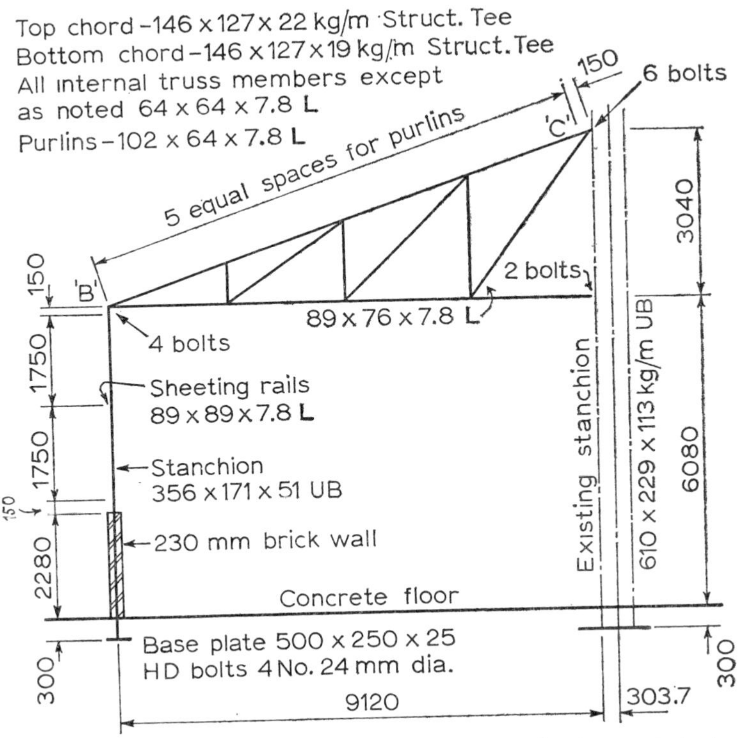

All bolts 20 mm dia. black bolts except where stated
All fillet weld 6 mm continuous

Figure 13.23 Steel-frame extension to an existing building

4 The part floor plan and elevation of the steel for a multi-storey building is shown in Figure 13.22. Make detail drawings for beams B1, B2 and B3 and stanchion Al. Scale: 1 in 25. All particulars are shown on the drawing.

5 A steel-frame extension to an existing building is shown in Figure 13.23. The frame dimensions, member sizes and numbers of bolts in the field joints are shown. Frames are at 5 m centres.

(a) Make a detail drawing of the frame. Scale: 1 in 25
(b) Detail joints A, B, C. Scale: 1 in 5

Appendix

Table 1. CLOSE-TOLERANCE BOLTS: TENSION, SHEAR, AND BEARING VALUES FOR ISO METRIC CLOSE-TOLERANCE TURNED BOLTS,‡ STRENGTH GRADE 4.6, COARSE PITCH SERIES. BS 4190:1967

Nominal size of bolt (mm)	*Area of shank* (mm^2)	*Area at root of thread* (mm^2)	*Tension value at 130* N/mm^2 (kN)	*Shear value at 95* N/mm^2		*Enclosed bearing value at 300* N/mm^2 (kN)† *Thickness of plate passed through or of enclosed plate* (mm)									
				Single shear (kN)	*Double shear* (kN)	6	($\frac{1}{4}$ in) *6.35*	($\frac{5}{16}$ in) *7.93*	*8*	($\frac{3}{8}$ in) *9.53*	*10*	*12*	($\frac{1}{2}$ in) *12.7*	($\frac{5}{8}$ in) *15.8*	*16*
M12	113	84.3	10.9	10.7	21.4	21.6	22.8	28.5	28.8	34.3	36	43.2	44.5	56.8	57.5
M16	201	157	20.4	19.1	38.2	28.8	30.4	38	38.4	45.7	48	57.6	59.3	75.7	76.6
M20	314	245	31.9	29.8	59.6	36	38	47.5	48	57.1	60	72	74.2	94.8	96
M22*	380	303	39.4	36.1	72.2	39.6	41.8	52.2	52.7	62.9	66	79.3	81.5	104	105.7
M24	453	353	45.9	43	86	43.1	45.6	57	57.5	68.5	72	86.5	89	113.5	115.2
M27*	573	459	59.6	54.4	108.8	48.5	51.4	64	64.7	77.2	81	97.3	100	128	129.7
M30	706	561	73	67	134	54	57	71.2	71.9	85.7	90	108	111.2	142	144
M33*	856	694	90.3	81.3	162.6	59.4	62.8	78.3	79	94.2	99	119	122.5	156	158.5
M36	1020	817	106.5	96.8	193.6	64.8	68.5	85.5	86.3	103	108	130	133.5	170.5	173

* Non-preferred sizes
† Single bearing value is 80% of enclosed bearing value.
‡ These are metric hexagon head bolts faced under head and turned on shank.

Table 2. BLACK BOLTS: TENSION SHEAR, AND BEARING VALUES FOR ISO METRIC BLACK HEXAGON BOLTS, STRENGTH GRADE 4.6, COARSE PITCH SERIES. BS 4190:1967.

Nominal size of bolt (mm)	*Area of shank* (mm^2)	*Area at root of thread* (mm^2)	*Tension value at 130* N/mm^2 (kN)	*Shear value at 80* N/mm^2: *Single shear* (kN)	*Double shear* (kN)	*Enclosed bearing value at 200* N/mm^2 (kN)† — *Thickness of plate passed through or of enclosed plate* (mm): *6*	($\frac{1}{4}$ in) *6.35*	($\frac{5}{16}$ in) *7.93*	*8*	($\frac{3}{8}$ in) *9.53*	*10*	*12*	($\frac{1}{2}$ in) *12.7*	($\frac{5}{8}$ in) *15.8*	*16*
M12	113	84.3	10.9	9.03	18.1	14.4	15.2	19	19.2	22.9	24	28.8	30.5	37.9	38.4
M16	201	157	20.4	16.1	32.2	19.2	20.3	25.4	25.6	30.5	32	38.4	40.7	50.5	51.2
M20	314	245	31.9	25.1	50.2	24.0	25.4	31.7	32	38.1	40	48	50.8	63.2	64
M22*	380	303	39.4	30.4	60.8	26.4	28	34.9	35.2	41.9	44	52.8	55.9	69.5	70.5
M24	453	353	45.9	36.3	72.6	28.8	30.5	38	38.4	45.7	48	57.6	61	75.8	76.8
M27*	573	459	59.6	45.8	91.6	32.4	34.3	42.8	43.2	51.5	54	64.9	68.6	85.3	86.5
M30	706	561	73	56.3	112.6	36.0	38.1	47.5	48	57.2	60	72	76.2	94.8	96
M33*	856	694	90.3	68.5	137	39.6	41.9	52.3	52.8	62.9	66	79.2	83.9	104.2	105.6
M36	1020	817	106.5	81.5	163	43.1	45.7	57	57.6	68.5	72	86.5	91.5	113.8	115

* Non-preferred sizes.

† Single bearing value is 80% of enclosed bearing value.

Table 3. HIGH-STRENGTH FRICTION-GRIP BOLTS
(Refer to BS 4395, 1969, Part 1: Metric Units)

Nominal diameter (mm)	*Proof load* (kN)	*Tension value* (kN)	*Shear value No wind* (kN)	*Shear value Incl. wind* (kN)
M12	49.4	29.6	15.9	18.5
M16	92.1	55.2	29.6	34.5
M20	144	86.4	46.2	54.0
M22	177	106.1	56.8	66.4
M24	207	124	66.5	77.5
M27	234	140	75.2	87.8
M30	286	172	92.0	107.2
M36	418	251	134.5	157

Table 4. STRENGTH OF FILLET WELDS. GRADE 43 STEEL
(Allowable stress = 115 N/mm², Strength = leg length × 0.7 × 115 N/mm)

Imperial sizes			*S.I. sizes*	
Leg length (in)	*Leg length* (mm)	*Strength* (N/mm)	*Leg length* (mm)	*Strength* (N/mm)
			3	242
$\frac{3}{16}$	4.77	384	4	322
$\frac{1}{4}$	6.35	511	5	402
$\frac{5}{16}$	7.93	638	6	483
$\frac{3}{8}$	9.53	766	8	644
$\frac{7}{16}$	11.1	894	10	805
$\frac{1}{2}$	12.7	1020	12	965
$\frac{9}{16}$	14.4	1160	14	1128
$\frac{5}{8}$	15.9	1280	16	1288
$\frac{11}{16}$	17.5	1410	18	1450
$\frac{3}{4}$	19.0	1530	20	1610

BRITISH STANDARDS AND CODES OF PRACTICE REFERRED TO IN THE TEXT:

BS 449, Part 2. *The use of structural steel in building* (1969)

CP3, Chapter V, *Loading*. Part 1. *Dead and imposed loads* (1967)

CP3, Chapter V, *Loading*. Part 2. *Wind loads* (1970)

BS 499, Part 2. *Welding terms and symbols* (1965)

BS 4190, *ISO metric black hexagon bolts, screws and nuts* (1967)

BS 4395, Part 1. *High-strength friction-grip bolts and associated nuts and washers for structural engineering. Metric series*. Part 1. *General grade* (1969)

BS 1856, *General requirements for the metal-arc welding of mild steel* (1964)

CP114, Part 2. *The structural use of reinforced concrete in buildings* (1969)

CP2008, *Protection of iron and steel structures from corrosion*

Bibliography

Handbooks

Handbook on Structural Steelwork, British Constructional Steelwork Association Ltd and the Constructional Steel Research and Development Organisation

Handbook for Welding Design, Vol. 1., *Fundamentals*, Welding Institute–Pitman, London (1967)

Structural hollow sections (3 parts): *Technical data; Jointing; Welding* British Steel Corporation, Tubes Division

General Textbooks on Design and Drawing

Bresler, B., and Lin, T. Y., *Design of Steel Structures*, Wiley, New York (1964)

Jude, D. V., *Civil Engineering Drawing*, McGraw-Hill, London (1971)

Lothers, J. E., *Design in Structural Steel*, Prentice Hall, New Jersey (1950)

Lothers, J. E., *Advanced Design in Structural Steel*, Prentice Hall, New Jersey (1950)

Robb, I., *Steel Frame Design Examples*, Cleaver–Hume London (1971)

Russell, P., and Dowell, G., *Competitive Design of Steel Structures* Chapman & Hall, London (1950)

Steel Designer's Manual 4th edn. (Metric) prepared for the Constructional Steel Research and Development Organisation, Crosby Lockwood, London (1972)

General Textbooks on Structural Mechanics

Morgan W., and Williams, D. T., *Structural Mechanics*, Pitman, London (1963)

Mills, G. M., *The Theory of Structures*, Macmillan, London (1965)

Index